Crafting Medicine

synthesis

A series in the history of chemistry, broadly construed, edited by Carin Berkowitz, Angela N. H. Creager, John E. Lesch, Lawrence M. Principe, Alan Rocke, and E. C. Spary, in partnership with the Science History Institute

Crafting Medicine

ARTISANS, KNOWLEDGE, AND THE COMMON MAN IN HIERONYMUS BRUNSCHWIG'S BOOKS ON SURGERY AND DISTILLATION

Tillmann Taape

The University of Chicago Press CHICAGO AND LONDON

The University of Chicago Press, Chicago 60637
The University of Chicago Press, Ltd., London

Published 2025
Printed in the United States of America

34 33 32 31 30 29 28 27 26 25   1 2 3 4 5

ISBN-13: 978-0-226-84058-1 (cloth)
ISBN-13: 978-0-226-84060-4 (paper)
ISBN-13: 978-0-226-84059-8 (e-book)
DOI: https://doi.org/10.7208/chicago/9780226840598.001.0001

Library of Congress Control Number: 2025004963

♾ This paper meets the requirements of ANSI/NISO Z39.48-1992 (Permanence of Paper).

For Qian

Contents

Part V. Crafting Text: Readers, Editors, Translators

Abbreviations and Conventions

AVCUS	Archives de la Ville et de la Communauté Urbaine de Strasbourg
BL	British Library, London
BSB	Bayerische Staatsbibliothek, Munich
CUL	Cambridge University Library
HAB	Herzog August Bibliothek, Wolfenbüttel
HBA	Harvard Botany Arnold Collection
HEH	Henry E. Huntington Library, Pasadena
HUL	Humboldt University Library, Berlin
JRL	John Rylands Library, Manchester
LOCR	Library of Congress Rosenwald Collection
STABI	Staatsbibliothek zu Berlin
WL	Wellcome Library, London

Individual copies of Brunschwig's books are referred to by the abbreviation of the holding institution (see above) followed by the classmark.

For manuscript sources, diplomatic transcriptions are provided, with abbreviations resolved in italics.

For printed sources, the transcription is lightly normalized for readability, and common abbreviations are resolved silently.

PART I

Crafting Medicine in the Early Age of Print

✷ INTRODUCTION ✷

Medicine, Artisans, Books, and Knowledge

In the years around 1500, a handful of books printed in German articulated a new kind of medical expertise for new kinds of practitioners and readers. They include a surgical manual, a plague treatise, and two books on pharmaceutical distillation, all published in Strasbourg by an otherwise obscure local craftsman. For Hieronymus Brunschwig (ca. 1450–ca. 1530), *artznei*—the business of healing—was first and foremost *kunst*—an art or craft—or even *handwerk*—literally "handwork." But he also read widely and wrote prolifically in his native Alsatian dialect. Drawing on text-based and practical traditions of knowledge, Brunschwig's writings represent a new kind of printed book. His surgical textbook marks the beginning of a wave of vernacular publications by craft surgeons. The two books on distillation spawned a new printed genre of *Destillierbücher* that rose to popularity over the course of the sixteenth century, especially with nonacademic readers. Contemporary translations into Latin, Dutch, and English testify to the wide appeal of these books, as do the traces of avid reading and annotating that survive in numerous copies. They fed an appetite for workable medical knowledge in the vernacular, and for new ways of knowing nature and the human body.

Hieronymus Brunschwig was a rather unlikely author at a time when the great majority of books were written in Latin by men with university degrees. His formal training was that of a craftsman, in the local guild of bathers and barbers. How this otherwise obscure artisan came to publish foundational medical books is not entirely clear. Born around 1450, he spent most his life living and working in Strasbourg.[1] No papers in his hand survive, and his name appears in the Strasbourg archives on only two occasions that neatly bookend his life. We first hear of him as young man in his twenties about to go to war for his native city: "Jeronimy Brunzwig—one battle axe" is listed on a small scrap of paper, alongside a handful of other local barbers drafted for military duty in the Burgundy campaigns in the 1470s.[2] In this mere archival snippet, Brunschwig appears above all as a member of one the local guilds that organized the predominantly artisanal population of a free imperial city. As he later records in his books, he returned from the Burgundy fields a more experienced surgeon. Brunschwig also developed a deep interest in pharmaceutical distillation, and whether or not he was an "apothecary," as some contemporary sources suggest, it is clear that he made and dispensed remedies from a shop near the Strasbourg fish market.[3] By the time Brunschwig's name turns up again in the archives, in 1537, he was already dead. Again, he is only mentioned incidentally, in a dispute over property, as "the experienced Hieronymus Brunschwig, our late citizen."[4] Brunschwig would not have objected to being remembered as an experienced craftsman. In his books, he proudly identified as a craft surgeon (*wundarzt*) and emphasized his practical skill and artisanal credentials. Indeed, I argue that these were central to his formulation of a way of knowing that anticipates the more general turn towards materials, sensory experience, and particulars in early modern natural inquiry.

But Brunschwig was also an avid reader of medical and alchemical texts. We should perhaps not take him at face value when he writes of "the books great and small and by good masters, which I have seen and read in many an old library, I may truthfully say about three thousand."[5] But he clearly had access to a range of texts on surgery, pharmacy, and medical alchemy, likely through German manuscripts that were in circulation from the fifteenth century.[6] While he borrowed freely from his sources, his was not a passive or "derivative" reception as has sometimes been claimed.[7] As I show, he reads with purpose and from his own distinctly artisanal perspective, distilling medical theories and alchemical ideas into workable practices in the workshop or at the sickbed. This combination of textual study and the work of the artisan is the driving force behind the impact of Brunschwig's books and what they offered—a new way of knowing nature and the body.

This larger project animating Brunschwig's works has not thus far been recognized in the historiography, even as Brunschwig has become something of a household name. His books—and especially their striking woodcut images—appear in histories of surgery and pharmacy, often to illustrate the early state of medical practice as a backdrop for later developments.[8] The lack of substantive engagement is no doubt in part owed to early scholarship on Brunschwig that closed down conversations through its positivist assessments. Hagiographic histories of surgery celebrated Brunschwig as a technically accomplished forebear of the discipline.[9] Conversely, traditional philological scholars of German medical literature dismissed Brunschwig's surgical writing as entirely derivative of medieval texts.[10] Robert Forbes's *Short History of the Art of Distillation* offers another disciplinary retrospective in which Brunschwig serves as a mere point of departure for later technical progress.[11] From these limited perspectives, the broader significance of Brunschwig and his works could hardly be appreciated.

Brunschwig has no doubt also been drowned out by louder and more controversial historical voices. Writing a few decades later, self-proclaimed iconoclasts such as Paracelsus or Leonardo Fioravanti deployed craft knowledge and "new" alchemical conceptions of medicine as a platform to challenge the learned establishment. Their polemic writings clamored for the attention of their contemporaries, and historians, too, have naturally been drawn to these controversial larger-than-life figures.[12] This may be one reason why Brunschwig has thus far barely featured in the recent wave of innovative scholarship on early modern cultures of practical knowledge, or "artisanal epistemology" in Pamela Smith's term.[13] Apart from my own work, the only recent exception is an essay by Alisha Rankin that identifies Brunschwig as a prime example of such artisan expertise.[14] My study recovers Brunschwig's chronologically early and remarkably self-assured articulation of practical knowledge making at the intersection of craft, medicine, and alchemy—by listening to the quieter voice of an actual artisan rather than the vociferous posturing of a Paracelsus or Fioravanti.

Crafting Medicine is the first sustained scholarly study of this remarkable artisan-turned-author and his milestone publications. It offers a reading of Brunschwig's books as a unified project of crafting new medical knowledge for new audiences, in particular non-learned practitioners. To make sense of Brunschwig's eclectic and multifaceted project, I combine recent approaches to artisanal and alchemical knowledge cultures with social histories of medical practice and methods for the study of material text. I read Brunschwig's books alongside medical and alchemical texts as well as the Strasbourg guild and hospital archives, to show how they shaped and were shaped by different cultures of text and practice. Chapter 1 sets the stage by tracing how Brunschwig's project was embedded in local artisanal, medical, and intellectual cultures. The chapters in part 2 examine his surgical work, while part 3 focuses on the distillation books. In part 4, I analyze Brunschwig's artisanal and experiential epistemologies across his works. Part 5 investigates Brunschwig's readers, imagined and real, and the transmission of his expertise in contemporary translations of his works into Dutch and English. Three main analytical themes run through the book—Brunschwig's vernacular books as agents of change, his expertise intersecting medicine, alchemy, and craft, and the codification of practical knowledge. The following sections introduce these themes, their main arguments, and their contributions to histories of medicine, science, and print.

Crafting Vernacular Print: Books as Agents of Change

Brunschwig's books mark a new beginning in medical print and the politics of knowledge. Historians of the German lands have charted the rise of vernacular medical books addressed to the "common man" over the course of the sixteenth century, but the contours of this new readership have remained somewhat vague.[15] Published at a seminal point in this vernacular wave, Brunschwig's books are remarkably explicit about the kinds of readers they intend to reach. They offer fresh insights into the emergence of new audiences and into social uses of medical print that were often highly politicized, and sometimes distinctly local.[16]

His first publication, the *Buch der Cirurgia* (1497), was the first printed book by a craft surgeon, and one of the earliest surgical texts ever to be published.[17] Issued as a copiously illustrated folio volume of over 250 pages, it marked the beginning of a fruitful collaboration with the Strasbourg printer Johann Grüninger. Writing in his forties, Brunschwig addresses young aspiring surgeons in the voice of a seasoned master, bolstering their training with instructions taken from the texts of medieval master surgeons as well his own experience. The book is organized into seven parts addressing bedside manner, treatments for external wounds, fractures and dislocations, and recipes for surgical plasters and poultices. The *Cirurgia* was a success in the print market. Within less than a year of its publication, an unauthorized version was printed in Augsburg, and over the course of the sixteenth century a total of six editions in German was published (see table 1).[18] By 1524, the *Cirurgia* had been translated into Dutch and from there to English.[19] As I show, Brunschwig's first book had a lasting impact on surgical print and practice, crafting new identities for surgeons as medical practitioners (see chapter 2) and valuable citizens of urban artisanal communities (chapter 3).

Three years later, in 1500, Brunschwig's *Liber pestilentialis* was published. Written in German (despite its Latin title), it was an illustrated work of around seventy pages discussing how medical practitioners and ordinary citizens could navigate plague epidemics. It offers the usual advice for protection against noxious miasmas as well as a regimen for the sick and instructions for lancing buboes. The bulk of the text is borrowed from a previous plague treatise by Heinrich Stainhöwel, town physician of Ulm, although Brunschwig adds his own instructions for pomanders and

TABLE 1. German editions of Brunschwig's books

YEAR 1ST ED.	TITLE	NOTES ON EDITION(S)	EDITIONS TOTAL
	Surgery textbook		
1497	*Dis ist das buch der Cirurgia. Hantwirckung der wund artzny . . .* (Strasbourg: Grüninger)	1497 unauthorized Augsburg edition. 1513 2nd Grüninger edition. 1518, 1534, 1589 further editions by other printers.	6
	Plague treatise		
1500	*Liber pestilentialis de venenis epidimie . . .* (Strasbourg: Grüninger)		1
	Small book of distillation		
1500	*Liber de arte distillandi de simplicibus: Das Buch der rechten Kunst zü distilieren die eintzigen Ding* (Strasbourg: Grüninger)	1538–1566 six subsequent editions by Frankfurt printers.	7
1505	*Medicinarius. Das buch der Gesuntheit . . .* (Strasbourg: Grüninger)	The *Small Book* bound with a partial translation of *De vita* by Marsilio Ficino and additional chapters on distillation by Brunschwig. Six subsequent editions by Grüninger: 1509, 1515, 1521, 1528, 1531, 1537.	7
	Large book of distillation		
1512	*Liber de arte Distillandi de Compositis . . .* (Strasbourg: Grüninger)	1519, 1532 editions by Grüninger. 1539–1614 six further editions by other printers.	9
	Partial editions		
1529	*Apoteck für den gemeinen man . . .* (Erfurt: Melchior Sachse)	Reprint of the medical recipes collection titled "Thesaurus pauperum" in the final section of the *Large Book*. Bound with Michael Puff von Schrick, *Von den ausgeprenten wassern.* 32 further editions between 1529 and 1599.	33

Continued on next page

TABLE 1. *(continued)*

YEAR 1ST ED.	TITLE	NOTES ON EDITION(S)	EDITIONS TOTAL
	Thesaurus Pauperum Einn fürtreffliche und volkomne Haußapoteck . . . (Frankfurt: Christian Egenolff)	Reprint of the medical recipes collection titled "Thesaurus pauperum" in the final section of the *Large Book.* 21 further editions between 1538 and 1664, often simply as *Hausapoteck.*	22
	Total German editions		85

distilled waters against the plague.[20] While this short and fairly conventional plague treatise does not tell us much about Brunschwig's project of crafting medicine, it does contribute to our understanding of his overall mission to empower vernacular readers.

Brunschwig's distillation manuals were also written in German, despite their rather awkward Latin–German titles. The first was published in 1500 as *Liber de arte distillandi de simplicibus,* with the self-translating subtitle *Das buch der rechten kunst zü distilieren die eintzigen ding* (Book on the true art of distilling simple things). Its heftier companion, published in 1512, was titled *Liber de arte distillandi de compositis,* or in German, *Das buch der waren kunst zü distillieren die Composita vnn simplicia* (Book of the true art of distilling composites and simples). To avoid confusion between these long and unwieldy titles, I will simply call the books—as Brunschwig did himself—the *Small Book of Distillation* and the *Large Book of Distillation,* respectively.

As I show in chapter 4, the *Small Book of Distillation* offers a detailed beginners' guide to the art. It opens with a step-by-step description of different techniques with numerous woodcut illustrations showing specialist glassware and other pieces of apparatus, followed by an extensive illustrated herbal with instructions for harvesting and distilling medicinal plants and the application of distilled waters for different diseases. To navigate the book, readers could refer to a table of contents and a set of indices for looking up specific herbs, diseases, and cures. These, however, were a little unwieldy to use and partly redundant, contributing significantly to the book's 460-page bulk. The problem was recognized and amended with a single index in the second edition of 1505. Here, the *Small Book* appears as part of a composite volume called *Medicinarius,* printed alongside a translation of the first two of the *Three Books on Life* (De vita libri tres) written in the 1480s by the Italian humanist and scholar of the occult Marsilio Ficino. Following this learned discussion of physical health and the prolongation of human life, the volume is rounded off with an additional section by Brunschwig, on *aqua vitae* and other more complex distilled remedies that might help achieve Ficino's lofty aims. This move towards more ambitious concepts and techniques came to full fruition in Brunschwig's final work.

Even as he was writing the *Small Book of Distillation*, Brunschwig was already envisioning his eventual magnum opus: "In good time," he promised his readers, "I will also publish the large book for learned people of this work."[21] Published more than a decade later and weighing in at just over seven hundred pages, the *Large Book of Distillation* reads like comprehensive companion volume to its smaller sibling. The range of specialist apparatus is expanded, including larger furnaces for running multiple distillations in parallel and a water-cooled still to distill strong spirits of wine. In the book's more theoretical passages, we can see Brunschwig grappling with alchemical ideas about matter that informed his own understanding of distillation, as I discuss in chapter 5.

Like the *Cirurgia*, Brunschwig's distillation books were the first of their kind. The only prior publication, Michael Puff von Schrick's *Büchlein von den ausgebrannten Wässern* (Booklet of distilled waters) of 1476, was a reference work for the use of distilled remedies and did not include any instructions for making them. As such, it was a successful and often reprinted volume, indicating that there was a market for the topic.[22] Brunschwig supplied that demand with the first printed works to describe the art of distillation in any technical and visual detail. They are some of the earliest examples of the emerging tradition of "how-to" literature and books of secrets that scholars have placed at the center of recent narratives in the history of early modern science and medicine.[23]

The distillation books were an even greater publishing success than the *Cirurgia* (see table 1). They were reprinted as complete works, but also in the form of small, cheap volumes sold as *Apoteck für den gemeynen man* (Pharmacy for the common man) or *Haußapoteck* (Home pharmacy). These contained the final section of the *Large Book*, a "poor man's treasure" of simple medical recipes, sometimes together with Schrick's work on distilled waters. As we shall see in chapter 4, Brunschwig's winning combination of practical distillation manual and illustrated herbal sparked a new printed genre of *Destillierbücher* or "distillation books," laying the foundations for (al)chemical approaches to matter and the body that drove major changes in science and medicine over the course of the sixteenth century.[24]

Overall, Brunschwig's works are remarkable for their deliberate cultivation of a new readership. While the *Cirurgia* is addressed primarily to practitioners of surgery already in formal training, the *Liber pestilentialis* opens up to a wider audience of ordinary citizens and householders. The *Small Book of Distillation,* too, invites nonspecialist readers, with its step-by-step instructions. The second edition adds more demanding techniques alongside Ficino's more theoretical work on long life. Finally, the *Large Book* explicitly includes learned readers in its intended audience and engages with challenging operations and complex matter theory. For all these technical and learned additions in some of his works, Brunschwig never loses sight of the "common man" as his core readership.

By exploring how Brunschwig brought medical expertise and agency to new readers, *Crafting Medicine* joins conversations about the politics and social function of printed books, such as Mary Fissell's work on the role of vernacular medical print in Restoration politics of reproduction,[25] and Hannah Murphy's analysis of the local impact of print in bringing about a new medical order in early modern Nuremberg.[26] Chapter 1 shows that Brunschwig's books were shaped by local artisanal, medical, and intellectual cultures, and were at least in part intended to intervene in those cultures, as I discuss in chapters 2, 3, and 4. Chapter 8 argues that publishing books for the "common man" was a pioneering move around 1500—especially in light of the debates about the education of laypeople that exercised local intellectuals long before the Reformation came to Strasbourg. Even reform-minded humanists had reservations about the social and intellectual ambitions of the new middling sort, embodied by the troublesome figure of the "striped layman." The illustrations in Brunschwig's books, I argue, turn this negatively charged striped figure into a model reader of medical craft. My investigation of actual readers' traces in surviving copies (chapter 10) suggests that Brunschwig succeeded in attracting readers on a spectrum from lay to learned, including a significant middle ground of "hybrid" readers. At the local level, this study of Brunschwig's books as agents of change offers a new understanding of the lively publishing landscape in Brunschwig's native city. In her landmark study of early Strasbourg print, Miriam Usher Chrisman concluded that there were two incommensurable "lay" and "learned" cultures, divided by the use of the German or Latin language.[27] I contend that, on the contrary, these boundaries were being debated as early as 1500, from both sides of the supposed two cultures, and that Brunschwig's books for the "common man" were instrumental in these negotiations. More generally, *Crafting Medicine* shows that the controversies around vernacular books for the layman were about natural as well as spiritual knowledge, and that printed books about medicine played a key part in these debates on the eve of the Reformation.

Artisans, Nature, and the Body

The potential of Brunschwig's books as agents of change sprang in large part from their articulation of a new kind of medical expertise. Throughout this book, I explore how that expertise was shaped by different traditions of knowledge—textual and practical, learned and vernacular, artisanal, medical, and alchemical. In parts two and three, I focus on the surgical work and distillation books, respectively, before turning to overarching questions about Brunschwig's epistemic engagement with craft knowledge and experience in part 4.

To investigate Brunschwig's surgical expertise, I read his *Cirurgia* alongside its local sibling publication, the *Feldtbuoch der Wundtartzney*, published in Strasbourg in 1517 by Brunschwig's slightly younger colleague, Hans von Gersdorff. Histories of medicine regularly gesture towards these books, sometimes rather grandly as a "Strasbourg school" of surgery,[28] but they rarely move past their striking woodcuts to engage with their project of codifying a new kind of surgical expertise.[29] As I show in chapter 2, Brunschwig and Gersdorff sought to elevate the status of surgery by drawing on what Michael McVaugh has called the "rational surgery" of the Middle Ages.[30] Influenced by the surgical writings of Guy de Chauliac, they revived the medieval ideal of surgery as a rational science as well as a craft. At the same time, they emphasized their experience in the Burgundy battlefields and the practical skill that sets them apart from physicians. In their confident articulations of this hybrid expertise, Brunschwig and Gersdorff subtly pushed against the boundaries of the traditional medical hierarchy in which learned physicians made a rational diagnosis and instructed "mere" craftsmen such as surgeons and apothecaries to let blood or compound remedies. As social historians of medicine have observed time and again, this ideal was constantly ignored, subverted, and renegotiated, sometimes leading to open conflict between physicians, regular artisanal healers, and unauthorized "irregular" practitioners.[31] As we shall see, there were efforts to establish learned control over medical practice in Strasbourg, but relations remained peaceful on the whole, and left room for Brunschwig and Gersdorff to push the boundaries of what surgeons could know and do.

Surgical expertise and identities were not only negotiated in relation to medical practice, but also within the social fabric of the city. Social historians have long been interested in surgeons and other "artisans of the body," in Sandra Cavallo's useful phrase, and explored how they negotiated their often precarious place in urban society.[32] In chapter 3, I read Brunschwig and Gersdorff alongside archival documents that speak to social life of body workers in Strasbourg, to show how they sought to carve out an honored and valued place for surgeons in the urban community. By implicating surgeons in official inspection panels and as forensic experts in court, Brunschwig and Gersdorff depicted the surgeon in the role of a guardians of artisans' bodies in their layered dimensions of health, labor, and honor. This contributes to social histories of "body workers," but also centers surgeons as medical experts in civic governance, an area of scholarship where the focus has thus far been on learned physicians as officeholders.[33]

Overall, my study of Brunschwig and Gersdorff in these chapters contributes to larger narratives in the history of surgery. Traditionally this has been a story of surgery's slow and gradual rise from a craft tradition to a recognized discipline in academic medicine, aided by the foundation of corporate professional bodies and by the establishment of an authoritative textual canon.[34] In this story, we usually see surgery being elevated by university-trained practitioners and humanist scholars. This book offers a counternarrative: as well as being pulled from above by learned men, surgery was also pushed from below by artisans crafting surgical expertise and identity, from the very beginning of surgical print.

Part 3 turns to the distillation books to investigate how Brunschwig's reading of medical and alchemical theory and his artisanal practice come together in a new understanding of material transformations inside and outside the body. In explaining the effect of his distilled remedies, as I show in chapter 4, Brunschwig imagines a body whose inner workings are driven by moving liquids and "spirits," reminiscent of the distillation process itself. I argue that this nascent idea of a "distilling body" resonated with a widespread view of the body in terms of "fluxes" and shaped subsequent Paracelsian and other "chemical" accounts, thus providing a missing link in histories of the body.[35]

In addition to Brunschwig's medical rationales, chapter 4 explores how he frames them as appealing commodities and discusses their production in terms of material and labor. By analyzing these economies of distillation, my study joins conversations about the "medical marketplace" as a category of historical analysis,[36] but also contributes to recent scholarship, notably by Alisha Rankin and Elaine Leong, on charitable and domestic cultures of medicine-making beyond the "market."[37] Brunschwig's books were themselves commodities, and their explicit turn towards new audiences (as discussed) showcase how to be successful in the early marketplace of print.[38] His account of distillation, I argue, shaped new agencies for non-learned makers and users—a market ploy that was imitated by later *Destillierbücher*.

Brunschwig's distilling expertise was intimately linked to alchemical ideas and techniques. As I show in chapter 5, his framing of distilled waters as safe and efficient remedies owed much to his reading of alchemical matter theories. Building on intellectual histories of these ideas, *Crafting Medicine* shows how they come to life in the workshop and the market, offering new frameworks for non-learned cultures of knowledge and medical care.[39] In so doing, my study joins a lively field of studies on early modern alchemy, including the work of Bruce Moran, Tara Nummedal, and Jenny Rampling.[40] These histories still leave something of a gap between learned medieval alchemy and Paracelsian or other "chymical" cultures from the second half of the sixteenth century onwards.[41] *Crafting Medicine* fills this lacuna with a much-needed account of the medical alchemy of the early *Destillierbücher*—a system of ideas and practices that worked seamlessly within traditional Galenic physic even as it articulated the artisanal processes and apparatus that would form the basis of Paracelsus's and others' challenges to that tradition.

Ultimately, *Crafting Medicine* asks how Brunschwig—and other artisans like him—understood the material world, the body, and how they could be manipulated by human art. In chapter 5, I explore how Brunschwig assembles an understanding of distillation as an epistemic encounter with matter, a close physical engagement in which the artificer must become nature's apprentice. Chapter 6 turns once more to the Strasbourg archives to explore artisans' notions of skill as a habit formed by practice, and shows how Brunschwig conveys this expertise of the body and the senses in print. This reading of Brunschwig links histories of medicine and alchemy with scholarship on practical cultures of knowledge making. Questions about artisans' contributions go back to the early Marxist histories of Edgar Zilsel and others,[42] but they have been pursued with renewed vigor and originality in more recent studies, including William Eamon's work on the tradition of secrets,[43] and scholarship on artisanal ways of knowing nature, notably the work of Pamela Long and Pamela Smith.[44] The recent digital edition and interdisciplinary study *Secrets of Art and Nature in Renaissance France*, undertaken by myself and my colleagues at the Making and Knowing Project under Smith's direction, investigates an anonymous French craftsman's "material imaginary"—his underlying understanding of the material world.[45] Building on this work, I characterize Brunschwig's material imaginary as a hybrid expertise, shaped by medical and alchemical theories, artisanal practice, and embodied experience.

In these historiographies of medicine, alchemy, and craft, the figure of Paracelsus looms large for his truculent rejection of Galenic medicine, his chemical approach to the body and medical treatment, and his advocacy of personal experience and artisanal practice as the most secure source of natural knowledge.[46] His self-consciously radical voice, amplified and adapted by contemporary as well as later writers, has become something of a singularity in the historical imagination, sometimes making it hard to view sixteenth-century alchemical medicine through anything but a Paracelsian lens. Brunschwig is an important early source for these intersecting knowledge cultures, and as Charles Webster has suggested, his influence on Paracelsus may be more significant than has thus far been realized.[47] It would be easy to offer a teleological reading of Brunschwig as an important proto-Paracelsian. This book aims to do the opposite. Rather than looking at Brunschwig through Paracelsus, I use Brunschwig as a lens to look past this towering and sometimes distorting figure to make sense of hitherto neglected understandings of matter and the body around the turn of the sixteenth century. This focuses historical attention on genuine craft expertise and on epistemic cultures that wove together learned theories with embodied practice rather than placing them in polemic opposition.

From Practice to Print and Beyond: Codifying "How-to" Knowledge

As well as exploring Brunschwig's artisanal way of knowing, *Crafting Medicine* shows how such knowledge was captured and transmitted, in the printed book and beyond. During the early modern period, artists, architects, chemists, metalworkers, and other experts increasingly took to the pen, and recent scholarship has shown how they grappled with the seemingly insurmountable problem of abstracting and codifying their practical knowledge.[48] The role of images in conveying natural knowledge, too, was renegotiated in the sixteenth century, as Sachiko Kusukawa has shown in her study of anatomical and botanical print.[49] As I argue in chapter 6, Brunschwig provides a remarkably lucid artisanal perspective on these issues. He reflects on the limits of words and images in capturing artisans' tacit skills and in conveying knowledge about plants and bodies. However, in collaboration with his publisher, Johann Grüninger, he rises to the challenge. Through cautionary tales, well-placed images, and instructions for the reader's own sensory awareness, Brunschwig's books provide something of a virtual apprenticeship in print.

Chapter 7 explores Brunschwig's complex relationship with experience as a source of knowledge, to place him in larger narratives about epistemic shifts towards "empirical" inquiry in science and medicine. Previous scholarship has tended to focus on the mathematical sciences[50] and on learned conceptions of observation and *historia*,[51] although a recent special issue on "Testing Drugs and Trying Cures" and Alisha Rankin's study of poison trials have begun to explore a greater diversity of early modern experimental and experiential cultures of knowing.[52] By characterizing what I call Brunschwig's "vernacular empiricism," my study reveals a nuanced and remarkably early impetus towards empirical knowledge making that unites artisanal, occult, and learned epistemologies as well as local conversations about the dangers of experiencing the material world.

Chapter 9 traces the subsequent movement of expertise across languages, in the translations of Brunschwig's books into Dutch and from there into English. While there is a growing body of scholarship on translation both from the history of science and early modern studies, translation between vernacular languages by non-learned experts is still under-studied.[53] The trajectories of Brunschwig's works, I argue, reveal cultures of what I call "vernacular text work" with distinctive strategies for transmitting terminology, technical detail, and experiential knowledge. The transmission of knowledge did not stop with the printed book, but continued into the material practices of readers, as chapter 10 shows by analyzing annotations and other traces in surviving copies of Brunschwig's works. Building on the history of reading, and especially William Sherman's study of early modern "book use," I explore epistemic and material practices from interventions signaling ownership to strategies for organizing and retrieving knowledge.[54] I show that the annotations bear out Brunschwig's aim to reach a wide readership, from Latinate students of medicine to vernacular surgeons and the "common man," with a majority of annotators reflecting a new middling or hybrid audience.

Overall, *Crafting Medicine* thus traces itineraries of knowledge transmission from artisanal know-how to printed text and image, from one vernacular to another, and from the printed book to the scribbling pen, exploring new relationships between practical text and textual practice. As the following chapter shows, this arc of transmission begins among the everyday lives of craftspeople, medical practitioners, and scholars of Brunschwig's native Strasbourg.

✷ 1 ✷

Artisans, Healers, Humanists

PRACTICING AND WRITING MEDICINE IN STRASBOURG, CA. 1500

Brunschwig lived and worked in Strasbourg all his life, and his books were shaped by the local environment of a free imperial city. In the decades around 1500, Strasbourg enjoyed political independence and prosperity. It was surrounded by fertile land and sat at the crossroads of the trade on the River Ill, the Rhine, and along the east–west route that passed across Strasbourg's pontoon bridge. One local publication claimed that its name derived from the German for "road" (*straße*) because "the city is a road to Lothringia, France, Burgundy, Lombardy, Switzerland, Breisgau, Swabia, the Low Countries, and everywhere, and it lies in the middle of a highway, both by land and by water."[1] Its main exports included cannon, a sturdy cloth known as "Strasbourg gray," good Rhenish wine from the region, and the brandy distilled from it by the local *weinbrenner* whose techniques and apparatus appear in Brunschwig's distillation books. Overall, there was no one dominant industry, and prosperity was distributed fairly equally. Strasbourg was part of a wealthy federation of small local powers, largely undisturbed by the ambitions of overreaching princes. The free imperial cities of the Upper Rhine were significant political players, and among them Strasbourg was second only to Basel in size and influence. During Brunschwig's lifetime, the peace of the region was threatened only once, by the military conflicts that saw our young author go to war armed with an axe. Rising in defense of their local autonomy against Charles the Bold in the Burgundy Wars of the 1470s, the Upper Rhenish powers joined forces with the Swiss confederacy in a resounding victory.[2]

Strasbourg housed the headquarters of several regional organizations of master craftsmen and journeymen that reached far into the empire. Within the city walls, too, political and professional life was ruled by artisanal corporations. The guilds organized citizens' military duties, taxes, health insurance, and court cases, as well as their professional training and social life. The bathers and barbers of Brunschwig's guild, together with a few apothecaries, likely represented a significant proportion of the medical services available in the city. It is only towards the end of Brunschwig's life that the city authorities began to take administrative control over medical treatment through municipal institutions and practitioners. Although there was no university during Brunschwig's lifetime, Strasbourg was a center of publishing and learned debate, especially about the value of practical knowledge and vernacular books for ordinary people. This chapter situates Brunschwig in his native city, by reading his books alongside the archives of the guilds and hospitals as well as local humanist publications. It shows how Brunschwig's project shaped and was shaped, socially and epistemically, by Strasbourg's intersecting artisanal, medical, and intellectual cultures, setting the local scene for the larger questions that are developed in subsequent chapters.

"Nobler than All Other Arts": The Craft of Healing in the City of Artisans

For Brunschwig, healing was a craft. He refers to surgery and pharmacy as *kunst*, meaning "art" in its wider early modern sense that covered a wide range of crafts and other practical knowledge, or as *handwerk*, which literally means "handwork," but was also used to refer to a specific trade or the guild that represented it.[3] In the *Small Book of Distillation*, Brunschwig emphasizes that a good medical practitioner was first and foremost a good "craftsman" (*werckman*), and that it would be foolish to consult anyone who was not:

> To such an untrained and short-witted healer you so lightly surrender your life, disregarding that not every craftsman is to be called a master of the craft which he has never learned, and you would not trust anyone with even the smallest bit of work in your house if he had never learned it: to make you a garment from good cloth or shoes from good leather, much less a goose pen from good wood. . . . . . The cloth, leather, or wood would be spoiled, and yet this can be bought again for little money, whereas you wager body and life, which cannot be bought for any amount of money or gold.[4]

The explicit comparison of healers with other trades identifies the art of healing as one craft among many others. The skill of a shoemaker or carpenter was essentially the same as that of a healer, even though they worked on very different materials. At the same time, medicine held a special place among the crafts because it worked with precious "body and life" rather than cloth or leather. As Brunschwig puts it in the second edition of the *Small Book*, "healing is nobler than all other arts: if a master falls sick, he cannot practice his art, therefore he desires medicine which will allow him to keep his art and livelihood."[5] For Brunschwig, medicine was integral to craft, but it played a distinct and crucial part in its shared economy of materials, bodies, and expertise.

Structurally, that shared culture was maintained by the city's corporations. There were two noble societies (*constoffeln*) and a range of craft guilds (*zünfte*). These were umbrella organizations that can be disaggregated into different aspects of city life. They were occupational associations representing a specific trade. Socially, they were confraternities that gathered in their own heated *trinkstube* or drinking hall, organized their communal religious life, and practiced a form of mutual social insurance. They were also military units that organized night watch duties and drafted members for the defense of a certain section of city wall, or indeed for campaigns such as the Burgundy Wars. As political bodies, the guilds elected representatives to the city council.[6] They collected members' taxes and judged many of their misdeeds in their own internal tribunals, and thus represented a significant portion of Strasbourg's administrative and juridical apparatus.[7] These aspects of a "guild" overlapped in complex ways. Brunschwig's confraternity of barbers, for example, was represented with a single seat in council by a political guild that also included the confraternity of bathers. The two groups maintained separate *trinkstuben* and identities, and fought over clients, as we shall see in chapter 3. Although a reform of the city council in the 1480s left the craft guilds with slightly fewer seats overall, artisans remained an important political and social presence in the city.[8]

Urban society was stratified along these corporate lines, and there was a clear hierarchy among guilds, visibly enacted several times a year in the marching order of ceremonial processions for feast days and the annual *schwörtag* (oath day). This hierarchy partly reflected the wealth and political power of a guild, but mainly dealt in the more symbolic capital of honor.[9] The goldsmiths' skill and honesty in working with precious metals brought them high status, whereas Brunschwig's colleagues, the barbers and bath masters, seem to have held a much lower rank—witness the dissolution of their joint guild and the loss of their only council seat. As I argue in chapter 3, one likely explanation lies in early modern artisans' ambivalent sensibilities surrounding the human body.[10] Many did not see "body and life" as precious materials, as Brunschwig did, but feared their dishonoring taint. Brunschwig's books grapple with these artisanal sensibilities, seeking to tackle surgery's stigma in both text and image.

Urban Networks of Materials and Expertise

Brunschwig's comparison of healing to carpentry and shoemaking places his medical project squarely within urban artisanal culture. Reading his books alongside archival records of local crafts, a network of materials and craft expertise emerges. The best honey for distilling sweet medicinal waters came from "behind the small place called Oberkirch in the diocese of Strasbourg,"[11] and the earthenware jugs made in the neighboring village of Haguenau were excellent for storing distilled remedies.[12] Better still were those made from "the earth of Syburg, a town in the diocese of Cologne,"[13] or in the nearby village called Sigen—provided one knew the local production schedule: "One must come around the day of Our Lady or fourteen days before, or around St. Margaret's day. They no longer work at other times of the year, for there are no more than twelve masters."[14] When it came to glass vessels for distillation, the best were "those made by a Bohemian glassblower or from Venetian glass."[15] Elaborate pieces such as the "pelican" could be hard to find because they were "difficult to make, and one seldom finds a glassblower who can make them" (see figure 4.8).[16] Close to home and further afield, Brunschwig knew the places, makers, and materials that produced the paraphernalia of his craft.

Brunschwig also borrowed the expertise and equipment of Strasbourg's other artisans. For some alchemical remedies, the ingredients had to be calcined at very high temperatures. Since these were routinely achieved by experts such as chalk burners and brickmakers, Brunschwig suggests using their kilns to calcine one's ingredients "for as long as the chalk and bricks are fired; here in Strasbourg that is twelve or fourteen days."[17] Distillation, on the other hand, required the low, steady fire of a purpose-built furnace of curved bricks (see figure 4.5). Brunschwig includes instructions for making these, complete with a double-page illustration that serves as a life-size template for a wooden mold (see figure 6.4). Readers can use it to make their own bricks, or, better still, "if you can get to a brickmaker, bring him the form and have the bricks made for you."[18] Archival records show that local brickmakers did indeed "make and keep individual molds, and all pieces shall be made from these."[19] To label flasks of distilled waters, Brunschwig refers to a different set of artisanal trades: the metal labels should be "inscribed with a stylus, engraved by a goldsmith, or embossed with the steel punches for making type at a printer's shop."[20] To make *aurum potabile* (potable gold), one should flatten gold florins "as the goldsmiths do when gilding,"[21] or leave the operation to the specialist skill of a "goldbeater" who knew best how to make delicate sheets of gold leaf.[22]

Such interdependencies between different artisans are also depicted in books of trade such as the well-known *Ständebuch* of 1528, with woodcuts by Jost Amman and poems by the shoemaker Hans Sachs describing different occupations and their place in society. The poem about the goldbeater, for example, is almost entirely a list of crafts that use his product, from illumination to embroidery. The instrument maker provides tools for engravers and woodturners, but also barbers' implements, and the coppersmith makes "helmets useful for distilling waters."[23] Medicine was enmeshed in a rich artisanal network that spanned the early modern city, and Brunschwig's own texts and practices were deeply embedded in this local culture of materials and expertise. To explore how that underlying expertise was understood by local craftspeople, we turn again to the archival record.

Crafting Werschafft*: Artisanal Standards of Workmanship*

As social historians have noted, artisans' collective notions of workmanship and skill are rarely spelled out, and we have to look for their written traces in regulations or accounts of inspections, examinations, and disputes.[24] The Strasbourg archives are patchy, but close reading can reveal underlying assumptions of what it meant to know one's craft. The goldsmiths, for example, were aware that their social and moral status was closely linked to the precious materials they worked with, and stipulated tight regulations that have survived in the archives in various iterations. A panel of sworn masters carried out fortnightly inspections to make sure that their members used high-quality gemstones and gold of at least 18 karats. Anyone found working with "dyed crystals" or gold-plated lesser metals faced hefty fines and could lose their guild membership.[25] The regulations further state that no piece of jewelry may be sold in the city unless it is stamped with the guild's seal, indicating that it had been "inspected by the master and the tribunal of the goldsmiths' guild and acknowledged to be *werschafft*."[26] The term *werschafft* denotes something like "sound product" with connotations of good quality, but also a sense of warranty for the buyer.[27]

Used as a verbal equivalent to the guild's seal of approval, *werschafft* emphasizes the quality and purity of raw materials, and it has been argued that this was the main function of artisanal hallmarks: to signal the craftsman's honesty more than his skill.[28] For the Strasbourg goldsmiths, however, skill did come into play when they judged whether a journeyman's work was *werschafft* enough to make him a master. The candidate's masterpiece must be made "with his own hand," and "whoever submits a piece that he has not made with his own hand . . . . . . shall have his shop closed down, and be barred from working in Strasbourg for a year and a day."[29] The goldsmith's *werschafft* signaled an ethos of manual skill as well as honesty in working with valuable materials.

The term recurs with some regularity in the Strasbourg archives wherever the question of sound craftsmanship is at stake. For example, a series of council records from the 1460s document a drawn-out dispute over a badly cast gun. The gunner claimed that "since the gun has withstood the first three shots, he has delivered *werschafft*."[30] The city authorities who had commissioned the gun disagreed. As proceedings dragged on, six expert witnesses were asked to assess whether the gun was *werschafft*. They concluded that it was dangerously flawed and would not "withstand the test of time or prove *werschafft* unless it were newly cast."[31]

The archives show that the medical trades were part of this artisanal culture, just as Brunschwig claimed. An ordinance for the regulation of pharmacy from around 1500 tells us that each apothecary shop in Strasbourg should be inspected twice a year, by "the city's two doctors and the oldest and most knowledgeable of the apothecaries," plus a trustworthy citizen who "understands the materials."[32] They were expected to assess the quality of the wares on offer and to "discard, pour out, or burn anything they do not find to be useful or suitable."[33] Perhaps in an effort to codify local standards, the inspectors were also tasked with setting down written guidelines to specify "how all materials, roots, herbs, waters and other things such as electuaries, syrups, and other mixtures should be kept in apothecary shops, and for how long each may remain *werschafft* and usable."[34] In the case of pharmaceuticals, the standards of *werschafft* were especially important—because citizens' health was at stake, and because many materia medica were unfamiliar substances from faraway places that lost their efficacy over time and could open the door to fraud. As Brunschwig warned in his *Small Book of Distillation*, "it often happens that the heathen or Saracens boil or macerate rhubarb or aloe wood, use the water, and then dry the rhubarb and sell it again as new—unbeknownst to the inexperienced apothecary."[35] Brunschwig does not claim to be an apothecary, but he was very experienced indeed at recognizing the properties and shelf life of medicines, as we shall see in chapter 4.

Patchy as they are, the Strasbourg archives show that there was just such a shared culture of artisanal expertise as Brunschwig implies in his comparison of craft and medicine. It coalesced around the term *werschafft* as a standard for good workmanship that involved sound materials, an ethos of honor, and manual skill. In chapter 6, I further explore the notion of skill, how it was acquired through the repeated practice of artisanal training, and how Brunschwig goes about conveying such embodied knowledge in print.

Healing and Healers in Strasbourg

As the pharmacy regulations show, the artisanal culture of *werschafft* intersected with the local landscape of medical practice, and in particular with municipal efforts to regulate it. Like other cities, Strasbourg had charitable institutions to take care of certain groups of poor sufferers. Records of the *Mehrere Spital* (Large Hospital) date back to the twelfth century, but like many of its kind, it originally provided general shelter and care for those in need, and only gradually took on a distinctly medical function. Outside the city, two religious foundations going back to the fifteenth century took in lepers, known as *sondersiechen* (special sufferers) or simply *guthleute* (good people)—contrary to persistent popular myths, sufferers of leprosy were not universally reviled or forcibly segregated in premodern Europe.[36] A more centralized system of care developed relatively late in Brunschwig's life. As the new "French disease," also known as pox or *blattern*, afflicted a growing number of citizens in Strasbourg and elsewhere, a dedicated *Blatterhaus* was founded in 1502.[37] The administration of the Large Hospital changed hands from the clergy to the city authorities during the fifteenth century, and in 1515—three years after Brunschwig had published his final book—it was reconstituted with a medical focus. The new ordinance appointed a pharmaceutical assistant, a surgeon, and a physician who was to supervise the manual work of the other two, as per the traditional medical hierarchy.[38]

Whether or not physicians in fact managed to exert this kind of control in practice, artisans usually did the bulk of the day-to-day medical work in early modern hospitals and cities. In Strasbourg as elsewhere, such work was so enmeshed in the landscape of crafts that distinctly "medical" practitioners can be hard to locate. The bathers (*bader*) and barbers (*scherer*) belonging to Brunschwig's guild took care of citizens' bodies by washing, shaving, and cutting hair as well as cupping or occasional bloodletting. The boundaries between medical or surgical treatment and other kinds of "body work" were fluid and contested, as we shall see in chapter 3. During the time Brunschwig was writing his books, only a few *bader* and *scherer* turn up in the archival record, and it is hard to know to what extent they worked as healers, except for the few who are explicitly identified as a *wundarzt*, a "wound healer" or surgeon, such as Brunschwig's younger colleague Hans von Gersdorff.[39] It is only after the city authorities centralized municipal health care in the 1520s that we see an increase in practitioners referred to as *wundarzt* or *blatterarzt* (pox healer), including two female healers described as *ärztin* working at the pox hospital. This increasing recognition of medical artisans may in part be due to Brunschwig's works. As we shall see in the following two chapters, he crafts a distinct identity for surgeons as medical practitioners, and emphasizes their service to the urban community—as guardians of artisans' bodies, and therefore, their livelihoods and even their honor.

Apothecaries barely appear in the archival record apart from the pharmacy regulations already discussed. The inspections were clearly taken seriously by the city authorities as a means of maintaining public health and order. For the apothecaries, they were perhaps something of a double-edged sword. On the one hand, they had their shops and wares scrutinized twice a year and might have their stock destroyed if it was found wanting. On the other hand, one of their own was involved in the assessment. This was somewhat unusual—elsewhere, this responsibility was exclusively entrusted to learned physicians. As well as being included in the inspection panel, Strasbourg apothecaries were given some leeway to prescribe medicines on their own account to close friends and family, which was strictly forbidden in other places.[40] Reading between the lines of the regulations, the apothecaries' expertise was apparently held in comparatively high esteem in the city.

The Strasbourg system of inspections was evidently so efficient that it became the envy of other major cities. In 1502, the Strasbourg council received a polite letter from the mayor of Freiburg. Since their apothecaries were increasingly "cheated with fraudulent materials," the Freiburg council had "resolved to inspect the pharmacy shops," and asked their colleagues in Strasbourg to send the "highly learned Fuchs, your physician" to assist with the inspections.[41] Nothing further is known about this Doctor Fuchs, but the letter suggests that he was part of the panel that policed the *werschafft* of Strasbourg pharmaceuticals. In the 1530s and '40s, the Strasbourg ordinance served as a model for new pharmacy regulations adopted in Freiburg and Bern.[42] As Hannah Murphy and others have shown, such regulations became an important and contested tool of municipal governance in free imperial cities.[43]

Physicians, too, were part of this administrative machinery. When he was appointed to the Large Hospital in 1515, Wendelin Hock became Strasbourg's first recorded *stadtarzt*, or municipal physician—a salaried office that was becoming increasingly common in imperial cities. Municipal physicians often had administrative responsibilities as well as treating the poor free of charge, and sometimes acquired considerable influence in city governance.[44] It was only towards the end of Brunschwig's life that the intellectual buzz of the Reformation attracted high-profile medical scholars such as Otto Brunfels, known today as one of the fathers of German botany. Brunfels taught medicine at a private school attached to the Carmelites, which later merged with the Latin *Gymnasium*, established by the reformer Johann Sturm in 1538.[45] Brunschwig did not live to see this new institution that offered classes in medical theory, taught among others by such luminaries as Guinter von Andernach, an early promoter of the new anatomy of the Renaissance and former teacher of Vesalius in Paris.[46] Learned medicine thus only gained an institutional foothold in the city long after Brunschwig had published his last book. Before 1512, only two physicians appear in the Strasbourg archival records. As we shall see in the following chapter, one of them expressed strong views against artisans practicing medicine without learned supervision, providing a learned foil against which we can read Brunschwig's confident articulation of surgical expertise. The physician Lorenz Fries was another important interlocutor. Before he moved to Strasbourg in 1519, he practiced in nearby Colmar, and clearly knew Brunschwig or at least his works—as we shall see in chapters 7 and 8, he endorsed and even continued Brunschwig's project of providing practical medical knowledge for vernacular readers.

Overall, the medical landscape of Strasbourg was fairly typical of a free imperial city. Its medical administration emerged gradually out of existing poor relief, but these structures were only consolidated towards the end of Brunschwig's life, and years after he had published his books. While he was writing them, he would have witnessed that apothecaries were given important civic responsibilities, and that their expertise was respected. Brunschwig dispensed medicines from his own shop and shows a remarkable command of pharmaceutical knowledge and techniques in his books, but he never claims to be an apothecary. Instead, he clearly identifies as a *wundarzt* or surgeon—a category of practitioner whose status in the urban society of artisans was evidently less secure. As a member of the guild of bathers and barbers, Brunschwig could not have failed to note its decline. Among his fellow barbers, there does not seem to have been a strong sense of a distinctly medical or surgical identity, and perhaps he realized that many of them did not possess the necessary know-how—as we shall see, he began his *Cirurgia* by complaining about the "ignorance" among the surgeons of his day. Brunschwig's books complement the patchy archival evidence to provide a clearer picture of early modern medical artisans, their expertise, and their place in the city and the landscape of medical practice.

Humanists and Printers

If Strasbourg appears on the historical map as a center of learned culture, it is usually in the context of the city's role during the Reformation in the 1520s and '30s, driven by such prominent figures as Martin Bucer and Johann Sturm.[47] The local intellectual climate of the preceding generation is less well known, but it was just as alive with learned debate, and already fomenting social and educational reform. Just as Brunschwig began publishing his works, the lawyer and satirist Sebastian Brant, by then widely known for his *Narrenschiff* (Ship of fools),[48] returned to his native Strasbourg to take up the prestigious office of *stadtschreiber*, the official secretary and chronicler of the city. Brant gathered around him a circle of like-minded humanist scholars, including the cleric and educator Jakob Wimpheling, the preacher Johann Geiler von Kaysersberg, who delivered a series of sermons on Brant's *Narrenschiff* in the Strasbourg cathedral, and Thomas Murner, who published satirical works inspired by Brant, as well as a legal manual.[49] The works of the Strasbourg humanists addressed sociopolitical issues of the day, often in the vernacular. Geiler, for example, was a fervent advocate of the sick poor and instrumental in the foundation of the *Blatterhaus* as well as the Major Hospital's conversion into a municipal institution of healing.[50] He was also aware of the concerns that animated his mostly artisan congregation, and he knew to speak their language to convey a spiritual message. From his pulpit in the cathedral, he preached a series of forty-one sermons, in plain German, on ants as a model both for individual duty and orderly communal life. Geiler extols the insects' industry and cooperation—although they "bustle about pell-mell, yet each one knows what it should do,"[51] just like specialist artisans who depend on each other. "A bather needs a shoemaker, and I, too, need a shoemaker,"[52] Geiler explains, promising to "trade spiritual for temporal goods."[53] In the anthill of urban society, the network of craft skills and materials that Brunschwig evoked in his books served as a guiding ideal of Christian communal life.

The topic that was most hotly debated by the early Strasbourg humanists was educational reform. Wimpheling laid out plans for a new education system, including public Latin schools, and squabbled with Murner over the pedagogical value of a card game for law students.[54] For all their reformist zeal, however, these intellectuals often felt acutely ambivalent about the education of laypeople, especially their unfettered access to the rising tide of vernacular books. While Brant and his circle were all for publishing illustrated volumes in German to instruct and even entertain, most of them drew the line at common citizens reading theological works, lest they turn around and question learned clerical authority based on their muddled understanding. The early Strasbourg humanists also had reservations about sensory experience, which they identified as the layman's way of knowing the world, as we shall see in chapter 7. In chapter 8, I show how these learned anxieties came to a head in a vivid trope coined by Geiler—the "striped layman" who embodied the intellectual and social ambitions of the middling sort, especially as a new kind of vernacular reader.

The intellectual culture of Brunschwig's Strasbourg was intimately linked to its thriving printing business, which measured up to such European publishing hubs as Basel, Lyon, and Venice. It was increasingly driven by polemical exchanges between Protestant and Catholic writers, works of medicine and natural philosophy, and new vernacular books for the "common man." During Brunschwig's lifetime, some twenty-seven printers were active in Strasbourg, and his publisher, Johann Grüninger, was one of the most productive.[55] At over 60 percent, Grüninger published by far the highest proportion of vernacular works of any printer in the city.[56] The press was known for its sloppy proofreading, but above all for illustrating its volumes with prime examples of the fine Alsatian school of woodcut illustration.[57] Grüninger's visual expertise and his innovations in print layout were crucial not only to Brunschwig's project of codifying practical knowledge, as I show in chapter 6, but also for the books' visual arguments that turned the "striped layman" of humanist discourse into a model reader of vernacular books and a new kind of artisan-healer (see chapter 8). From the early humanists to the Reformation, Strasbourg was a hotbed of negotiation and debate about the politics of knowledge and print. Throughout this book, I show that Brunschwig's works were tuned into these conversations and made significant claims about what artisans and the "common man" could and should know about medicine.

Conclusion

This chapter has begun to map the social and epistemic cultures—artisanal, medical, and intellectual—that Brunschwig's books engage. Each will be explored in more depth in the chapters that follow and placed in conversation with larger written and practical traditions of medicine, alchemy, and how-to knowledge. Brunschwig's books not only offer a unique window into these local and wider cultures, but also represent important interventions in the knowledge politics of the day. His goal is not just to describe the craft of medicine, but to supply a perceived lack of expertise and stem the tide of "untrained, short-witted" healers he bemoans as the outset of his comparison of medicine to other crafts. He bolsters the education of surgeons with authoritative texts as well as his own experience, and subtly crafts a stronger identity for the surgeon as an honorable medical artisan and model citizen. In the distillation books, he opens up pharmaceutical knowledge and techniques to the "common man." While promoting the individual agency of trained practitioners and laypeople, Brunschwig also emphasizes the benefit of his project to his native city and to artisanal communities in general. While the printed book is often hailed as a universalizing innovation that created larger and larger audiences and cultures of knowledge, Hannah Murphy has pointed out that medical print was often distinctly local. Brunschwig's books were not only shaped by local cultures, but also intervened in local as well as larger conversations about the place of craft in medicine and medicine in craft.

PART II

Crafting Surgical Identities

2

Between Craft and Learning

NEGOTIATING THE PLACE OF SURGERY IN THE MEDICAL LANDSCAPE

By training and guild membership, Brunschwig was first and foremost a surgeon, and already quite an experienced one by the time he published his first book, the *Buch der Cirurgia* (Book of surgery) in 1497. It was the first time an artisan-surgeon codified his expertise in print. The *Cirurgia* was not only the first published book on surgery in German, but also one of the earliest printed surgeries in Europe, preceded only by a couple of Italian and French editions of medieval learned texts.[1] At the hands of the capable Strasbourg printer Johann Grüninger, the *Cirurgia* was published as a handsome folio volume of over 250 pages, with woodcut illustrations of surgical instruments and practitioners at work. Its seven "tracts" deal with the appropriate conduct for a surgeon, general principles of wound treatment, specific wounds in different parts of the body, and an antidotary of recipes for ointments and plasters.

The *Cirurgia* had lasting appeal in the print market. Within less than a year, it was copied by the Schönsperger press in Augsburg, spurring Grüninger to commission two new woodcuts and additional chapters for an augmented second edition—or rather multiple versions that combined sheets of the original print run with new material in different ways, making for a rather tangled publication history.[2] Before the end of the sixteenth century, three other printers published versions of Brunschwig's book that closely followed the original German, and there were translations into Low German, Dutch, and English, all before 1550, which are discussed in chapter 9.[3] No prior book on surgery had achieved such a wide reach. A number of learned medieval treatises were in circulation, including vernacular translations.[4] Artisan-surgeons, too, had composed manuscript treatises, but these were generally intended for a narrower circle of the author's immediate family and apprentices or potential patrons.[5] With his printed book, Brunschwig addressed "all who want to learn surgery," especially "young prospective masters and journeymen of barbers and surgeons."[6] Writing the first textbook for a wide audience was an opportunity to redefine what surgery was and what a surgeon should be. This chapter explores what Brunschwig made of this opportunity.

Any study of the *Cirurgia* and its local context would be incomplete without its natural companion, the *Feldtbuoch der Wundtartzney* (Field book of surgery) by Hans von Gersdorff, another Strasbourg surgeon a few years younger than Brunschwig. First published in 1517, the *Feldtbuoch* went through many print runs and enjoyed great popular success.[7] At around two hundred pages, it was slightly shorter than Brunschwig's work, but it covered additional topics, especially skin diseases such as St. Anthony's fire and leprosy, which Gersdorff got to know well during his years working at the St. Anthony Hospital. This book, too, was lavishly illustrated, with woodcuts showing dramatic surgical interventions and impressive instruments, some perhaps of Gersdorff's own invention. Some editors and translators clearly found these images more impressive than those in the *Cirurgia* and combined images from Gersdorff with Brunschwig's text (see chapter 9).

Published in the same place and by contemporary authors, the *Cirurgia* and the *Feldtbuoch* are often cited together as seminal works of surgical print, or even as a "Strasbourg school" of surgery, but they rarely receive more than cursory treatment.[8] Older scholarship alternately hails them as harbingers of technical progress or dismisses them as textually derivative.[9] More recently, Melanie Panse has offered a thorough study of the *Feldtbuoch* and Heidi Hausse has provided a careful and illuminating reading of Gersdorff in her history of amputation and prosthetics.[10] A reappraisal of Brunschwig's *Cirurgia* is long overdue. In this chapter and the next, I read Brunschwig and Gersdorff alongside archival records that document the medical and artisanal worlds of Strasbourg around the turn of the sixteenth century. By locating these printed books in the city, I argue that the *Cirurgia* and the *Feldtbuoch*, for all their indebtedness to existing textual traditions, represent an innovative bid to craft new epistemic and social identities for surgery.

These books are a unique opportunity to explore an area of medical practice that is still less well understood than learned physic, even though it touched the lives of a far greater proportion of the population in medieval and early modern Europe. In Tudor London, for example, surgeons were the largest group of medical practitioners.[11] In German towns, too, surgical practitioners were present in significant numbers, with on average one surgeon per four hundred to five hundred inhabitants.[12] For Strasbourg, existing figures for the period 1550 to 1630 suggest one of the highest densities, one surgeon per 470 inhabitants.[13] More numerous than physicians, and often considerably cheaper, surgical practitioners were the only medical professionals most people ever consulted over the course of their lives.[14] Surgery was the medical service of the emerging middling classes that were to become, over the course of the sixteenth century, an increasingly health-conscious demographic consuming medical services, remedies, and books.

The overarching narrative in the history of surgery in the medieval and early modern periods is that of a manual craft trade that was gradually assimilated into academic medicine. Traditionally, that history begins in earnest with Enlightenment efforts to rationalize and professionalize.[15] This is in part because modern surgery, at least as understood in the Western world, offers a clearly defined body of knowledge and practitioners that readily sustain unifying historical narratives.[16] In the premodern period, there was no such stable entity—what we would now call surgery was written and practiced by a range of artisans and scholars across Europe, with different ideas as to what the discipline should look like, and how it might be consolidated.

A number of more recent scholarly efforts to illuminate this fragmented picture have focused on textual traditions. Collectively, these studies identify an important common thread in histories of surgery in Europe: the medieval articulation of a learned surgery and various early modern efforts to revive this ideal. The "rational surgery" of the Middle Ages, as Michael McVaugh has called it, began in the medical schools of northern Italy, where surgical writers complemented practical instructions with learned concepts from ancient Greek and Arabic sources, and reached its epitome in in the 1360s with the *Chirurgia magna* of the French physician Guy de Chauliac.[17] These texts were still copied, read, and translated well into the sixteenth century, and formed the core canon for Brunschwig and Gersdorff. That canon grew significantly through Renaissance efforts to mine ancient texts for new knowledge. Alongside instructions for surgical practice, they also recovered from Galen's writings the ideal of a unified medicine that encompassed both surgery and physic. Some humanist scholars worked towards this ideal by publishing vernacular translations of the classics for the benefit of ordinary surgeons, including affordable French editions of Galenic texts published in 1530s Lyon that were, in turn, translated into English.[18] In the German lands, surgical texts by Chauliac and other medieval masters circulated in the vernacular, but there was less of a culture of surgical humanism. German surgeons were arguably better served by the vernacular craft tradition of surgical writing that Brunschwig had started.[19]

Social and institutional histories have highlighted the different ways in which surgeons negotiated their position in the medical landscape, especially with respect to learned physicians. A number of Italian universities included surgery in their curricula from at least the fourteenth century. Graduate surgeons formed an elite group that was closer to physicians than to workaday barber-surgeons, but it was not closed to ambitious artisans.[20] In the corporate medical landscape of early modern France, surgeons were often in direct conflict with physicians in their attempts to forge a strong collective identity. The surgical college of St. Cosmas in Paris, for example, set up as an academic institution to rival the Faculty of Medicine, complete with MA degrees, gowns, and mortarboards, temporarily shaking off the physicians' control at various points in the sixteenth and seventeenth centuries.[21] In England, too, surgeons sought greater corporate and academic recognition. In 1540, they merged with the barbers' guild to form the Barber-Surgeons' Company.[22] With the support of the physician John Caius, the surgeons formed an elite group apart from the company's "mere" barbers, aiming for an expertise and status closer to that of the physicians.[23]

As the example of the plucky Parisian surgeons shows, such social and occupational wrangling sometimes brought on conflicts and boundary disputes with learned physicians. In the traditional medical hierarchy, rational physicians determined the causes of diseases in individual humoral bodies and instructed the manual work of surgeons and apothecaries. In practice, surgeons and other non-learned practitioners often managed to resist or subvert learned control, to the extent that the hierarchy existed more as a theoretical norm than a practical reality.[24] Nevertheless, it provided one important background against which surgeons fashioned their status and identity.

Histories of surgery have done much to explore textual and social efforts towards Galen's ideal of a unified medicine in which surgery and physic complemented each other side by side. Thus far, however, they are skewed towards a narrative in which surgery is being pulled up from above by physicians, humanist scholars, and increasingly elite learned surgeons consolidating a shared body of knowledge. My study of Brunschwig and Gersdorff offers a hitherto overlooked counternarrative. As well as being pulled from above, surgery was pushed from below, by literate artisans or "vernacular surgeons," to use Heidi Hausse's term.[25] From the very beginning of the print tradition, their confident articulations of identity and expertise shaped surgical cultures just as significantly as later and more learned writers.

The chapter begins with an exploration of the medical landscape and prevailing attitudes towards surgery in Strasbourg. I then show how Brunschwig and Gersdorff renegotiate the traditional hierarchy by emphasizing surgeons' manual skill, especially on the battlefield, and by reviving the medieval ideal of the learned surgeon. In their treatments of human anatomy, I argue, these artisan-authors articulate concerns about the relationship between practical, rational, and visual knowledge that exercised generations of learned experts. The final section discusses how, based on their view of surgical expertise, Brunschwig and Gersdorff pushed the boundaries of the surgeon's legitimate field of practice with respect to the physician.

Between Learned Physicians and Ignorant Empirics: Surgeons in Strasbourg's Medical Landscape

As far as we can tell from the patchy archives, Strasbourg housed a fairly typical cast of medical characters, bearing in mind that there was no university, and that efforts to centralize the local medical administration only happened late in the lives of our authors (see chapter 1). Brunschwig published the *Cirurgia* before the appointment of a municipal physician, surgeon, and apothecary at the Large Hospital in 1515, and Gersdorff's *Feldtbuoch* appeared only two years after these changes. Both lived through the decline and dissolution of the joint guild of bathers and barbers in the 1480s. A handful of contemporary *bader* and *scherer* are mentioned in the archives prior to this date, but in absence of the more specific designation of *wundarzt*, we do not know to what extent they did surgical work.[26] The exception is one Conrad Hornung who appears next to Gersdorff in a list of conscripts for the Burgundy campaigns, and later as one of three medical experts petitioning the council to have Gersdorff join their panel for the inspection of suspected lepers.[27]

Surgeons with official appointments appear more frequently in the archival record as the city authorities consolidated their control over public health in the 1520s. Lamprecht Röderer, a specialist stonecutter and oculist, offered his services to the authorities for free in 1524 and was hired two years later in a permanent capacity, for a salary of fifty-two florins per annum. Around the same time, the surgeon Lienhard Reinlin was hired explicitly as a as a *blatterarzt* (a specialist in treating the venereal *blattern* or pox) in 1526, for a payment of three shillings per patient per week, plus a bonus of two gold florins for each successful cure.[28] Two decades later, the contract appointing the surgeon Adolf Meng to the *Blatterhaus* states that "none other shall be admitted to treat the poor in the house whose diseases are within his knowledge and experience."[29] While Meng was free to consult outside help as necessary, he was given full responsibility for one of the city's charitable institutions. It is tempting to speculate that the appointment of surgeons to such important posts owed something to Brunschwig's and Gersdorff's publications.

We know little of learned physicians in Strasbourg at the time when Brunschwig and Gersdorff were writing their books, before the Reformation and the foundation of the *Gymnasium* drew in such high-profile scholars as Otto Brunfels and Guinther von Andernach. From the 1460s to the early 1500s, only two physicians appear by name in the city archives, Thoman von Hochberg and Johann Widmann, who shared the unusual qualification of holding degrees in surgery as well as physic. Widmann offers a rare glimpse of the local medical landscape in a letter he sent to the city council in 1483, in which he speaks of himself as a "future doctor here in Strasbourg."[30] He had come to the city to treat the mother of a friend, the humanist scholar Peter Schott. Widmann purchased citizenship and perhaps intended to settle, but before he did so he thoughtfully shared in his letter a long list of improvements to the council's management of healthcare. First and foremost, he questioned "whether it is good and useful that one should allow anyone here to dispense remedies and other things belonging to the doctor of medicine, including baptized Jews, barbers, old or foolish women, all called *emperici* in Latin, and even the executioner, as I am reliably told is the case here in Strasbourg."[31] Widmann describes the usual spectrum of practitioners that could be found in most cities, from barbers to neighborhood wisewomen and other "irregular" practitioners.[32] To assert his learned authority, he lumps them all together as "empirics" whose knowledge was based on experience rather than rational principles. Regulation of such practitioners, he implies, was too lax. Strasbourg was behind the times compared to "large cities" where, for example, midwives had to be sworn in by a physician.[33] He submits that "the practice of medicine, that is physic, should be allowed to none except doctors of medicine, and not to an empiric," to avoid the "great damage that arises from their ignorance."[34] Barbers in particular should "perform nothing other than venesection and do other things that the doctors command."[35] Like many other physicians, Widmann emphasizes a medical order in which surgeons were subordinate to physicians' learned authority in all but the most straightforward external injuries.

Widmann was perhaps exaggerating for rhetorical effect, but it is possible that Strasbourg, in the absence of a medical faculty or other body of physicians, was indeed more permissive than other places. The archival record is too thin to provide a clear picture. Nor do we know whether the council acted on the complaints made by Widmann, who in any case only stayed in the city for another couple of years. It is not until 1515, when the Major Hospital first employed an in-house physician, that we see any formal statement of learned authority over medical practice. According to the new regulations, the physician should "advise and help the hospital's barber" and "instruct him truly as to when and how he should bleed each [patient], or apply cupping, baths, or leeches, such that none shall be neglected or damaged out of ignorance."[36] The barber could be trusted to perform key medical techniques, but the when and how was up to the physician's learned opinion.

Neither Widmann's letter nor the hospital regulations should be read as a straightforward reflection of local conditions. Physicians commonly adopted a rhetorical stance against nonacademic healers while at the same time collecting the cures and treatments of these same "empirics" in their private notebooks—because they worked, and because they were popular with paying clients.[37] In his bid for control over medical practitioners, Widmann likely exaggerated his disdain for barbers and other "empirics." He certainly was not opposed to surgical work per se, since he proudly called himself *doctor medicinarum*, a doctor of both medicines, that is to say physic and surgery.[38] The hospital regulations represent a desired norm rather than everyday reality, and are in any case worded rather mildly, in terms of advising, helping, and instructing rather than surveillance. There is nothing in the archives to suggest any real antagonism between surgeons and physicians, and as we shall see, both Brunschwig and Gersdorff speak of coexistence, and even collaboration, rather than medical turf wars.

While the rigid rhetoric in Widmann's letter and the hospital ordinances hardly reflect the realities of medical practice, they do map out the stereotypes with which surgeons had to contend. Widmann draws a clear dichotomy between his own rational command of learned theory and the unstructured knowledge of "empirics" who knew their cures by experience alone and had no understanding of the underlying causes. Widmann and the hospital document place surgical practitioners in the latter category; both use the word *unwissenheit* (ignorance) to explain why barbers should be directed by a knowledgeable academic. Even though this strict hierarchy of knowledge and practice was honored more often in the breach than in the observance, surgeons' identity struggles necessarily involved renegotiating these boundaries of legitimate practice.[39]

In the *Cirurgia*, Brunschwig mounts a spirited defense against this image of surgery as ignorant and purely empirical. (As I show in chapter 7, he also defends the value of empirical knowledge itself.) He begins, however, by acknowledging that far too many surgeons were indeed woefully ignorant: "Many people are damaged by the young prospective masters of barbers and surgeons who propose to do what they have not learned, and of which they know neither beginning, middle nor end."[40] Throughout the book, he complains more than ten times about "several foolish surgeons [whose] ignorance impedes healing" and other such inept practitioners.[41] Using the same word as Widmann and the hospital regulations, Brunschwig identifies "ignorance" as a key problem. However, rather than subjecting surgeons' work to learned supervision, Brunschwig proposes to empower them with the requisite knowledge. Both he and Gersdorff set out to provide aspiring masters and journeymen with an education that tackled the prevailing *unwissenheit* at the root. To combat surgery's image as learned medicine's grubby, irrational younger sister, Brunschwig and Gersdorff developed a twofold strategy already present in the writings of the medieval surgeons: they revalued the manual skill and "empirical" expertise of their trade, but also demonstrated that surgery had its own basis in reason.

"Handwork" and Field Surgery

Claims to specialist manual expertise were a powerful way of setting surgery apart from learned physic and other healing work. Both authors open their description of the surgeon's role with the etymology of the word *cirurgicus*, explaining its origin in the Greek *chiros* meaning "hand," since "his work is done with the hand, which the physician does not do."[42] This etymologically informed distinction of physic and surgery was a standard feature of medieval Latin surgeries, but Brunschwig's definition of the "surgeon as a handworker" rings all the more powerful in the vernacular: *handwerk*, literally "handwork," designated skilled manual labor, but also a craft or even the guild representing that craft in the fabric of urban social life.[43] As we saw in chapter 1, these institutions articulated artisanal expertise in terms of handwork and skilled practice. This language echoes in Brunschwig's book, and Gersdorff, too, describes his own expertise as "exercised in practice and with the hand."[44]

In addition to their practical training in the guild system, our surgeons' expertise came largely from their experience on the battlefield. They write of what they did and learned as they participated in military campaigns against Charles the Bold during the Burgundy Wars in the 1470s.[45] The very title of the *Feldtbuoch* makes it out as a "field book," an impression underscored by a dramatic woodcut of a surgeon extracting an arrow while the battle is still raging in the background (figure 2.1). While the images in Brunschwig's book are more sedate, his text includes vivid accounts of similar treatments.

FIGURE 2.1. A surgeon extracting an arrow on the battlefield. Hans von Gersdorff, *Feldtbuoch der Wundtartzney* (1517), fol. 38r. Credit: Herzog August Bibliothek, Wolfenbüttel, A: 153.5 Quod. 2° (5).

> In the Burgundy Wars [in the battle] of Blomunt, it so happened that someone was shot with an arrow, and the iron [head] remained inside while the wooden shaft came out. Many good masters were called to him and proposed to extract the iron, but it would not give way. Then Hans Meier, a barber and surgeon from Strasbourg, was called. Seeing these futile attempts, he took the tongs into his hand and realized that the iron had barbed hooks, so he turned the tongs around such that the direction of the barb was opposite to the way it had entered, and immediately the arrowhead came out. He was greatly praised by all the other masters, knights, and vassals, but they did not know how he had got it out.[46]

Brunschwig highlights Meier's association with Strasbourg and the nature of his extraordinary skill: As soon as his hands grasped the instrument that held the obstinate projectile, he realized what was amiss in the depth of the wound. By the joint touch of hand and instrument, he knew what no one else could see.[47] Accounts of successful cures were common enough in the medieval surgical tradition, where they served as learned case narratives. Brunschwig's spectacular anecdote, on the other hand, celebrates the sensory knowledge and embodied expertise of the artisan-surgeon.

This skilled union of hand and instrument also runs through the images. Both title pages evoke the material culture of injury in a traditional "wound man" image, depicting the variety of weapons that might gash, crush, bruise, or penetrate the body. Inside the books, these are more than matched by woodcuts showing the variety of surgical instruments used to heal such injuries. In the *Cirurgia,* they are impressively arrayed in a decorated cabinet, from small knives and probes to the shears and forceps for extracting bullets or arrows (figure 2.2). Gersdorff's *Feldtbuoch* unpacks the surgeon's instrument chest in detailed, full-page woodcuts showing instruments for treating gunshot wounds, (figure 2.3), devices for the delicate operation of treating depressed skull fractures (figure 2.4), and a series of contraptions for straightening fractures and dislocations, perhaps of Gersdorff's own invention (figure 2.5). Their captions name the implements and describe their purpose, while disembodied human hands demonstrate their skilled use. Like the story about Meier's knowing touch, these images locate the surgeon's expertise in the interplay of specialist tools and skilled hands.

Hands-on field surgery was also an opportunity for surgeons to develop their art beyond what their predecessors had set down in writing. Emergencies sometimes called for improvisation, and this played a part in Brunschwig's epistemology, as we shall see in chapter 7. But there was also a whole new genre of injury. Gunshot wounds were unknown to the

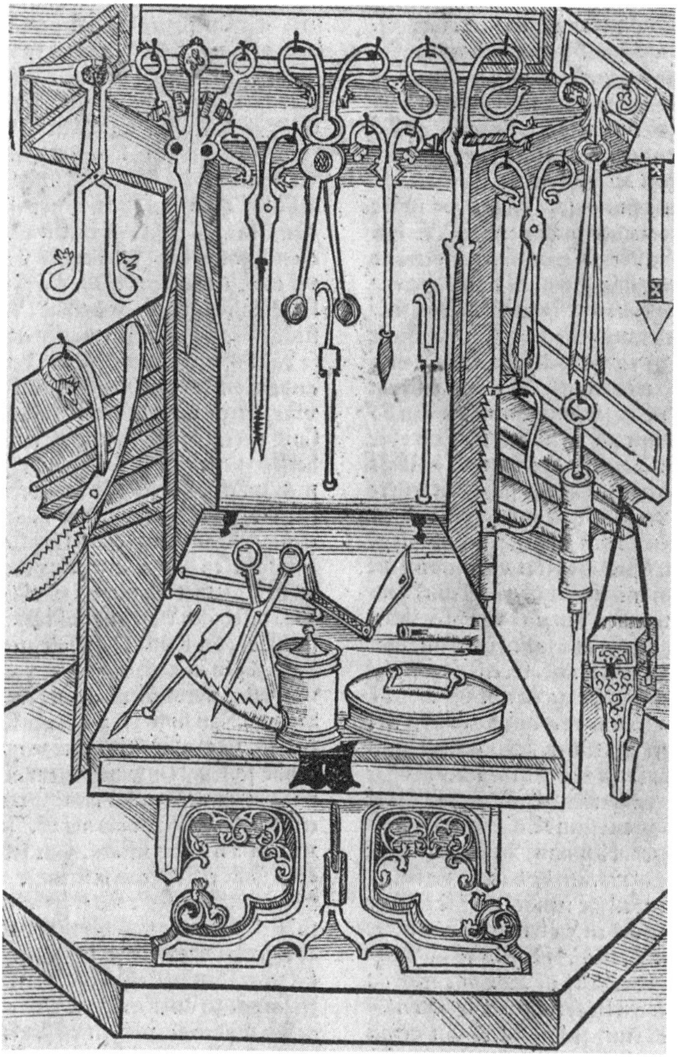

FIGURE 2.2. Cabinet of surgical instruments. Brunschwig, *Cirurgia* (Strasbourg, 1497), fol. 19r. Credit: The Huntington Library, San Marino, California, 85770.

medieval tradition, and challenged late medieval and early modern practitioners to devise new surgical cures and rationales. In fact, Brunschwig and Gersdorff leaned towards different explanations that were common at the time.[48] While Brunschwig writes that the "powder has poisoned the wound" and needs to be removed, Gersdorff insists that one needs to "extinguish the burn" caused by the powder.[49] Their treatments, however,

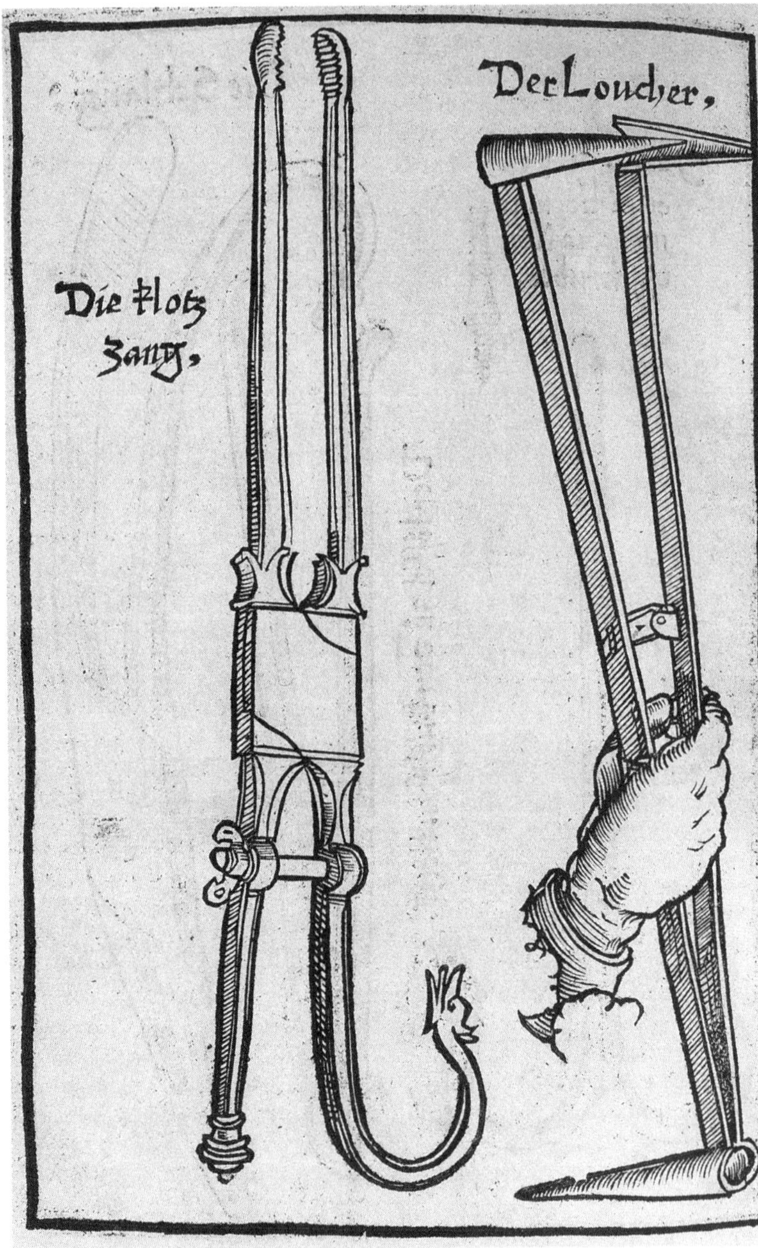

FIGURE 2.3. Bullet extractor and instrument for widening the wound canal. Gersdorff, *Feldtbuoch* (Strasbourg, 1517), fol. 51v. Credit: Herzog August Bibliothek, Wolfenbüttel, A: 153.5 Quod. 2° (5).

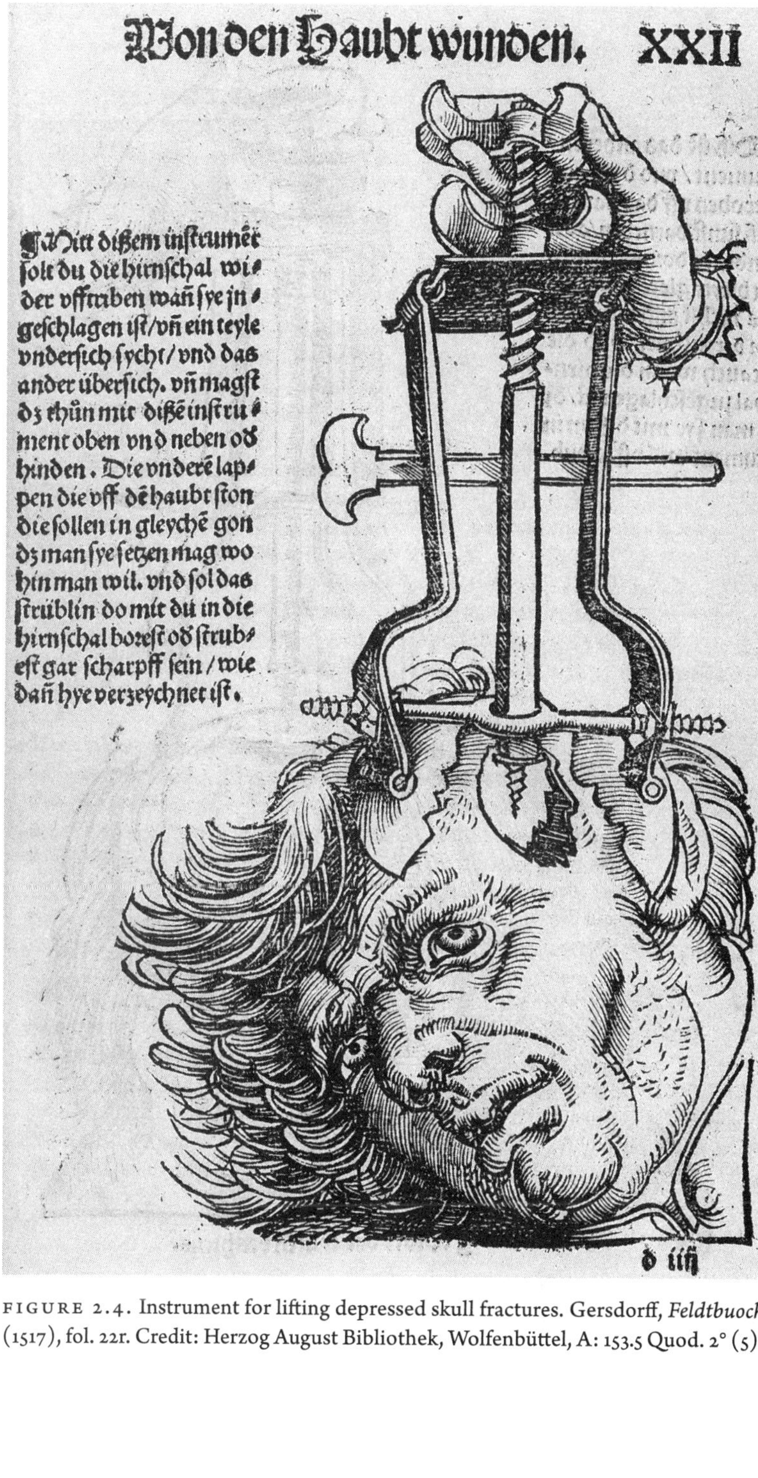

Von den Haubt wunden. XXII

¶ Mitt dißem inſtrumẽt ſolt du die hirnſchal wi- der vffttriben wañ ſye jn- geſchlagen iſt/vñ ein teyle vnderſich ſycht/vnd das ander überſich. vñ magſt dz thůn mit dißẽ inſtru- ment oben vnd neben od' hinden. Die vnderẽ lap- pen die vff dẽ haubt ſton die ſollen in gleychẽ gon dz man ſye ſetzen mag wo hin man wil. vnd ſol das ſtrüblin do mit du in die hirnſchal boreſt od' ſtrub- eſt gar ſcharpff ſein/wie dañ hye verzeychnet iſt.

d iiij

FIGURE 2.4. Instrument for lifting depressed skull fractures. Gersdorff, *Feldtbuoch* (1517), fol. 22r. Credit: Herzog August Bibliothek, Wolfenbüttel, A: 153.5 Quod. 2° (5).

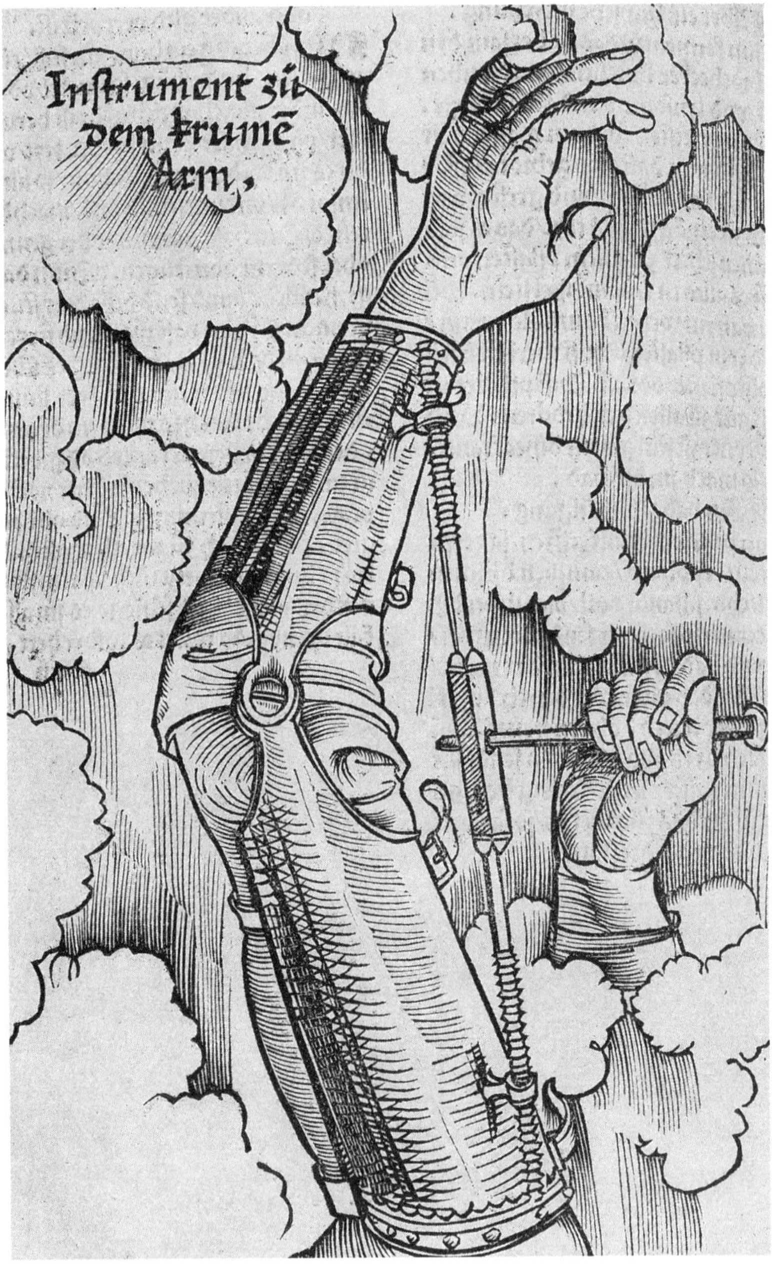

FIGURE 2.5. Adjustable splint to straighten a crooked arm. Gersdorff, *Feldtbuoch* (1517), fol. 34v. Credit: Herzog August Bibliothek, Wolfenbüttel, A: 153.5 Quod. 2° (5).

are not too dissimilar. Both recommend cleaning the wound with oil and keeping it open with a small wedge of lard to promote the formation of pus, and both include images of instruments for extracting shot. Gersdorff writes of his treatment that he has "used it often during the Burgundy Wars with my master called Mulartzt, surgeon to Duke Sigmund of Austria."[50] This master, whom he also calls "master Nicklaus," was probably Klaus von Matrei, who owed his position at the Tyrol court at least in part to a presentation manuscript that included a section on field surgery.[51] Writing of their wartime experiences, both surgeons identify the battlefield as a place where surgeons honed their embodied expertise of hand and instrument, developed new cures, and made a name for themselves.

A Science and a Craft: Reviving the Medieval Ideal of the Learned Surgeon

While handwork was central to surgical expertise and its distinction from physic, a good surgeon also needed theoretical knowledge. The learned writers of the medieval tradition made surgery rational precisely so that it would stand above other crafts that relied on manual skill alone. As Guy de Chauliac put it, if healers were not required to master geometry, astronomy, or any other ancient teachings, then "tanners, carpenters, smiths, and others would give up their crafts in a heartbeat and rush to medicine to become doctors."[52] Brunschwig, too, insists that a surgeon should know "not only medicine, but also other natural arts of philosophy."[53] He should "read about it and study often, for something which is read ten times over while paying diligent attention is still only just comprehended by reason."[54] As well as practicing their handwork, aspiring surgeons ought to exercise their rational minds by studying texts.

Medieval learned surgeons such as Guy de Chauliac provided successful models for presenting surgical knowledge as rational knowledge. For a start, it had to be organized in a way that showed how the surgeon's various techniques and treatments derived from an overarching logical framework. The medieval masters consolidated a typology of general principles of surgical treatment. These different *intentiones*—cleaning the wound, staunching bleeding, appropriate medication, phlebotomy, purging, regimen—were set out as rational categories with their own sections in surgical texts, separate from instructions for treating particular wounds in various parts of the body.[55] Another rational way of structuring surgical works was the distinction of wounds into two categories based on ancient theories of generation—what we would now call embryonic development. According to Galen, some parts of the body, such as flesh and muscles, were generated from blood, while others, such as bones and tendons, arose from seed. The medieval masters translated this learned dichotomy into distinct rationales for wound healing: ruptures in the blood-derived parts—mere flesh wounds—could heal through what they called first intention, knitting together with the original tissue. The so-called spermatic body parts, however, were deemed incapable of such regeneration. Bone fractures, for example, could only heal by second intention, where the rupture was filled in with a different type of bodily matter.[56] Surgical texts such as Guy's *Chirurgia magna* reflect this distinction by dealing with flesh wounds and bone fractures in separate *tractatus*.

The significance of rational ordering principles was not lost on Brunschwig. His table of contents clearly sets out the structure of his book. General principles of treatment are covered in the second *tractat* or treatise, whereas specific techniques pertaining to particular wounds or body parts are dealt with in later treatises. These are clearly grouped according to first and second intention, with flesh wounds in the third treatise, followed by bone fractures and dislocations in treatises five and six. Gersdorff's somewhat more jumbled instructions show no clear separation of bony and fleshy parts, but his table of contents does flag up sections dealing with general learned principles such as "good rules and teachings of phlebotomy in general," or "the workings and nature of the twelve heavenly signs."[57] For both authors, the medieval tradition provided a rational scaffold on which to display and unfold their expertise.

Part of that expertise was a mastery of the more theoretical aspects of learned humoral pathology. As the medieval masters had pointed out, no two cases were ever the same, and surgeons needed to assess the attendant *contingentia* for each treatment.[58] Brunschwig insists that they should know "the power, virtue and property of each healing thing, since that which heals soon in one country heals slowly in another, or it heals in one limb and not another, in one complexion but not another."[59] As we shall see in more detail in chapter 7, Brunschwig had little time for surgeons who acted on a purely empirical basis, smearing the same plaster on every wound. Only through a rational understanding of the body could a surgeon hope to navigate the contingencies of location, body parts, and the patient's "complexion" or humoral balance.

Crafting the Rational Body: Anatomy and Images

In addition to a body's internal principles, surgeons also needed to understand its structure, and they came to play a key part in the revival of anatomical knowledge. The medieval masters had readily co-opted anatomy as the rational organizing principle for surgical knowledge, and early Renaissance writers followed suit,[60] even though some would later criticize anatomical dissection as corrupting surgeons' skill and morals.[61] Brunschwig, too, insists that a surgeon ought to learn anatomy, "so that he knows where to cut or cauterize a person" without causing "lethal damage."[62] It is not until the 1513 edition of the *Cirurgia,* however, that the subject is covered in its own fifteen-page chapter with a labeled woodcut of a skeleton. Even then, Brunschwig seems to doubt the image's ability to convey knowledge. Anatomy, he writes, had to be "seen and experienced," and while text and image could convey the general idea, ultimately "the skin, flesh, and veins cannot be revealed by way of the figure; they are seen by way of inspection of dead bodies through cutting, boiling, or corroding with quicklime."[63] Like surgery as a whole, anatomy was a hybrid knowledge that could not be learned from books alone; it had to be read within the body itself, exposed and experienced layer by layer through skilled handwork.

Published four years after Brunschwig added his anatomical treatise to the *Cirurgia,* Gersdorff's *Feldtbuoch* shows a more confident use of anatomical images. The whole-page woodcut of a dissected body illustrates the visceral organs discussed in Gersdorff's chapter on anatomy, and at the same time functions as a "phlebotomy man," indicating the most important sites for letting blood, as described in the subsequent chapter (figure 2.6). At the same time as the *Feldtbuoch,* the printer Johann Schott also issued two broadsheet images of a visceral anatomy (figure 2.7) and a skeleton (figure 2.8), which point to the *Feldtbuoch* for further reading and were sometimes bound with it. Gersdorff writes that both visceral anatomies (the broadsheet and the "phlebotomy man") were based on the dissection of a hanged criminal that same year, "in the presence of several learned and distinguished physicians, doctors, surgeons, and barbers, . . . artfully expounded by the experienced and highly learned doctor of medicine, Wendelin Hock of Brackenau, and at the same time faithfully drawn with all its shape, color, and true indication, as you find it in the following figure."[64] By emphasizing the presence of learned and artisanal medical experts, the credentials of the physician in charge, and the immediacy of the artistic rendering, Gersdorff bridges the distance between the dissection and the printed image. The phlebotomy man is confidently titled *Augenschinliche Anatomy*—an anatomy "evident to the eye" of the reader, even one who had never experienced a dissection firsthand.[65]

Brunschwig and Gersdorff's concerns about the relationship between manual experience, medical learning, and visual representation cut to the heart of anatomy's historical role as a site for encounters and exchange

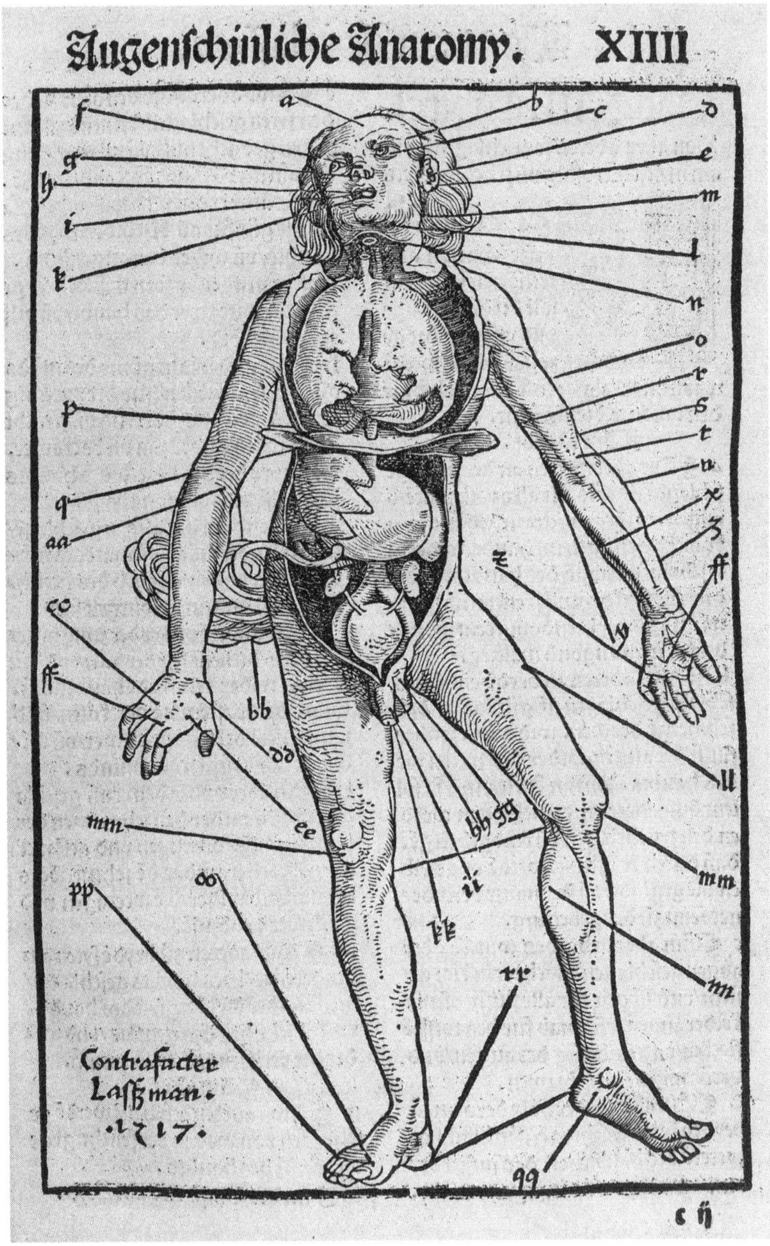

FIGURE 2.6. Phlebotomy man with visceral anatomy. Gersdorff, *Feldtbuoch* (1517), fol. 14r. Credit: Herzog August Bibliothek, Wolfenbüttel, A: 153.5 Quod. 2° (5).

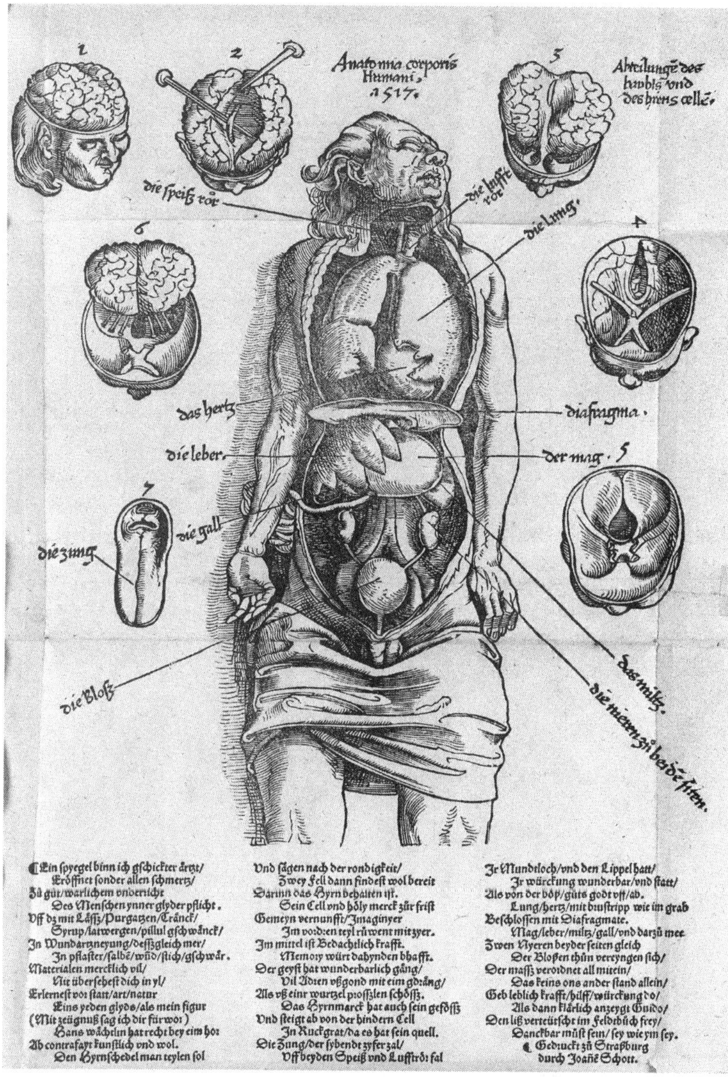

FIGURE 2.7. Printed broadsheet showing visceral anatomy (Schott, 1517). Credit: Wikimedia Commons.

between surgeons and physicians. To most physicians before the late Middle Ages, anatomy was first and foremost a work of scholarship, tracing words and concepts in ancient texts with philological precision. Knowing the body with knife and hand was the purview of barbers and other body workers, but also experts of the dead body such as the nuns and priests who embalmed the bodies of saints, looked for bodily signs of holiness, or cut unborn babies from dead mothers' wombs for an emergency

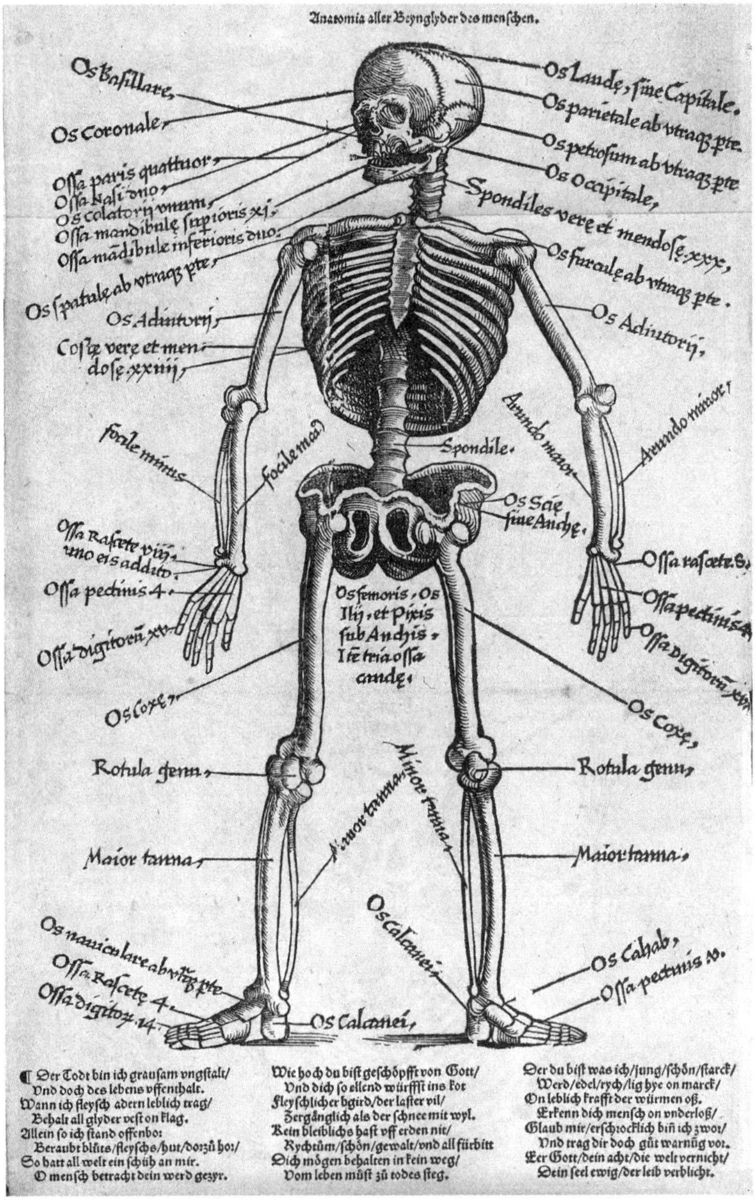

FIGURE 2.8. Printed broadsheet showing skeletal anatomy (Schott, 1517). Credit: Wikimedia Commons.

baptism.[66] When Italian universities began using dissection as a teaching tool, text-based learning and the skilled manipulation of real bodies were brought together in the same room, but they were not allowed to mingle: the professor lectured on the ancient writings of Galen while the *prosector*, usually a surgeon, dissected the relevant structures.[67] There were clear hierarchies of knowledge and practice: dissection was gory manual labor, and dissected bodies served not to gain new insights, but to illustrate ancient learning.

This practice was slow to spread north of the Alps, and the Strasbourg dissection of a hanged criminal in 1517 was clearly an extraordinary event, documented in printed images and Gersdorff's book. His description of the event reflects the traditional hierarchy: the physician Wendelin Hock "expounded" on the anatomical structures while the actual dissecting was likely left to one or more of the attending surgeons. Brunschwig and Gersdorff were well aware that surgeons traditionally stood on one side of the dissection table, but they staunchly disputed that theirs was in any sense the wrong side. For Brunschwig, the skilled handwork of dissecting, etching, and boiling was ultimately the only way to learn anatomy. For Gersdorff, the anatomical images achieved visual self-evidence precisely because they drew together the physician's learned expertise, the draftsman's skill with shapes and colors, and the surgeon's and barber's with knives and probes.

In their treatment of anatomical images, Brunschwig and Gersdorff raised concerns that became central to the visual practices of humanist physicians several decades later, notably the botanical works of Leonhard Fuchs and Conrad Gessner, and the anatomical woodcuts of Andreas Vesalius, illuminatingly studied by Sachiko Kusukawa.[68] These scholars pointed out that *autopsia* (seeing for oneself) was a key element of ancient science, and developed learned rationales for their programs of visualization. Their epistemic arguments about images drawn *contra factum* (from a real observed object) or *ad vivum* (from life) echo Brunschwig's insistence on firsthand experience and the visual claims of Gersdorff's "evident" anatomy and "counterfeit" phlebotomy man. Fuchs even included in his book the portraits of the artisans who produced his woodcuts, drawing attention to the skilled process of image-making in the same way Gersdorff highlighted the ability of the draftsman present at the dissection.

The gradual drawing together of manual skill and learned knowledge in the field of anatomy is nowhere illustrated more clearly than in the works of the Flemish physician Andreas Vesalius. Working in the renowned medical faculties of Paris, Padua, and Bologna, he deplored that the ancient ideal of a unified medicine had been fragmented, mainly because physicians refused to perform surgery and anatomical dissection. Medical students were better off learning from butchers, studying the bones in the Paris ossuaries, or performing dissections themselves. Since this was not always possible, Vesalius presented the detailed woodcut images in his *De humani corporis fabrica* (1543) as reliable tools for learning anatomy, constructed from many instances of personal *autopsia*. As Kusukawa has pointed out, the author portrait in the *Fabrica* shows Vesalius dissecting a human hand and forearm on a workbench cluttered with surgical tools, making playful allusion to the anatomist's own skilled hand. Like Brunschwig and Gersdorff before him, Vesalius located his skill in the interplay of hands and instruments.[69]

As the example of Vesalius shows, Brunschwig and Gersdorff anticipated the central issues of Renaissance anatomy: the relationship between firsthand dissection, printed images, and professional identities uniting skill and learning. While Vesalius's call to restore the ancient unity of surgery and physic was taken up by other medical humanists, it was only very gradually adopted in practice. Even in Italy, it is only towards the end of the sixteenth century that anatomical dissection became anything like the site of exchange and mobility between surgery and physic that Vesalius envisioned.[70] Several decades before Vesalius, Brunschwig and Gersdorff already point towards the same ideal of unity and collaboration between surgeons and physicians. But since they were writing from the ostensibly inferior position of the surgeon rather than the academic elite, they had to renegotiate the traditional disciplinary boundaries with care.

Negotiating the Place of Surgery at the Bedside and in the City

Brunschwig and Gersdorff knew that providing surgeons with learned knowledge of complexions, phlebotomy, and anatomy was a controversial move that was likely to rattle (however quietly) the partitions of the medical hierarchy. Brunschwig's tone is deferential towards "learned and experienced people" and asks them to "accept this [book] from me in good will as I desire to continue learning from you."[71] He does not question the primacy of physicians in certain cases; if a wound is complicated by a fever, the surgeon should seek learned advice "since it belongs more to the physician than the surgeon."[72]

In other cases, the emphasis on artisanal handwork provides an opportunity to renegotiate the boundaries of what a surgeon ought to do, or what a surgical author ought to write about. Poisonous bites, for example, fall within the physician's domain insofar as the poison acts internally and needs to be purged by a regimen. Nonetheless, Brunschwig feels that he "ought to write about it, and all outward things which can be felt and learned with the hand" since this falls to the surgeon as a handworker.[73] He suggests cooperation: a physician should be consulted "to stop the poison and protect the heart and strength of the person," but the treatment of the external wound should be left to the surgeon.[74] Given this overlap in their practice, Brunschwig concludes that "no one can be a good surgeon unless he is somewhat learned in physic. Similarly, the physician should also be learned in surgery, since one is often faced with that which is appropriate for the other to do."[75] A surgeon might have to act quickly even in cases which fall under the remit of physic, since "before you have sent for a physician the person may be half dead."[76] Brunschwig often cites the difficulty of reaching a physician, or indeed any practitioner, especially in rural areas.[77] While this may be rhetoric calculated to advertise and legitimate his books, it was not unknown for rural surgeons to obtain special dispensation to practice physic in the absence of a physician.[78] In any case, Brunschwig clearly intends to provide learned knowledge that could obviate the need for a physician in some cases, and subtly extends the remit of surgical practice into learned territory.

Gersdorff, too, defers to the physician's domain of expertise in some cases while pushing the boundary in others. In his chapter on astrology, he makes sure to state that he is only covering the subject "as far as regards [the surgeon], for to search further for the aspects of the planets and the administration of other medicines belongs to the astrologers and physicians."[79] Elsewhere, Gersdorff acknowledges that he received the recipe for a wound plaster from a physician, but makes a point of stating that "he and I have used it together," suggesting close collaboration with a learned colleague. In a poignant (although probably coincidental) twist, the physician in question was none other than "Doctor Widmann from Tübingen," the very man wrote to the Strasbourg council to demand stricter control of surgical practice.[80]

Gersdorff is even less cautious of traditional boundaries when it comes to his own specialist expertise. He worked at St. Anthony's Hospital in Strasbourg, a small institution dedicated to treating the severe inflammations occasioned by a skin disease known as St. Anthony's fire. Clearly speaking as an expert on the matter, Gersdorff describes how to recognize the symptoms—which, he explains, were difficult to tell apart from those of other skin conditions—and provides recipes for a number of plasters and ointments. If these did not quench the "fire," the limb often had to be amputated: Gersdorff himself claims to have "cut off one or two hundred limbs at St. Anthony's in Strasbourg."[81] At the conclusion of his detailed instructions for treatment and recovery, he notes that ultimately "every barber can recognize for himself" what needs to be done and "find good advice in the aptitude of his understanding."[82] Confident in surgeons' judgment, he makes no mention of consulting a physician. His confidence was not misplaced: as we shall see in the following chapter, the city's medical examiners were keen to recruit Gersdorff as an official inspector to judge cases of leprosy, another skin disease with even greater implications for municipal public health.

Conclusion

As craftsmen seeking to reaffirm the place of surgery in the landscape of medical practice, Brunschwig and Gersdorff had to contend with the idea of a medical hierarchy that, although far from universally enforced in practice, considered surgeons as second-rate manual laborers doing the physician's bidding. As we have seen, part of their strategy was to codify their manual expertise in print and to emphasize the true value of handwork, artisanal training, and war experience. Building on the medieval model of the rational surgeon, they also insisted on the theoretical foundations of surgery, a body of knowledge that overlapped with learned physic.

Their books are remarkable early efforts in drawing together manual skill and medical learning in printed text and image. Particularly in the field of anatomy, Brunschwig and Gersdorff's grappling with the relationship between firsthand experience, theoretical knowledge of structures, and the printed image hit upon concerns that would shape scholarly conversations for decades to come, culminating in Vesalius's call for a renewed unity of medical knowledge.

These claims to learning as well as skill entail careful renegotiation of the traditional boundaries between surgeons and physicians. Brunschwig gently softens that boundary to create a relatively broad area of overlap, based on the surgeon's claim to anything involving handwork, and on pragmatic arguments about physicians' availability in an emergency or in rural areas. Two decades later, Gersdorff makes more decisive claims to independent diagnosis and treatment, in particular where his specialist expertise is concerned. Decades before learned anatomists and humanists began to elevate the image of surgery from above, artisan-writers had prepared the ground by pushing it from below. While these efforts were essential to craft a new identity for surgery with respect to learned physic and the landscape of medical practice, it was just as important to situate surgical work with respect to other crafts and the social landscape of the early modern city, as we shall see in the following chapter.

3

Body, Honor, Health, and Handwork

SURGEONS IN THE CITY OF ARTISANS

In the previous chapter, we have seen how Brunschwig and his local colleague Gersdorff crafted a new image for surgery. Writing from below—as artisans rather than learned physicians—they emphasized surgeons' manual skill and rational learning to push for their rightful place in the medical landscape. But this was only half the battle. As this chapter shows, the Strasbourg surgeons also negotiated surgery's identity within the social life of the city. While authors and publishers usually intended printed texts to have a wide geographic and cultural reach, it is easy to overestimate the universality of print. As Hannah Murphy has shown in her study of municipal physicians, some printed medical books were designed for distinctly local impact.[1] By reading Brunschwig's and Gersdorff's surgical books alongside records of urban life from the Strasbourg archives, I show how they speak to the concerns of the artisan population in a free imperial city, in particular their sensibilities surrounding the body as a source of labor, identity, and honor.

These kinds of tensions and sensibilities were important determinants of surgeons' status and identity across Europe, although the outcomes vary widely from place to place. For a start, practitioners of surgery are rarely visible in the historical record as a clearly delineated group with a strong medical identity. They usually belonged to a broader social milieu of "artisans of the body," to use Sandra Cavallo's useful phrase, which might include tailors, upholsterers, and bathhouse attendants as well as barbers and surgeons, blurring the boundaries between medical treatment and other crafts concerned with well-being, appearance, and polite urban sociability.[2]

In German cities, too, we can usefully locate surgical practice within larger cultures of artisanal "body work." A *bader* (bather or bath master) looked after his clients' cleanliness and comfort, but might add healing and soothing herbs to a bath, give massages, or perform cupping to improve health. A *scherer* (barber or barber-surgeon) trimmed beards and hair as well as letting blood, stitching up wounds, or setting bones. The term *wundarzt,* literally "wound healer" or "wound doctor," has more distinctly medical connotations, and distinguishes a surgeon from a *leibarzt* (body doctor), i.e., a physician. A *wundarzt* was sometimes hired as *stadtarzt* (town doctor), alongside or instead of a municipal physician.[3] While some surgical careers progressed in clear steps from *bader* to *scherer* and then *wundarzt,* with an examination at each step,[4] in other contexts these labels are interchangeable.[5] Where formal distinctions did exist, the boundaries of their practice and competence were always subject to negotiation and, frequently, a source of conflict, especially between bathers and barbers, as we shall see.[6]

If surgical practitioners (and their historians) struggled to negotiate a clear occupational identity, their status in society could be even more ambiguous. Across the Holy Roman Empire, surgeons managed to obtain enviable positions at court, and in some cities, they held important medical or administrative posts and were respected guild members.[7] In many other places, surgeons and other body workers occupied some of the lowest rungs of the social ladder.[8] Brunschwig and Gersdorff, I argue, were keenly aware of surgery's tenuous position in the social fabric of the city and took care to position their expertise as an asset to the urban community. I begin by exploring that community and the place surgical practitioners had in it. I then show how Brunschwig and Gersdorff emphasize surgeons' role in the governance of the city, and as guardians of artisans' bodies, overseeing not only their health, but also their labor and their honor.

Body Work in the City of Artisans

Early modern practitioners of surgery were often hairdressers and barbers as well as healers, blending into a large and diffuse milieu of "body work." Strasbourg was no exception. The barbers were the closest thing to a recognized "surgical" trade, since they regularly performed phlebotomy and other medical services. As of the fourteenth century, they shared a political corporation and a council seat with the bathers. Perhaps due to this close proximity, the two groups could never quite agree where one trade ended and the other began. City council records from 1435 to 1513 document recurring squabbles that always played out the same way. The barbers complained that "the bathers . . . have employed journeymen barbers in their homes and bathhouses and shave many people there, and thus interfere with the barbers' craft."[9] The bathers countered that the barbers regularly washed their customers, "which after all is half the [bathers'] craft."[10] Furthermore, the bathers went on, it was common practice for bathhouses to employ journeymen barbers. This, the barbers countered, was only the case in "those towns which are not organized in guilds."[11] Whatever mayhem reigned outside of guild structures, the barbers felt that in any orderly city a clear distinction from the bathers' trade was a matter of artisanal propriety. The exasperated council ruled that the bathers should not shave and the barbers should not offer bathing or cupping—except for "any sick our wounded people they are currently advising."[12] The repeated rulings do not seem to have done much to settle this boundary dispute once and for all. It is significant, however, that the council grants the barbers the better share of medical body work.

This was symptomatic of the increasing divergence of barbers and bathers during the latter half of the fifteenth century. By 1470, the bathers could no longer afford to pay the rent and heating fuel for their *trinkstube* (each trade maintained such a communal drinking hall), which was consequently dissolved by the city council, and the bathers had to find refuge in the barbers' hall, losing their guild seal, banner, and identity. This decline was compounded as citizens increasingly preferred to have heated baths installed in their homes rather than using the public houses. The barbers fared better and their numbers increased, mopping up some of the bathers' lost trade and offering a wider range of services, from washing and trimming hair to various kinds of medical treatment.[13]

On the whole, however, neither the barbers' nor the bathers' social standing was by any means prestigious. As we have seen, Strasbourg citizens—artisans, merchants, and even wealthy patricians—were organized in guilds that formed a clear hierarchy, based in part on wealth, but also a deeply held sense of honor associated with different kinds of work.[14] A guild's status was marked by its number of representatives in council and in the marching order of annual processions through the city.[15] From wealthy goldsmiths to smelly tanners, citizens' civic, moral, and occupational identities were inextricably linked in this artisanal economy of honor. Surgical practitioners in Brunschwig's day started from a weak position. In the annual *Kreuzgang* procession, they marched furthest from the crucifix. As a military contingent, they did not pull much weight. In the Burgundy Wars, as we have seen, Brunschwig and his fellow barbers joined the fray armed with battle axes—a poor man's weapon at a time when other guilds could increasingly afford portable firearms.[16] Their joint guild never had more than one seat in the council. When the structure of the council was overhauled in 1482, their seat was one of the casualties, and it was resolved that their guild should be disbanded altogether. There were plans to divide its members equally among the other guilds, but it is not clear that these plans were carried out.[17] More likely, the bathers were subsumed wholesale in the smiths' guild, and the barbers with the grain merchants—this is in any case where they resurface in the archival record during the seventeenth century.[18] What is clear is that the corporate structures of late fifteenth-century Strasbourg did not provide surgical practitioners with the strong professional identity that Brunschwig and Gersdorff aimed to forge in their writings. The Strasbourg archives yield no explicit explanation of surgical practitioners' lowly status, but it was likely due to their association with other kinds of body work that were deemed dishonorable.

The sense of honor that was the chief social currency of urban communities was not an abstract idea, but a physical phenomenon. Honor was a question of the body: it could be tainted by a disfiguring wound or the mere touch of a dishonorable person.[19] Perhaps paradoxically, the very trades that were skilled at taking care of the honorable body were often themselves held in low esteem. In some cities, bath house keepers were not even allowed to bear arms, to join guilds, or to hold public offices.[20] The bathhouse was a place where people could be tempted to discard, along with their clothes, the niceties of civic order and sexual propriety. According to—often exaggerated—cultural stereotypes, it was a place where prostitution, theft, and gambling were rampant. Those who worked there were seen as complicit in this moral gray area, especially if they allowed it to spill into the streets: working in a hot and steamy environment, bathhouse attendants were usually scantily clad, and sometimes did not bother to cover up before leaving the premises, to the outrage of decent citizens.[21] The Strasbourg bathers were not technically considered dishonorable and barred from official life as was the case elsewhere. They could be full citizens and were represented, however minimally, in council. Nevertheless, they were not held in high regard and were de facto excluded from certain functions—the watchman of the cathedral tower, for example, could not be a bather.[22] This could reflect poorly on barbers and surgeons who were often part of the same guild as the bathers, and did similar work—so much so that there were frequent squabbles about who should do what, as we have seen in the Strasbourg brotherhood.

Surgeons' social standing could also suffer from a perceived association with another, more profoundly reviled occupation: that of the executioner. His lowly duties included torturing and executing criminal bodies, supervising prostitutes, emptying outhouses, and driving lepers out of town. As Kathy Stuart has argued, artisans in particular feared the physical taint of the executioner's dishonor if they so much as touched him, and these anxieties often extended to surgeons and others whose work with human bodies, dead or alive, was just a little too close to the executioner's vile trade.[23] The Strasbourg archives suggest that precisely such a disreputable aura surrounded the local hangman: In an ordinance of 1500, he was ordered to stand in a separate place in church, not to touch any food in the market, and not to approach honorable people in public—in short, to avoid any contact that might pollute the embodied honor of decent citizens.[24] It is not clear to what extent this penumbra of dishonor affected other local body workers, but it may be one reason for the bathers' and barbers' weak position in the guild hierarchy.

Overall, the barbers' position was unenviable, but not desperate. Their trade was not especially lucrative, but it received some recognition from the council, who ruled in their favor in their dispute with the bathers on the grounds that they provided medical services. They were more prosperous than the bathers and held in slightly better regard. They ranked low in the hierarchy of crafts, perhaps due to an aura of dishonor that attached to all body work, but they were not officially dishonorable and had full civic rights. From Brunschwig's point of view, the occupational and social identity of surgical practitioners must have looked paltry, but redeemable.

Indeed, across the Holy Roman Empire, body work could also open up avenues of surprising social mobility. While surgeons, barbers, and bathers remained at the fringes of honorable society in many cities across the empire, in others they established successful practices and held important official posts.[25] Gersdorff's master, Klaus von Matrei, parlayed his surgical expertise into a lucrative court position. The career of his contemporary, Hans Seyff, documents an even more remarkable trajectory. Born into serfdom, he passed his examinations as a bather and then a surgeon with flying colors and was freed to become personal surgeon to the Duke of Bavaria. He traveled widely and held a number of official positions with salaries higher than those of many learned physicians.[26] At the height of his career, he was summoned to participate in the most high-profile operation of his day. When the leg of Emperor Friedrich III became so gangrenous that it could not be saved, Seyff was the practitioner chosen to amputate one of the noblest limbs in Europe.[27] Later in the sixteenth century and south of the Alps, where surgery generally had a firmer footing, the Bolognese surgeon Giovanni Battista Cortesi managed an astonishing rise from a mere *stufaiolo* or bath attendant—the lowest kind of body work on the local scene—to a professor of medicine.[28]

This kind of surgical social mobility was undergirded by writing as well as practical proficiency: both Matrei and Seyff authored noted manuscripts that helped secure their advancement. These were not public documents but private presentation copies or practical instructions for a select circle of sons and professional successors. Brunschwig and later Gersdorff were the first artisan-surgeons to realize the potential of the print medium for reaching much wider audiences and crafting new surgical identities at a larger scale. As we have seen in the previous chapter, they did so in large part by inscribing their books into the medieval tradition and its ideal of surgery as a rational science as well as a skilled art. However, as I aim to show here, they never lost sight of the more immediate concerns of practicing surgery in a local community of artisans whose identities were shaped by bodily labor and embodied honor.

Surgeons of the Collective Body: Expert Testimony and City Governance

To establish surgeons' place in the city, Brunschwig and Gersdorff emphasized their civic status and function. As we have seen in the previous chapter, Gersdorff based his skill in treating St. Anthony's fire and other skin diseases on his long years of service in one of the city's public hospitals. He also wrote an entire chapter on leprosy, another skin disease that was an even greater public health issue. In Strasbourg as in many other German cities, lepers were entitled to charitable donations or medical care. This made the annual *Siechenschau,* the official "inspection of the sick," a significant event in the city's calendar. For the inspectors, it was a welcome opportunity to display their medical authority without having to cure anyone—a privilege that physicians usually claimed for themselves.[29] Johann Widmann, the physician who wrote to the Strasbourg council to suggest tighter medical control (see chapter 2), specifically questions "whether barbers should inspect leprosy and be allowed to pronounce judgment on a par with doctors," although he grudgingly admits that they "have some knowledge of it."[30] For Gersdorff, however, leprosy was surgeons' business. He confidently explains his approach to diagnosis, visually underscored by a full-page woodcut of a surgeon—perhaps Gersdorff himself—inspecting a patient (figure 3.1).

The Strasbourg archives suggest that the local authorities agreed with Gersdorff's claim to expertise. According to regulations that can be dated back at least to the mid-fifteenth century, the inspection of lepers fell to a team of two physicians and two surgeons.[31] In an undated document, the appointed inspectors, whose number had dropped to three, reported that they struggled to cope with the large number of cases, and requested an additional expert. After all, they claimed, "it is common in other cities and lands [to have] a doctor and two or three surgeons," just as the original

FIGURE 3.1. A surgeon, perhaps Gersdorff, inspecting a patient for signs of leprosy. Gersdorff, *Feldtbuoch* (1517), fol. 77v. Credit: Herzog August Bibliothek, Wolfenbüttel, A: 153.5 Quod. 2° (5).

Strasbourg ordinance stated. "While no doctor desires to do this," they pointed out, "it is not necessary to have a doctor, but rather a surgeon to whom such a disease is obvious."[32] In fact, they had a specific surgeon in mind for the job—none other than "Hans von Gersdorff, surgeon behind the cathedral, who is well-versed and experienced in this matter." While the surgeon's expertise lay in the "perception of external more than internal lesions," the inspectors (including at least one physician) were confident in Gersdorff's ability to read these external signs of a disease that affected the whole body.[33]

Brunschwig speaks to a different civic function of surgical practitioners, as expert witnesses in legal proceedings. Archival documents show that barbers were regularly appointed as *wunden beseher* (wound inspectors) to provide official assessments of the gravity of an injury that had become the object of a court case. City law classified injuries into distinct categories: small or large *wunden* (wounds) and small or large *blutrunsen*, a term referring to lacerated and bleeding skin (I shall use "laceration" as the closest English noun).[34] In 1528, the council made plans to improve the existing system. Up to that point, the five inspector positions had simply rotated among the city's barbers, including some who "had little understanding of such things."[35] In the future, the council resolved, they would handpick "those whom they deem most skilled."[36] This mattered because, as the council acknowledge, the classification of injuries could "not be properly set down" a priori—wounds could be "much more damaging in one place [on the body] than in another," and the assessment in each case was trusted to the "insight of the wound inspectors."[37] Their pronouncements were legally binding and consequential: a large wound carried a punishment of up to five years' banishment from the city and a hefty fine.

Brunschwig does not state explicitly that he acted as a wound inspector for the city, but it seems clear that he must have done so. For one thing, if the office rotated among the local barbers prior to 1528, it is difficult to imagine that he repeatedly missed his turn. For another, he includes an entire chapter dedicated to the distinction of different injuries, using precisely the categories of large and small wounds and lacerations spelled out in the council document. He also discusses the additional categories of large and small "loss of limb," paying attention to the significance of different body parts for different kinds of labor.[38] In his assessment of an injury, Brunschwig takes into account not only size and location of the wound, but also to what extent it impairs an artisan's ability to work. Only a slight difference should be made, he writes, between someone losing their entire hand and someone merely losing their fingers "which are, after all, quite necessary to his *hantierung*."[39] A plaintiff's *hantierung* (which could be translated as "handling," but really encompasses anything done with the hand) further depended on their "craft or trade":[40] "Blacksmiths, cartwrights, carpenters, and those who work with heavy hammers or tools, and also those who work with the entire arm, require free movement of the axle and the uppermost joint [i.e., shoulder]; if this is injured, it is a great paralysis to them and a loss of a limb."[41] As far as these craftsmen's livelihood was concerned, Brunschwig suggests, a bust shoulder amounted to robbing them of the entire arm, and should be reckoned accordingly in court. To a tailor, furrier, or cobbler, on the other hand, the loss of a working elbow would be the most devastating injury, and should be counted as a "loss of limb." While a mangled left hand was somewhat less debilitating for many trades than an injury to the right hand, to a weaver or embroiderer it was just as bad. In his capacity as a legal witness, Brunschwig translated physical injuries into real-life implications for *hantierung* and livelihood, displaying a fine-grained understanding of his fellow citizens' working bodies.

The severity of an injury also depended on the quality of the subsequent surgical treatment. While inspecting wounds in the throat or the shoulder joint, he observed that "if the surgeon is good, it is a small wound, but if he is inexperienced, it is a large wound."[42] The inspector's attention to the shoulders, arms, elbows, and fingers of working artisans translated to the surgeon's healing practice. In a chapter of instructions for proper bandaging, Brunschwig reiterates that particular care should be taken with the arm, since "this limb is much used and necessary to man, especially for craftsmen such as blacksmiths, cartwrights, carpenters and such like who work with the whole arm."[43] As we have seen, he would go on to claim in his *Small Book of Distillation* that "healing is nobler than all other arts: if a master falls sick, he cannot practice his art, therefore he desires medicine which will allow him to keep his art and livelihood."[44] As expert witnesses or as healers, surgeons played a key part in keeping the collective and individual bodies of artisans in good working order.

Surgeons of the Honorable Body

Labor and livelihood were not the only things at stake when it came to assessing and restoring bodily integrity. Brunschwig writes that surgeon-inspectors must bear in mind that an artisan's body was not only the instrument of his *hantierung* and the seat of his professional identity, but also the site and public face of his honor. If a man had his nose injured, this constituted a small wound; for a woman, however, it was "a wound, and also an eternal disgrace, since it may be suspected that it was cut off due to adultery."[45] Piercing someone's cheek with a triangular blade literally added insult to injury (or, in Brunschwig's words, "disgrace" to "wound") because the victim "may be suspected of having been branded on the cheek"—a common punishment for thieves.[46] Some injuries cut deeper than flesh and bone, marking the body in ways that undeservedly signaled crime, punishment, or dishonor.

Brunschwig's discussion of these types of injuries speaks to a nexus of appearance, honor, and urban social life that runs like a thread through premodern surgical literature—a thread that historians are only beginning to unravel.[47] The medieval rational surgeons knew from ancient physiognomy that a person's appearance reflected not only external beauty, but also inner moral virtues such as knowledge and honesty. They discussed where to draw the line between proper restoration of a dignified, natural appearance and artificial cosmetic enhancement, especially as urbanization increasingly drew people together at close quarters, giving rise to new kinds of polite society and closer mutual scrutiny for external signs of the moral qualities that made a good citizen.[48] Strasbourg was not the only city where the moral implications of physical injuries were taken seriously in the court of law. In fifteenth-century Bologna, anyone who drew blood from another citizen's face or damaged their eyes faced increased penalties. In fourteenth-century Lucca, citizens could obtain an official certificate stating that their mangled nose was due to an illness rather than a dishonoring punishment.[49] By the later sixteenth century, as Paolo Savoia has argued, Italian surgeons gained fame and success by operating within a "culture of the face" driven by urban honor and male violence. Losing one's nose in a duel was so devastating a blow that men were willing to undergo as long and painful a treatment as Gaspare Tagliacozzi's famous reconstructive nose surgery to restore their physical and moral integrity.[50]

Brunschwig locates the surgeon's role in the urban cultures of honor and bodily integrity of his own local context. Writing decades before Tagliacozzi, in an imperial city dominated by the concerns and institutions of craftspeople rather than a noble elite, he describes the surgeon as a crucial asset to the life of the city and the individual artisan, be it as a healer or an official inspector. His was not merely an expertise of the body's surface, as the traditional medical hierarchy would have it, but one that concerned a citizen's whole and many-layered person—as Brunschwig puts it, "his body, his honor, his health, and his *hantierung*."[51] This was a powerful counterimage to common views of surgery as an unclean, dishonorable sort of body work. Instead, Brunschwig portrays surgeons as guardians of civic order and individual labor, identity, and honor.

"The Honor of All Surgeons"

Such a positive image of surgery could only prevail if surgeons as a group lived up to it. Speaking again from his experience as an inspector, Brunschwig writes, "Oh how often have I seen an inexperienced surgeon bandage someone whom I then had to declare 'wounded.' If a different, more experienced surgeon had bandaged him in the first place, it would have been a 'laceration' [*bluotrunß*], and not a large one at that."[52] Brunschwig clearly worries less about the legal implications of the botched bandage than how this reflects on surgery in general. He implores "each and every surgeon" in such a case to "think on your honor and the honor of all surgeons" and leave the sick person "in the care of someone who knows the art."[53] In addition to the well-being of the patient, the individual practitioner's duty was to the collective honor of surgery.

Strasbourg's body workers took their collective reputation seriously, as becomes clear in the archival documentation of a dispute taking place in 1517, the same year Gersdorff's book was published. In an attempt to curb violent brawling, the council decreed that "all barbers and their journeymen shall henceforth be bound by their oaths to report to the alderman any such wounded or bleeding people who come to them or are brought in, as well as those who inflicted this upon them, with their name and surname, if they can ascertain it."[54] The barbers objected that their trade was built on a confidential relationship with their paying clients—indeed, "in relation to a patient or wounded person, the barber should be thought of as an innkeeper with respect to his guest," who "trusts him and seeks shelter." Instead of reporting a client with a suspicious injury, a barber "should rather be honor-bound to help the perpetrator escape."[55] To break that trust would be bad for business, since the barbers relied on "the favor of the common man."[56] In any case, once it became known that the barbers reported to the council, injured brawlers would simply take their business elsewhere, resorting to "pox doctors and old women" or other "unskilled healing."[57]

Moreover, the barbers feared that an obligation to report wounds would mean that their "journeymen barbers may be held in contempt in foreign parts where such reporting may be considered immoral."[58] This would have been a serious setback. During their traveling years, artisans relied on well-connected networks of journeymen's associations to find work and accommodation, but members of a guild with a mark of dishonor in the journeymen's "black book" were likely to be turned away.[59] At home and abroad, image was everything, and if the council compromised their reputation, the barbers said, they might as well "secretly snatch the bread from our mouths."[60]

Against this backdrop of artisanal anxieties, concerns about "the honor of all surgeons" gain a new urgency, even though at first glance Brunschwig's and Gersdorff's treatments of the issue seem closely modeled on their medieval antecedents' discussion of the ideal surgeon: he should be serious but mild-mannered, talk soothingly to patients and not too much; he should treat the poor for free but insist on a respectable salary from others.[61] At a closer look, however, our surgeons deviate from their favorite late medieval models such as Guy de Chauliac, who had reformulated the issue in ethical terms of individual responsibility and virtue. Instead, the Strasbourg surgeons shift the emphasis towards the surgeon's reputation. One should not attempt hopeless cures for money's sake, Gersdorff writes, because this would "arouse much clamor and give him a bad name," and Brunschwig, too, cautions that it would result in a "bad reputation" and "disgrace and opprobrium."[62] By refocusing the medieval catalog of surgical virtues on the issue of the practitioner's good name, Brunschwig and Gersdorff adapt their learned predecessors' writings to the concerns of artisans in the honor society of a free imperial city.

Surgery in Print and in the City

In their printed surgeries, Brunschwig and Gersdorff locate surgery firmly in the fabric of urban artisanal life. As healers, legal witnesses, and public health inspectors, surgeons were experts and guardians of artisans' bodies, both individual and collective, in all their many-layered complexity of health, labor, and honor. Implicitly writing against stereotypes of dishonor associated with body work, Brunschwig and Gersdorff portray surgeons as honorable and indispensable members of the artisanal community, and instruct surgeons themselves to earn and maintain the good opinion of their fellow citizens. In speaking to the concerns of local practitioners documented in the archives, they hit upon a fruitful strategy for promoting the image of surgery in print that resurfaces in subsequent publications, in Strasbourg and beyond.

The *Cirurgia* marks the beginning of a steady trickle of local publications that continued to emphasize the place of surgery in the city. In the dedicatory letter to his 1528 translation of a text by the medieval surgeon Lanfranco, local humanist Otto Brunfels notes, "when our ancient forebears wrote about statecraft [*policeien*], . . . they said: if a city has good authorities, good clergy, doctors, schoolmasters and midwives, then public order [*policei*] is provided."[63] However, according to Brunfels, one key function was missing in this traditional enumeration: "In my opinion, to these should be added the surgeons" since they were equally indispensable to "the common good; not only in times of war, but also in many other diseases of man."[64] Another Strasbourg print, Walther Hermann Ryff's *Kleiner Chirurgi* of 1542, reminds readers that surgery deals with the majority of accidents and diseases, so surely anyone could see "how useful and highly necessary surgery . . . is for upholding a proper and orderly rule [*pollicey*]."[65] From Brunschwig's *Cirurgia* onwards, surgical expertise became a matter of civic order or *policei*. This implicates the surgeon in the medical governance of early modern cities, where the focus of historical scholarship has thus far been on physicians.[66]

While Brunschwig's and Gersdorff's books were shaped by the local artisanal community, as I have argued here, they were of course produced for a broader audience and were widely disseminated and reprinted in other major cities.[67] One edition of Gersdorff's work, published in Frankfurt in 1551, was edited by Ryff, whose own book underscored surgery's role in city governance—thus it was perhaps his hand that changed Gersdorff's title to *Stat und Feldbuch Bewerter Wundarznei*—the "field book" had become a "city and field book." Surgery had come home from the battlefield and the rough terrain of the journeyman's travels to take its rightful place in the early modern city.

PART III

Distilling Knowledge

✷ 4 ✷

Distilling Medicine

SUBTLE REMEDIES FOR FLUID BODIES

Three years after the *Cirurgia,* Brunschwig revealed another string to his bow as a medical artisan-author by publishing the first of two books on distillation. This was not a departure from his training as a surgeon, but a natural extension. To a greater extent than is generally acknowledged, surgeons' expertise and practice consisted in making and applying topical remedies. Candidates for mastership were often examined not on their knife skills but their ability to concoct plasters and ointments.[1] The technique of distillation in particular was keenly adopted by the rational surgeons from the thirteenth century onwards.[2] To be sure, Brunschwig probably took his pharmaceutical interest further than most surgeons. He writes about a range of materia medica, including their Latin names, explains pharmaceutical jargon, weights, and measures, and describes techniques for preparing remedies, including complex distillations. Two contemporary commentators even refer to Brunschwig as an "apothecary." As we shall see in chapter 7, the physician Lorenz Fries from nearby Colmar alludes to Brunschwig in his discussion of apothecaries in a publication from 1518. Fries writes that people come running to his house for medicines "when they want to buy fish," suggesting that he ran some kind of shop or dispensary near the Strasbourg fish market.[3] In a 1532 edition of the *Large Book,* Bartholomäus Grüninger, the son and successor of Brunschwig's publisher, refers to the author as a "surgeon and apothecary." Brunschwig, however, never refers to himself in those terms. Even in his final work on distillation, he self-identifies as a *wundarzt* rather than *apotheker*. Indeed, there is no evidence that he had official license to practice this strictly regulated trade.[4] We should therefore be cautious about labeling Brunschwig an "apothecary," as historians have sometimes done.[5] It is perhaps more likely that he simply developed an unusually deep interest in a set of skills that was always a key part of surgery, even though it is often overlooked.

In his books on distillation, Brunschwig discussed a technique that was by no means unknown. Local regulations list distilled waters among the typical remedies for sale in an apothecary's shop,[6] and allude to "those who distill waters in the hospital."[7] Brunschwig himself mentions the distilling practices of "several nuns . . . as I have seen myself in Strasbourg"[8] and describes the equipment of local brandy distillers.[9] A high density of archaeological finds of distillation equipment bear out this lively distilling activity in the Strasbourg area and the Alsace region.[10]

Brunschwig was, however, the first to describe this widespread technique in any practical detail. Like the *Cirurgia*, his distillation books were written entirely in German, despite their partially Latin titles. The *Small Book of Distillation*, published in 1500, opens with a discussion of what distillation is and describes different methods and equipment, accompanied by a wealth of images showing flasks, alembics, and furnaces. Brunschwig's instructions are detailed and assume little prior knowledge, beginning with the bricks and mortar for furnaces and ending with the appropriate storage of distilled waters. Complementing the technical section, there is a substantial illustrated herbal providing basic botanical information, including the German, Latin, Greek, and sometimes Arabic names of plants or animals, as well as a description of their appearance. It tells readers which part of the plant to use, when to collect or distill it, and how to administer the resulting remedy. In addition to a table of contents, the first edition had two well-meant but unwieldy indices.

In its second edition (1505), the *Small Book* was improved with a more usable index and published under the title *Medicinarius* together with a partial German translation of Marsilio Ficino's *De vita*, juxtaposing Brunschwig's practical medical distillations with Ficino's ideas about human longevity, informed by Hermetism and other occult traditions. The *Large Book of Distillation* was published in 1512. True to its name at over seven hundred pages, it was conceived as something of a companion volume to the *Small Book*. It does not repeat the systematic treatment of basic distilling techniques and medicinal herbs of the *Small Book*, but rather expands the repertoire of techniques and equipment, adding more sophisticated furnaces and glassware. It also includes more in-depth theoretical discussion of alchemical concepts—including the notion of "quintessence," which was a formative influence on Brunschwig's understanding of distillation, as we shall see in chapter 5. In contrast to Brunschwig's other works, the *Large Book* is explicitly addressed to learned readers as well as the common man.

The books were a success. Brunschwig's publisher Grüninger produced six versions of the *Small Book* between 1509 and 1537, and six further editions were published by Frankfurt printers between 1538 and 1556 (see table 1). The *Large Book* as a whole was reissued twice by Grüninger, in 1519 and 1532, and in six further editions by other printers between 1539 and 1614. By far the largest print runs were achieved by two stand-alone versions of the final section of the *Large Book,* a collection of simple medical recipes headed "Thesaurus pauperum" (Poor man's treasure) after a medieval work of the same name.[11] The first independent reprint of these recipes, titled *Apoteck für den gemeynen man* (Common man's apothecary), combined them with Michael Puff von Schrick's *Booklet of Distilled Waters,* apparently to great commercial success: there were thirty-three editions between 1529 and 1599. The second spin-off only contained Brunschwig's recipes, alternatively titled *Thesaurus pauperum* and *Hausapoteck* (Home apothecary). It was printed over twenty times, and well into the seventeenth century. As we shall see, Brunschwig's idea of a "distillation book" was taken up by other authors, sparking a new printed genre.

As this proliferation of printed books indicates, distillation captured the attention of early moderns to an extent that has thus far not been recognized in historical scholarship. This contemporary interest is illustrated in the frontispiece to Jan van der Straet's collection of engravings entitled *Nova reperta.* A distilling furnace and alembic are prominently depicted alongside other "new inventions" that fueled the age's new impulses to understand and exploit nature: the "discovery" of the Americas, the compass, mechanical clocks, the printing press, gunpowder, and guaiac wood, a new-world remedy for the new disease of the age, the ("French") pox or *blattern* (for which there was a dedicated hospital in Strasbourg as in many cities) (figure 4.1). The seventh plate of the cycle is dedicated entirely to the art of *distillatio,* illustrating how the technique generated new knowledge, new books, new medicines, and advances in manufacturing (figure 4.2). Distillation was part of a sixteenth-century obsession with novelty, even though the technique was not truly a Renaissance invention. It had been practiced since antiquity, developed in the Middle East, and eagerly adopted by medieval European writers on medicine and alchemy in the Middle Ages.[12] As we have seen, by the time Brunschwig wrote his seminal books, it was already being used in various ways and for different purposes. But in an important sense, the sixteenth century really was a new age of distillation. As I will show through the lens of Brunschwig's books, distillation not only spawned new genres of books, but also shaped new cultures of understanding and working with natural matter and the human body, at the intersection of medicine, alchemy, and craft.

In medicine, it was the bread and butter of professional distillers such as apothecaries or *wasserbrenner,* literally "distillers of waters," a trade documented in many imperial cities since the Middle Ages and often practiced

FIGURE 4.1. Jan van der Straet, *Nova reperta* (c. 1590), frontispiece. Credit: Metropolitan Museum of Art.

FIGURE 4.2. Jan van der Straet, *Nova reperta* (c. 1590), plate 7, "Distillatio." Credit: Metropolitan Museum of Art.

by women.[13] In the domestic sphere, too, women were often charged with distilling remedies for the household, from the middling classes to high-born princesses.[14] Distillation became deeply implicated in medical theory as it shaped the central tenets of new "chemical" medicines. At the hands of heterodox writers such as Paracelsus or Leonardo Fioravanti, distillation provided powerful new remedies, but also new understandings of the body's inner workings.[15] Paracelsians, alchemists, and other students of nature formulated new questions around distillation, about what matter was fundamentally made of and how its "essence" could be extracted.[16] For many and perhaps most practitioners, distillation was a tool of their trade and a way to earn a living, from the *weinbrenner* of Brunschwig's Alsace who made sought-after Rhenish brandy, to miners, metalworkers, and artists who distilled strong mineral acids to assay metallic ores and etch printing plates.[17] As we shall see, Brunschwig worked at the intersection of these overlapping cultures of distillation, fully aware of their stakes and challenges.

Surprisingly, we have as yet no history of distillation as the epistemically and materially transformative process that it clearly was. This is in part owed to disciplinary and methodological fault lines in historical inquiry. There are intellectual histories of texts and ideas that show distillation's significance in premodern matter theories, and the quest for universal medicines, and metallic transmutation.[18] Technical histories, on the other hand, tend to treat distillation as a self-evident mechanical process that can be understood historically by tracing developments in the apparatus used, and pay little attention to the ways in which the underlying material transformations were understood by practitioners.[19] We need a history that takes account of ideas, practices, and material culture to understand how distillation worked fundamental material change through human art.

In this chapter and the next, I show that Brunschwig's books provide unique insights into a crucial part of that history. They combine unprecedented technical detail and hands-on experience with alchemical and medical theories, each enlivening the other in a rich new expertise of natural matter and the human body. I argue that Brunschwig understood distillation as a technology that responded to three key challenges—medical, cosmological, and artisanal—of working with matter. This chapter explores distillation as the medical project that it primarily was for Brunschwig. It asks how he understood the process as an intervention in the materialities of medicinal substances and the body. This in turn allows us to explain distillation's impact on medical theory and economies of healing. The following chapter explores how Brunschwig articulates a more general concept of matter through distillation, shaped by traditional metaphysics, alchemical thought, and the artisan's close physical encounter with materials.

To dispel some common anachronisms in our understanding of distillation, this chapter begins by exploring Brunschwig's concept of the process as a separation of "subtle" and "coarse" components of matter. Exploring how such "subtle" medicines acted, I then show how Brunschwig framed distilled remedies as safe and efficacious treatments for common folk, at the same time paving the way for new "chemical" understandings of the body in health and disease. This in turn changed the dynamics in economies of healing, creating new agencies especially for nonacademic practitioners. Last but not least, distillation had a significant impact on the sixteenth-century book market, where Brunschwig's works marked the rise of the *Destillierbuch* (distillation book) as a new kind of publication.

What Is Distillation?

Today, distillation is, in principle, quite a straightforward process that separates substances with different boiling points. A mixture of such substances is heated to the specific temperature at which one of the substances evaporates; the vapors then condense on a cooled surface and can be collected in a separate vessel. While the same basic principle is in operation in many early modern instances of "distillation," the term had much wider significance and covered a broader range of material processes. It is thus worth exploring how these processes were understood, and what exactly Brunschwig means by "distillation."

In the first chapter of the *Small Book,* Brunschwig writes that the purpose of distillation is "to separate the subtle from the coarse, and the coarse from the subtle."[20] At a glance, this basic principle of separating material substances corresponds to our modern-day definition. Many of the technical setups in Brunschwig's text and images could perform what we now mean by "distillation," and some are used to this day in more or less the same manner. However, a closer look at the range of techniques in Brunschwig's books reveals a much broader concept of distillation and how it worked to separate coarse from subtle matter.

In the *Small Book,* Brunschwig sets up a basic taxonomy of ten types of distillation, distinguishing five techniques "without cost," and five that can only be performed "at a cost and with a fire" because they require a special furnace.[21] The first group begins with what Brunschwig designates, in shaky Latin, as *per filtri distillatio.* This is the simplest way of separating messy aggregates. A long triangular piece of felt is placed in a pan

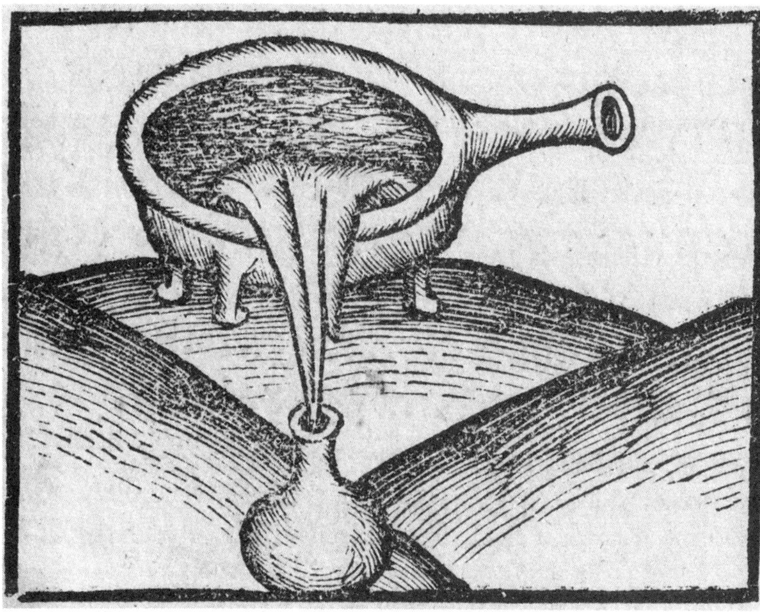

FIGURE 4.3. Distillation through a filter. Brunschwig, *Small Book*, fol. 7r. Credit: The Huntington Library, San Marino, California, 85769.

filled with liquid, with its narrow end in a receptacle placed lower than the first, such that the liquid will be drawn through the filter (figure 4.3). This technique can be used to clarify "water, wine or juice which one wants to separate from its murky and turbid substance,"[22] or to "rectify" distilled waters, removing impurities to make them last longer.[23] For what Brunschwig calls *distillatio per solem*, a glass flask filled with flowers is exposed to the heat of the sun; the liquid thus drawn from the flowers is caught in a second vessel. Neither technique fits our modern-day definition of distillation, but they fall rather neatly within the literal meaning of the Latin *destillare*, since both operate by "dripping down." With the remaining three techniques of this first group, it is less obvious how a separation of matter might be taking place. In each case, plant matter is heated in a single sealed glass flask, using a baker's oven (*per panem*), where the flask is wrapped in dough and baked alongside the usual loaves, fermenting horse manure (*per fimi equi*), or even a live anthill (*per formice*).

The second, more demanding group comprises techniques which modern readers might more easily recognize as distillations. The equipment varies in size, shape, and material according to purpose and availability, from prized Venetian glass to simple lead pots, but they have in common a basic setup consisting of a purpose-built furnace, a distilling vessel sometimes known as "cucurbit" topped with an alembic or "helmet" in which the liquid condenses (figure 4.4), and a receptacle for the distillate, assembled as shown in figure 4.5. These five techniques differ in the way in which the heat is controlled and mediated, ranging from a gentle water bath (*balneum mariae*) to a water bath with added horse manure (*per ventrem equinum*), to a bath of ashes (*per cineres*) or sand (*per arenam*), and finally direct exposure to the fire (*per ignem*), where the vessel is protected by nothing but a *lutum*, a paste of clay and straw to prevent the glass from cracking in the heat.

The repertoire of the *Small Book* is expanded in the *Large Book*, including stills of greater sophistication or output, such as a large two-story water bath furnace permitting several distillations to be run in parallel (figure 4.6). One significant modification to the process of distilling through an alembic is the addition of a water-cooling mechanism for distilling strong spirit of wine, or *aqua ardens* (figure 4.7).

This wide range of different operations and apparatuses raises questions about precisely what kind of material separation of "coarse" and "subtle" things they are supposed to perform, and how they are understood to achieve this. Existing technical histories of distillation tend to fall back on modern-day categories, distinguishing "open" setups, which use an alembic and are able to perform a separation of substances as we now understand it, and "closed" techniques operating in a sealed system, which are not distillations at all from a modern point of view, and can at best be called decoctions or infusions.[24] Such presentist interpretations ignore the historical reality in which Brunschwig and his contemporaries clearly *did* understand such processes as distillations, so the real question is *how* they did so.

The answer lies in the technical detail of Brunschwig's instructions and their underlying rationale. They show that "closed" techniques using a sealed flask in fact serve two distinct purposes. The first is a process that the *Small Book* terms "rectification" of already-distilled waters. Inside a sealed flask placed in warm horse manure, liquids may "rise and fall, by which means they are rectified, and improved in every way."[25] This rising and falling can be achieved in a flask topped with a "blind" alembic, an hourglass-shaped *circulatorium* made in one piece, or better still, the intricate two-armed flask known as a pelican (figure 4.8). In the *Large Book*, this technique is elaborated into a distinct category called "circular distillation" whereby the vapors "always rise up in the middle of the flask, running back through the arms, or from one vessel to the other; thus it is rendered beautiful and of greater power."[26]

The other type of process that could occur in a single, gently heated

FIGURE 4.4. Alembics made from glass (above) and copper (below). The latter was known as a *rosenhut* (rose helmet), reflecting its use in the distillation of rose water. Brunschwig, *Small Book*, fol. 1v, 2r. Credit: The Huntington Library, San Marino, California, 85769.

FIGURE 4.5. Complete still with furnace, a copper alembic known as a *rosenhut* (rose helmet), and a receptacle. Brunschwig, *Small Book*, fol. 4r. Credit: The Huntington Library, San Marino, California, 85769.

FIGURE 4.6. Large water bath furnace of the kind allegedly owned by the bishop of Trier. Brunschwig, *Large Book*, fol. 22v. Credit: Wellcome Library, London, EPB/D/1114.

flask is called "putrefying and digesting."[27] It served to soften particularly tough materials prior to distillation, as I discuss in more detail in chapter 5. Digestion is explicitly distinguished from circular distillation in a chapter on the "difference between digesting and circulating."[28] Both processes may use the gentle heat of fermenting dung and even the same type of

FIGURE 4.7. Water-cooled still. Brunschwig, *Large Book*, fol. 21r. Credit: Wellcome Library, London, EPB/D/1114.

glass vessel; the difference is that, while the vessel is completely immersed during digestion, for circular distillation the vessel should be left exposed at the top, "such that a third of the glass is outside, and that which has risen up through the heat will descend again due to the cool air."[29] This technique served to further purify already-distilled substances, clearly a very different material process from the preliminary digestion of tough ingredients. It is not the type of vessel used that distinguishes a process that is a kind of distillation from one that is not; it is the rising of vapors and falling of condensed liquid.

If two such conceptually distinct processes as digestion and circular distillation can occur in the same piece of glassware, what is the relationship between distillation through a regular alembic and circular distillation, whose "open" and "closed" apparatus suggest such fundamentally different

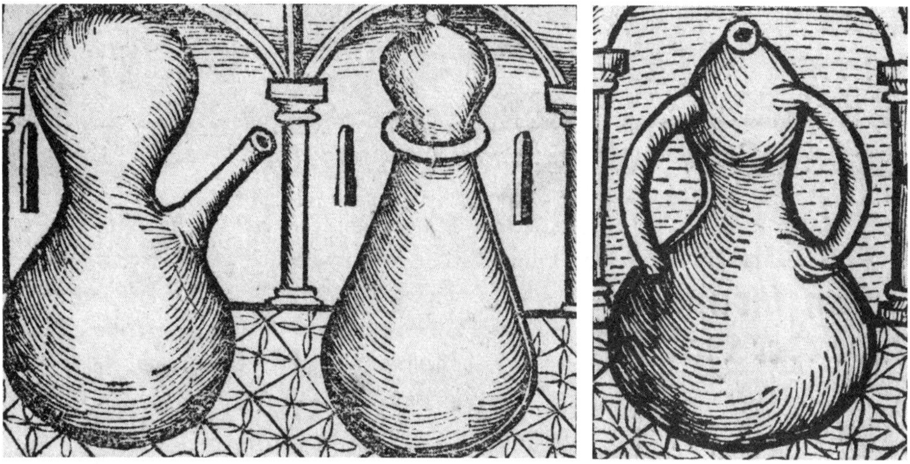

FIGURE 4.8. Vessels for circular distillation: *circulatorium* (left), "blind alembic" (middle), and "pelican" (right). Brunschwig, *Small Book*, fol. 1v–2r. Credit: The Huntington Library, San Marino, California, 85769.

processes to modern technical sensibilities? This can be discerned in a chapter of the *Large Book* that deals with the distilling of the "quintessence of wine." The passages are in large part borrowed from a medieval text, *On the Consideration of the Quintessence* by the Franciscan alchemist John of Rupescissa.[30] Rupescissa's ideas were formative in shaping Brunschwig's understanding of distillation, even though the quintessence of wine was something of a special case. The instructions involve multiple consecutive distillations *per alembicum*, and Brunschwig writes that "the more it is distilled, the more it is rendered subtle, spirit-like, and immaterial, such that it becomes more excellent."[31] Subsequently, circular distillation can be used to refine the distilled matter still further. "Since one can see that through one rising and falling of distillation the coarse and impure is separated and made more subtle, . . . what would happen, then, if it were to move up and down the *distillatorium* a thousand times? No doubt it would achieve the highest worth . . . as it is separated from the aggregate."[32] With Rupescissa, Brunschwig suggests that the repeated up and down of circular distillation in a single vessel is essentially the same as the rising and falling of a single distillation in an alembic, continued in a cyclical manner.

Although he describes distillation as fundamentally a process of separation of subtle from base matter, for Brunschwig the reach of the term distillation was not limited to processes that yielded physically separated fractions in the way our modern definition demands. Whether single and "open" or circular and "closed," the rising and falling of vapors and liquids determines distillation's transformative power, by separating pure from impure, and by rendering matter more "subtle."

This does not entirely resolve the questions raised by the range of techniques Brunschwig refers to at one time or another as "distillation." The filtration and distillation *per solem* do not seem to involve an up-and-down of evaporation and condensation, and are better served by the etymological description of "dripping down," while it is not at all clear what is happening during distillation *per panem*. These processes, however, remain fairly marginal in Brunschwig's works, whereas the rising and falling of distillation provide a working definition of what he most consistently thinks of as distillation. In this understanding of the process, technical details such as the right type of vessel and heat source are intertwined with scholastic and alchemical accounts of "subtle" matter. Chapter 5 explores in more detail how distillation intervened in the overarching cosmology and matter theory of Brunschwig's day. In the following section, I show how Brunschwig conceptualized subtle and coarse parts of medicines in relation to the human body, and why it was so important to separate them.

The Matter of Medicines

In Brunschwig's manuals, distillation is first and foremost a medical matter. The first chapter of the *Small Book* specifies what is at stake in making and administering medicines. Distilled waters are "separated from the gross, as from the *corpus* the virtue in its soul is drawn out from each thing, the better to send it to the place where it may bring greater benefit and healing, as I will show hereafter."[33] The healing virtue or "soul" of medicinal simples can be isolated from their physical shell or corpus, which smooths its passage through the body. As promised, the following chapter, "Why Distillation Was Thought up and Found," elaborates why this is desirable. The corpus of a medicine is not only something of an inert material prison which encloses and weighs down the active healing principle, but it can actually cause harm when ingested. Laxatives, for example, are especially problematic, because "material parts get stuck in various parts of the human body, such as the stomach, liver, or intestines, constricting and abrading them, or making them soft and weak, or causing pain in various ways."[34] Brunschwig discusses several procedures for mitigating such damage, but judges that none of them sufficiently eliminates the troubling corpus.

The method of the ancients was to chase a laxative with large volumes of a cleansing draft to flush out any dangerous substances. Brunschwig argues, however, that this leaves them plenty of time to do significant damage before leaving the body. It would be much safer to prevent the corpus from entering the body in the first place, for example by macerating medicinal herbs in water of endive. Although this watery extraction is already much safer, "one must still worry that its substance remains mingled in it."[35] Wrapping the simples in a cloth, much like a tea bag, is an improvement, but does not entirely prevent their substance from seeping through. After all, "such a macerating liquid takes on the color of many medicines. Therefore it takes on some of the substance of the corpus, since color is a physical property."[36] Boiling or seething medicinal simples is not much different; if anything, "more of the corpus of the medicine is mingled in a decoction."[37] Brunschwig concludes that "however much it is cleansed by the aforementioned methods to prevent damage to the human body caused by the substance of the corpus, the words of Avicenna and Hippocrates hold true: that medicine weakens and ages even if it is rectified, fermented, and put through maceration."[38] Dismissing one alternative technique after another as insufficient, Brunschwig works towards the only real solution, "a way to separate the coarse and earthy substance from the subtle, through this work which is called the handwork of distillation."[39] As in his surgical writings, Brunschwig emphasizes the importance of skilled handwork in dealing with the material aspects of medicine, but he thereby also ties artisanal expertise back into a larger theoretical discourse about the inner workings of the body.

Different people had different models and imaginaries of the body, but there was often considerable overlap even between learned physicians and common folk. Questions about what kind of matter should and should not enter the body were central to virtually all accounts. By framing these questions in terms of materials' passage through the system (and potentially causing accidents or blockages in the healthful traffic of bodily matter), Brunschwig speaks to a widespread understanding of the human body in terms of fluids and motion. Learned medicine was influenced by the Hippocratic tradition of rationalizing health and disease in terms of the flow of bodily humors, good or bad, and their correct evacuation, deviations, or blockages.[40] In his reading of the Rabelaisian grotesque body, the twentieth-century critic Mikhail Bakhtin suggests that these Hippocratic ideas shaped a view of the body as open and porous, freely exchanging with the environment all sorts of physical matter in the form of food, drink, sweat, tears, and excrement.[41] Following this idea, historians have uncovered an early modern understanding of the healthy and normal body as one whose fluids were moving unobstructed within and across its permeable boundaries.[42] Bodily liquids and their movements were seen as causes and expressions of the emotional, spiritual, and physiological integrity of individuals and communities; dryness and stagnation meant death, and a woman whose heart literally could not be "moved" to stir the bodily flow of genuine tears was surely a witch.[43] For laypeople in particular, images of fluid movements or fluxes inside their bodies were much more important in experiencing and articulating health and disease than the topography and function of distinct solid organs. Fears of blockages and misdirections of vital flows were among the most serious medical concerns, and it was not uncommon for learned practitioners to couch their diagnoses in these terms for the benefit of their paying customers.[44]

Within the framework of a body governed by moving fluids, Brunschwig's claims about distillation carry considerable force. While the corpus of medicines threatens to clog or disrupt vital passages, their distilled "spirit" may "through its subtlety more easily and quickly enter and penetrate the human body . . . with its virtue and power."[45] Moving through the body freely and without causing damage, the isolated medicinal virtue is more efficiently delivered, as we saw, "to the place where it may bring greater benefit and healing." This not only speaks to the widespread understanding of the body in terms of moving fluids rather than solid topography, but even implies that a detailed learned understanding of the latter is not always necessary to treat internal ailments. The orally administered purified *spiritus* of an appropriate herb finds its own way to the place where it is needed and causes no damage on the way. Although this account of distilled remedies' action is not explicitly offered to legitimate their use by common folk and medical self-helpers, it implicitly makes a strong case for distilled waters as efficacious, self-directing, and foolproof remedies that could safely be used by just such an audience.

It would be tempting to read this as Brunschwig's siding with a school of learned medical thought that best suited his aims. Two such paradigms became widely known during the early modern period. Gianna Pomata has shown that patients in seventeenth-century Bologna preferred the so-called Hippocratic idiom, which privileged flowing liquids. This placed sufferers in a horizontal power relationship with their practitioner, whose role was merely to aid the body's natural, self-healing evacuations. In contrast, the Aristotelian–Galenic idiom of the body as a well-ordered polis of discrete anatomical structures cast the learned physician in a vertical relationship to the patient, one of authoritative governance.[46] It is not clear, however, that Brunschwig would have been aware of this distinction. For one thing, medieval and Renaissance scholars of medicine were by and large working to reconcile Hippocratic and Galenic traditions rather than teasing them apart.[47] For another, Brunschwig was no such scholar, but an eclectic reader and compiler. His writings are certainly not dismissive of the anatomy of solid parts, especially in surgery. Whether or not Brunschwig was aware of the Hippocratic undercurrent of his defense of distillation, he chose a framework that had the clout of learned medicine and appealed to a popular understanding of the body.

Distilling Bodies

Brunschwig deliberately frames distillation as a safe and efficient mediator between the materialities of the body and the outside world. But an even deeper relationship was developing between the body and the distillation process, as they increasingly became analogous models of each other. One traditional view of the body, going back to Aristotle, understood human physiology in analogy to the process of cooking, with the stomach, liver, and other organs concocting ingested matter into blood and other vital substances. In the medieval Islamic tradition, distillation was practiced much earlier than in the Latin West, and here it began to rival the cooking paradigm as a model of the body. Medical writers such as Avicenna and Al-Razi explained that the stomach functioned like a heated alembic, causing vapors to rise into the head, where they condensed as in a cooling helmet, and were often discharged like a snotty distillate from the nose, as if from the beak or *solen* of an alembic.[48]

As distillation practices and literature became widespread in the wake of Brunschwig's publications, this analogy gained traction. Many of the subsequent *Destillierbücher* evoke the cyclical, transformative nature of distillation as a powerful model for bodily and natural processes. Walther Herrmann Ryff's *New Large Book of Distillation* (1545) borrows freely from Brunschwig, but its added preface relates distillation to the rising mists and falling rain of the macrocosmic water cycle, and to the microcosm of the human body, where "rising vapors" condense in the brain to form "snot or phlegm" which is extruded "through the nostrils."[49] Adam Lonicer's *Kreuterbuoch* makes the analogy more explicit: "The pan or vat of the stomach is heated by the hearth of the liver, and the inner vapors . . . rise into the head and skull as an alembic, and through its coldness are driven together to snot or phlegm into the nose as the beak of the alembic and the neck, [whence they] fall back again and must be swept out by sniffling, coughing and clearing of the throat."[50] In the early *Destillierbuch* tradition, the fluids and vapors of the human body could help understand the up and down of distillation, and vice versa.

For some medical writers and practitioners, the analogy became much closer, to the point where distillation and human physiology were seen as practically the same. The idea of the *homo alembicus*, of the human body as a distilling alembic, had been raised by Avicenna, but it came to its most literal fruition in the self-proclaimed new medicines that sprang up around distillation and other "chemical" practices over the course of the sixteenth century, overhauling or explicitly challenging traditional physic. The most well-known of these heterodox writers is Theophrastus von Hohenheim, known as Paracelsus. Doing away with the ancient concept of the four humors, Paracelsus explained the microcosm of the human body in terms of distillation and other chemical processes, operated by an inner alchemist, the *archeus*.[51] If the inner alchemy of the body was disturbed, the alchemist-physician could use external, artificial distillation to produce the necessary remedies.

The relationship between distillation and the body became more explicit in the context of urine analysis. Paracelsus criticized the traditional practice of uroscopy, whereby physicians detected in the appearance and sometimes the taste of a patient's urine traces of the humoral imbalances that ailed them. While he did not doubt that traces of disease were present in the urine, he insisted that they could not be seen with the naked eye. Just as silver dissolved in *aqua fortis* is invisible until it is precipitated, so the material traces of the body's alchemy had to be made visible by chemical means such as distillation. The matter is taken further in a text attributed to Paracelsus but likely not by his own pen, provocatively titled *Anatomia corporum adhuc viventium* (Anatomy of bodies that are still alive) to suggest that the distillation of urine allows one to know the living body in the way the anatomist can only ever know the dead body—a sort of chemical dissection by proxy. This involved a tripartite furnace built to exacting specifications and a bespoke alembic whose graduated subdivisions corresponded to regions of the body (figure 4.9).[52]

This approach was taken up by the goldsmith, alchemist, and medical practitioner Leonhard Thurneisser zum Thurn (1530–1595). He explicitly conceived of the body as a system of fluids whose health and emotions were governed by movements and evacuations—to him, a "sinking heart" or a "weeping eye" were not mere emotional rhetoric, but real physical phenomena.[53] The distillation of urine could replicate and make visible the material processes of disease in the body. Using a similar type of graduated alembic that represented the human body, he would observe, for example, a cloud of green fumes forming at the location corresponding to his female patient's uterus, then rising to the top of the alembic where they became thick and turbid without condensing. According to Thurneisser, this indicated that there were vapors permeating his patient's liver ducts and heart cavities, ultimately threatening to attack her brain. The distilla-

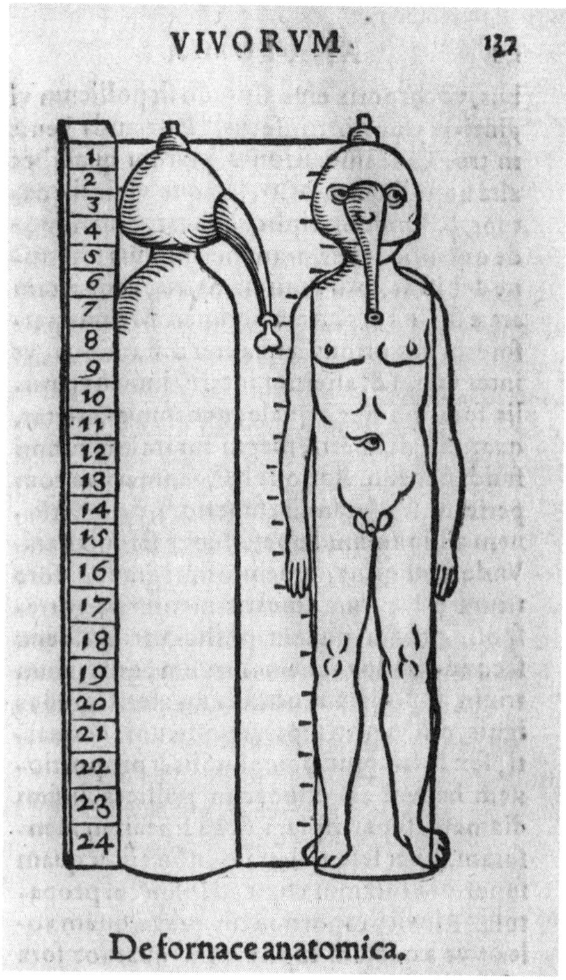

FIGURE 4.9. Alembic for urine analysis, with gradations corresponding to the human body. Gerhard Dorn, *Aurora Thesaurusque Philosophorum Theophrasti Paracelsi* (Basel, 1577), 137. Credit: Basel Universitätsbibliothek, UBH Le VIII 30:2.

tion of urine thus not only precipitated invisible disease matter, but could function as a kind of early modern equivalent of a time-lapse weather map, with bodily processes playing out dynamically in the artificial body of the alembic.[54]

Later writers on medicine and natural philosophy continued to use distillation as a model for their accounts of bodily processes. The French Paracelsian Joseph Duchesne argued that human physiology involved the gradual rarefaction of bodily matter into vapors and vital spirits, just as a series of distillations could produce the quintessence of wine: the mouth functions like the winepress, crushing and extracting the base matter, which then ferments in the stomach as it would in a wine cask. Bodily distillations separate the coarse *faeces* from the subtle components, which are further rarefied through circular distillation, driven by the heat of the heart and liver, into the body's own quintessence in the form of rarefied blood and vital spirit. Francis Bacon explained the aging process in analogy with a brandy still that is running too hot, thus burning up excess vital energy. The human body required artificial cooling (of the kind illustrated in the *Destillierbücher*) to make the process run efficiently. For Bacon, this meant that actual distilled *aqua ardens* was not a good remedy, as it too closely resembled humans' hot vital spirit and would thus further excite its combustion.[55] Using the same model for the body's inner workings, Bacon's reasoning thus made a significant departure from Brunschwig and many other distillers who, as we shall see, were influenced by an alchemical tradition that identified the quintessence of wine as a universal remedy.

Analogies between the alembic and the body run like a common thread through the emerging distilling literature—from Brunschwig's articulation of a material relationship between distillation, the subtle "souls" of remedies, and the healthful flow of matter within the human body, to Paracelsian chemical medicine and experimental natural philosophy. A complete history of the sixteenth-century emergence of what we might call the "distilling body" as a widespread model has yet to be written. While scholars have noted that Brunschwig's books likely had a significant influence on Paracelsus and his followers, it remains to be seen precisely how and to what extent Brunschwig paved the way for these new understandings of medicine and the human body.[56] It is in any case clear that he was the first to provide a widely read technical and theoretical rationale for distilled waters as safe and efficient medicines for a fluid material body as understood by broad swathes of the population.

Pleasant and New: Economies of Healing

Brunschwig's understanding of distillation not only explained why distilled remedies worked better than others, but also why they were more pleasant and appealing to patients and paying clients. By separating subtle healing powers from coarse matter, distillation makes "the disagreeable more agreeable."[57] Distilled waters could be mixed into medicinal syrups or electuaries to serve them up "in an agreeable, drinkable form."[58] Taken by themselves, they appeal to "those who take more pleasure in a potion or water than in things that are eaten or any other medicine."[59] This was no trivial matter since early modern sufferers had a greater say in their treatment than we might expect. Even in the Strasbourg hospital, where the poor received charitable care, the regulations stated that if a patient showed "revulsion or dread" for a particular treatment and could not be "gently persuaded" to take it, the doctor was "bound by his oath not to coerce them."[60] Those who did not rely on municipal charity could be even more discerning when shopping around for practitioners and remedies, in what historians of medicine have called a "medical marketplace."[61] The concept has sometimes been applied indiscriminately, reducing complex historical realities to laissez-faire economics, which threatens to obscure important areas of medical practice at charitable institutions (such as the Strasbourg hospital) and within families and neighborhoods.[62] Nonetheless, it can be a useful analytic lens that brings into focus the often underestimated agency of patients.[63] In an increasingly competitive market, Brunschwig promised clean and pleasant remedies that appealed to clients' tastes.

Distilled remedies rose to popularity in part because they played into early moderns' obsession with the new. As Stradanus's *Nova reperta* illustrates, distillation was part of a sixteenth-century culture of discovery and innovation. Although it had been practiced in one form or another for centuries, sixteenth-century writers such as the mining expert Vannoccio Biringuccio found a "pleasing novelty" in distillation and other chemical operations.[64] This was especially true in medicine where, as we have seen, distillation inspired new theoretical models of health and disease as well as new pharmaceutical methods and products.

These were appealing to many people who felt that traditional Galenic medicine could not keep up with the challenges of the age.[65] As Walther Ryff wrote in his *New Large Book of Distillation* (1545), "such artful extracts made by distillation were also in part invented so that one could cure the venom and devastating evil of newly arising diseases with new-found medicines," notably "the French ailment."[66] As we saw in Stradanus's *Nova reperta*, this "French disease," "pox," or "syphilis" was thought to be a new disease brought back to Europe from the New World. It continued to exercise Paracelsians and other "chemical" practitioners, who cornered the market by offering specific remedies instead of the Galenists' lengthy diets and regimens.[67] In his translation of a Paracelsian work, the London distiller and translator John Hester, whom we will meet again in chapter 9, criticized those who sought an answer in ancient authorities, since "the disease of the French Pocks was neyther knowne to them, nor to theyr successors." Instead, one should look to "men's industries" such as his own thriving distillery at St. Paul's Wharf.[68] Many of Hester's works were based on the writings of the Italian "professor of secrets" Leonardo Fioravanti, who cultivated personal fame in the medical market through his novel and powerful chemical medicines.[69] Pleasant, fashionable, and with credible rationales for their specific effects on the body, "new" distilled remedies frequently outcompeted traditional medicine's sluggish and unsatisfactory response to the new health threats of the age.

Distillation also played an important part in economies of healing beyond a transactional "marketplace." As social structures shifted in the century of the Reformation, responsibility for spiritual and physical welfare increasingly fell on the men and women in charge of an individual household. Such *hausväter* and *hausmütter* were an important part of the readership of vernacular medical works such as Brunschwig's.[70] Distillation produced not only medicines for the family, but also chicken soup and other home goods.[71] These often gendered domestic practices gave rise to rich cultures of collecting and exchanging recipes within family networks, studied by Elaine Leong.[72] English memoirist Elizabeth Freke, for example, copied recipes for distilled medicines from George Hartmann's *Family Physician* (1696) and kept "stilled waters" in her medicine cabinet.[73]

Noblewomen were by no means free from the gendered role of household caregiver; in fact, they were often seen as a *landesmutter*, a mother to a whole territory. As Alisha Rankin has shown, German princesses such as Countess Dorothea of Mansfeld took on responsibility for the sick poor, handing out homemade remedies, sometimes to hundreds of people at a time. An expert distiller herself, Dorothea also helped other princesses to establish their own court distilleries, among them the electress Anna of Saxony.[74] Anna owned at least three *Destillierbücher* (the herbals by Röslin and Lonicer, and Ryff's *New groß Distillier Buoch*), and her pharmaceutical expertise came to be widely recognized among the European nobility. When the Infanta Juana of Habsburg, the princess of Portugal, was plagued with painful and embarrassing hemorrhoids, Anna sent her a much-valued gift of a water distilled from fresh stag horns, along with a copy of the recipe.[75] Women like Dorothea and Anna did not hawk their wares in the marketplace, but they operated within important economies of medical expertise, charitable care, and the kinds of courtly gift exchanges that kept the European political machinery running smoothly.

Of course, printed books were themselves a key part of the business of distillation, and Brunschwig's works mark the beginning of a successful new genre to which we now turn.

Destillierbücher—*Brunschwig and the Print Culture of Distillation*

Clunky though Brunschwig's titles were, the idea of a "book of distillation" stuck. In the form of the snappier compound noun *Destillierbuch*, it appeared in the titles of a number of subsequent works, marking the rise of a new printed genre over the sixteenth century.[76] In 1533, the physician Eucharius Röslin, better known for his writings on childbirth, published a version of a popular herbal known as the *Hortus sanitatis* (Garden of health), which he deemed a "noble book for the common man." To this "home pharmacy," he decided to add "the *Destillierbuch* of Hieronymus Brunschwig, which is worthy of no little praise."[77]

Upon the death of Röslin's publisher, Christian Egenolff, much of the business fell to his son-in-law, Adam Lonicer. Also a physician by training, Lonicer decided to produce a completely reworked edition of Röslin's herbal. The prolific medical writer Walther Hermann Ryff published his *New Large Book of Distillation* in 1545, closely modeled on Brunschwig's *Small Book* with some added material and more complex apparatus. These editing projects were driven by business decisions and a good nose for what would sell in the marketplace of print. Röslin thought it worth investing considerable time and labor into his volume, "reading it over this winter in addition to my other business," and Lonicer in his turn felt justified in giving the book a complete editorial overhaul, including expensive new woodcuts, because "this book has been very pleasing and sellable to the common, simple man, such that it was often reissued in print."[78] Ryff claims that his "manifold writing work" served the "common good," but he was known even at the time as a liberal compiler of others' writings with a keen eye for marketable copy.[79] As distilled remedies became increasingly popular commodities, so did "pleasing and sellable" books that described how to make them.

The *Destillierbuch* phenomenon was part of a wider "how-to" literature that took off during the sixteenth century. Early modern readers developed a voracious appetite for books that explained how things were done or made. Manuscript collections of practical medicine and miscellanies of craft techniques, alchemy, and natural magic had already circulated in manuscript form in the Middle Ages, but with the advent of print, the trickle became a veritable flood. "Books of secrets" became one of the most popular genres, read by artisans and aspiring practitioners, but also by collectors and consumers of luxury goods seeking to enhance their connoisseurship.[80] Medical and alchemical topics were a mainstay of the literature of secrets, or *Kunstbüchlein* (booklets of art) as they were often called in the German lands. From such early examples as the *Rechter*

Gebrauch d'Alchimei (True use of alchemy) to the staggeringly successful *Secreti* published under the pseudonym Alessio Piemontese, significant portions of these secrets were dedicated to the art of distillation. With their practical instructions aimed at a wide readership, Brunschwig's books anticipated the literary appetite of a broadening reading public for medical, alchemical, and artisanal practice. Distillation lay squarely at the intersection of these interests.

The tradition of the dedicated *Destillierbuch* found its apogee in a comprehensive treatment by the naturalist and humanist Conrad Gessner. First published in Latin in 1552 under a pseudonym as *Thesaurus Euonymus Philiatri de remediis secretis* (Euonymus Philiater's treasure of secret remedies), it was translated into German in 1555. A well-read and bibliographically minded scholar, Gessner included a list of authors he had consulted. Brunschwig, Gessner notes, had not exactly invented the art of distillation, as many claimed, but he deserved credit as the writer by whom it was "first put into words and divulged in our German language."[81] He also mentions Walther Ryff and Adam Lonicer, who had "more briefly captured the writings of Brunschwig and Ryff."[82]

In Gessner's list, these *Destillierbuch* authors were in good company with medieval alchemical writers such as Ramon Llull and John of Rupescissa, but also contemporary scholars who had shown an interest in distillation. Gessner quotes entire passages from *De Subtilitate* (On subtlety), the opus magnum of Milanese physician and natural philosopher Girolamo Cardano, published a mere two years before Gessner's *Thesaurus*. At several points in over a thousand pages, Cardano turns to distillation as an art uniquely suited for producing and thinking about "subtlety" in a material sense: the "thinness" or rarefied nature of things like air or the spirit of wine.[83] If Cardano provided key ideas for Gessner's project of rationalizing the art of distillation, Brunschwig was Gessner's main source of practical instructions. He is cited as an authority on when to harvest fragrant herbs, how much to distill at once, and how to manage a steady temperature: "Brunschwig judges the heat of horse dung, through the admixture of quicklime, to exceed that of a bain-marie by half a step."[84] In Gessner's book, when it came to the nitty-gritty of artisanal practice, Brunschwig was the man to ask.

Conclusion

Brunschwig's works mark the beginning of a new age of distillation by providing the first detailed how-to instructions as well as material, medical, and economic rationales for distilled medicines. His conception of distillation as a process of cyclical separation and purification speaks to a rich underlying understanding of matter that is explored more fully in the following chapter. As a medical technology, distillation mediated between the materialities of medicinal substances and the human body, extracting from the former only the subtle parts that could safely and efficiently penetrate the latter. This material mediation grew into a deeper hermeneutic relationship between body and distillation, from explanatory analogy to physiological model and diagnostic tool. In the marketplace, distillation books and distilled products appealed to an appetite for novelty, practical knowledge, and pleasant remedies that operated outside the exclusionary and stayed framework of learned physic. Brunschwig's treatment of distillation thus fostered more independent agencies for patients, practitioners, and purveyors of medicines in different economies of healing. It also spelled out the agencies of artisans as knowers and workers of nature, as we shall see in the following chapter.

5

Working with Matter

In the previous chapter, we have seen how distilled waters functioned within the human body and in the market, offering a solution to what I have characterized as a medical challenge of working with matter. This chapter argues that Brunschwig's discussion of distillation responds to two further challenges, cosmological and artisanal. Brunschwig's account of his remedies as subtle, self-targeting medicinal substances raises questions about their materiality: What kinds of substances were they, and how was distillation thought to bring about their special properties? By exploring these questions, I show how distillation informed Brunschwig's underlying understanding of matter and how it could be worked by human art—what Pamela Smith has called a "material imaginary" in her studies of early modern artisans' ways of knowing.[1]

I begin by showing how Brunschwig's imaginary was shaped by longstanding intellectual traditions of cosmology and matter theory. In the Aristotelian cosmos, all matter on earth was made up of some combination of the four elements, whereas the heavens were a completely separate material sphere, maintained in unchanging perfection by an incorruptible fifth element. This scholastic view was well established among medieval scholars in Europe, but it had competition from ancient Hermetic and Stoic ideas that were introduced to the Latin West, mostly through Arabic sources, around the twelfth century. These include the concept of *pneuma,* a prime matter more fundamental still than the four elements, a subtle soul or spirit that was akin to celestial matter and formed the core of all earthly stuff.[2] In the medieval context, distillation was increasingly understood as a technique that could not only separate the four elements but also extract the underlying prime matter. Distilling the soul or "quintessence" of matter became a quest for the elixir, a universal medicine to cure all diseases, and this idea was eagerly developed by fourteenth-century European alchemical writers, notably Ramon Llull, Arnald of Villanova, and John of Rupescissa.[3] Brunschwig mentions Llull and Villanova as authorities, and as we shall see, he was a keen student of Rupescissa, borrowing from his writings and adapting the idea of a distillable quintessence to his own practical purposes. In particular, I argue, he draws on this textual tradition to frame distillation as a solution to a cosmological challenge of working with matter: its inherent instability among the vagaries of the sublunar world.

Through his practical, artisanal working through of learned matter theory, Brunschwig provides a unique opportunity to broaden existing intellectual histories of alchemical traditions and see those ideas come to life in the early modern workshop, as I show in the second section. This brings up a third, artisanal challenge of physically working with invisible, subtle, and sometimes stubborn materials. The final section of the chapter explores the sensory and embodied expertise required to make, store, and assess distilled waters. Distillation thus provides a unique lens into artisans' intimate encounters with and understandings of matter, offering new perspectives in recent historical scholarship on practical, artisanal, and embodied forms of knowledge.[4] I argue that Brunschwig understood distillation as a unique tool for making difficult, "obstinate" materials yield to human art, but at the same time as a kind of apprenticeship in which the practitioner became nature's patient pupil.

The Alchemy of Natural Matter

Medical remedies were, first and foremost, material substances, and as such they were inherently problematic. In the previous chapter, we have seen that the coarse, earthy corpus of materia medica was a danger to the body. From his daily workshop practice, Brunschwig knew that medicinal substances were also a danger to themselves, "for if you mash up herbs, roots or other substances, it becomes unpleasant, and also nondurable and corruptible, since its coarse and uncleansed substance soon causes foul corruption."[5] Like all earthly matter, remedies were prone to become corrupt and do more harm than good. Making, keeping, and using medicines involved grappling with this unstable materiality. Once again, Brunschwig presents distillation as the solution—through this technology alone, he writes, "the unkeepable may be kept, and the coarse cleansed from its putrefaction and destruction."[6] These were big claims, and they spoke to fundamental concerns about the instability of natural matter that went well beyond medicines or foodstuffs.

In the early modern cosmos, the heavens were maintained in timeless perfection by the regular and perfect movements of the celestial spheres, but the sublunar world—the domain of nature and man—was governed by messy cycles of generation, change, and decay. Brunschwig explains why this is so in a section "on the four elements," added to the second edition of the *Small Book*. These basic constituents of all earthly matter (fire, air, water, and earth) possess hot, cold, wet or dry qualities that are individually destructive. That is, if they are unevenly matched in any natural thing, those qualities in excess cause decay and corruption, much the same as might be inflicted by the heat of summer or a freezing winter. Natural matter is truly stable only when the four elements are in perfect balance, "as you can see in May when it is neither too dry nor too humid, neither too warm nor too cold."[7] The same is true, Brunschwig says, of people; "one who has the four elements in equal measure is healthy."[8] According to Galenic medicine, the four humors acted as the vehicles of elemental qualities in the human body and maintained it in good health as long as they were balanced according to the individual's innate complexion—a state that was more an ideal to be aimed for than a reality for most people.[9] Inside and out, the human body was part of a material world governed by the fluctuations of the elements in cycles of generation and degeneration.

Brunschwig's reference to those few perfectly balanced days in May places the fleeting nature of earthly matter within a discourse of passing time and changing seasons. In early modern thinking, the seasons provided a metaphor for the four stages of the human life cycle that associated youth, health, vigor, senescence, and death with the flourishing and decay of nature throughout the year, and with the cosmos at large.[10] While its perfect celestial revolutions mark the cycles of the seasons and human life, they are but imperfectly reflected on earth. Comparing the problem of elemental imbalance and corruption to the seasons thus throws into relief the time-bound, transient, and fluctuating nature of sublunar matter in contrast with the timeless and perfect regularity of the heavens and their motions. On the whole, the cosmology of Brunschwig's time offered a rather glum view of natural matter as unstable and unpredictable.

In his reading of alchemical texts, Brunschwig found ideas and practices that promised mastery over these vicissitudes of matter. By far his most important interlocutor was the fourteenth-century friar and alchemist John of Rupescissa. As I have shown in detail elsewhere, Brunschwig borrowed freely from Rupescissa and drew on his concept of "quintessence" to frame his distilled waters as reliable remedies in an otherwise unpredictable material world.[11] In the *Liber de consideratione quintae essentiae* (ca. 1350), Rupescissa describes a quest for a universal medicine that would preserve him and his fellow evangelical believers through the approaching apocalypse. His starting point was the ancient Stoic idea of a *spiritus mundi* or *pneuma*, the primal material principle that was present in all natural things and determined their properties.[12] Rupescissa explains how this "quintessence" could be extracted by distillation and used for medicinal purposes. One of these distilled remedies—the quintessence of wine—was the marvelous panacea he had been looking for. This strong alcoholic spirit, won through a series of sophisticated distillations, was especially worthy of the name *quinta essentia* or "fifth essence"—Rupescissa also calls it "fifth being" or "man's heaven," all of which allude to the substance of the heavens in the Aristotelian cosmos. This "fifth element" was perfectly balanced, unlike the four sublunar elements that represented the extremes of materials' properties (hot or cold, wet or dry). Free of these destructive qualities, the fifth element was itself incorruptible, but also maintained the heavenly spheres in changeless perfection.

Likewise, Rupescissa believed, the distilled quintessence of wine could not itself become corrupt, and could halt the corruption of natural matter, most importantly the human body. Other medicines, including the distilled quintessences of things other than wine, always affected a patient's humoral balance in the same direction through their specific properties—hot and dry, say. In contrast, the quintessence of wine "cuts off any excess of any one quality, and whichever quality is lacking it shall restore."[13] Acting as a true panacea, it adjusts any and all disease-causing humoral imbalances.[14] Although not literally a piece of the celestial spheres of the cosmos, "man's heaven" was the closest thing to material perfection humans could hope to create on earth. As Rupescissa scholar Leah DeVun has suggested, distilling the quintessence of wine amounted to "perfecting nature" by lending it near-celestial properties.[15]

Rupescissa's claims about the powers of distillation resonate deeply in Brunschwig's understanding of the process. Already at the beginning of the *Small Book*, he describes distillation in Rupescissan terms, as making "elemental things more heavenly."[16] In the *Large Book*, Brunschwig engages explicitly with the concept of quintessence, citing Rupescissa as an authority, and borrowing entire passages from the *Liber de consideratione*, perhaps via one of the German translations that circulated in manuscript form.[17] The description of circular distillation, for example, is part of a passage corresponding closely to the Rupescissan original. While a single cycle of "the rising and falling of distillation" noticeably cleanses the distilled matter of the elements' corruptive qualities, if the matter were allowed to "rise and fall a thousand times and more through the distillatorium," it would no longer be "mingled with the destructible" at all and thus become "like heaven."[18] Brunschwig eagerly latched on to Rupescissa's idea that distillation could channel some of the heavens' element-defying permanence onto earthly things.

While Brunschwig's engagement with Rupescissa's quintessence throws into relief his own concept of distillation, it is important to note that producing this marvelous panacea is not the main concern of his *Destillierbücher*. The great majority of remedies in Brunschwig's books are much more pedestrian. The *Small Book*, for example, does not mention the quintessence of wine except in an addendum to the second edition, and instead describes distillations of simple waters that are technically much less demanding and do not involve alcohol. Even in the *Large Book*, the number of alcoholic remedies is marginal compared to aqueous distillates.[19] This raises questions as to how these remedies, the staples of Brunschwig's works, are situated with respect to the quintessence ideal. How does Brunschwig's reading of Rupescissa frame these more pedestrian waters, and what does he think distillation achieves in their production?

As we have seen, although the *Liber de consideratione* is mainly concerned with the quintessence of wine, Rupescissa's starting point is the idea that "quintessence is in all things" in nature.[20] This general sense of quintessence as the material manifestation of the innate qualities of a thing is particularly clear in a short chapter on "extracting quintessence from fruit, roots, herbs and leaves."[21] Paraphrasing Rupescissa, Brunschwig explains that this ordinary kind of quintessence contains "the power and virtue of the herb, flower, or root."[22] In other words, Rupescissa's quintessence in the general sense corresponds to the healing virtue or "soul" that Brunschwig proposes to extract from the material corpus of natural things through distillation.

This would seem to imply that all distilled remedies are "quintessences," and indeed Brunschwig anticipates that readers might want to question just this point: "It might be asked what quintessence is, since it is no more than just another distilled water."[23] He replies, "True, it is no different in form and aspect than any other water, but nobler and more subtle in its power that is hidden in it, and invisibly united in a small amount of substance. This is because its phlegmatic or watery moisture has been consumed by many a digestion, distillation, and circulation."[24] A larger number of repeated distillations makes this quintessence of a herb more subtle and concentrates its healing power in a smaller amount of liquid compared to an ordinary distilled water. While simple waters have not been refined to the point that the liquid as a whole deserves to be called a "quintessence," they nonetheless contain the herb's medicinal virtue (its quintessence in the general sense), albeit in less concentrated and purified form. Simple waters and "quintessences," in other words, are not different in kind, but merely in their degree of perfection through distillation.

Most of Brunschwig's distilled waters, then, are not the purified quintessences of medicinal simples, much less the all-curing quintessence of wine, but rather more pedestrian versions of the quintessence concept. For one thing, they are not incorruptible material objects—they last considerably longer compared to mashed-up herbs and other such crude stuff, but as we shall see, they have a shelf life that Brunschwig discusses at some length. For another, distilled waters are not panaceas like the quintessence of wine, free of elemental qualities. On the shelves of his shop, Brunschwig arranges them according to such distinctly elemental categories as "herbs of a hot and dry substance" or "roots of a moist and phlegmatic complexion."[25] According to these qualities, distilled waters have a specific rather than a tempering effect on the body's humors and could be dangerous in the wrong hands. Recently distilled water of water lilies, for example, should not be taken internally "because it is half poisonous through its great cold."[26] Unlike a panacea, distilled waters had the potential to harm as well as heal.

Nonetheless, even when it comes to distilling simples, the material effects of the process are significant. Although the waters still contain residual elemental qualities that spell their eventual decay, this occurs in a regular and predictable manner. If the distilled cooling virtue of water lilies is initially dangerously powerful, after one year on the shelf it has faded sufficiently to be taken internally, and "in the third year its cooling effect is tempered."[27] Water of wild roses even changes from "cooling" to "comforting" as its properties decline.[28] Brunschwig is confident that the medicinal virtues of his distilled waters diminish by a specific and knowable amount each year. Properly managed, they are reliable remedies, at least for a time, and changes in their properties can be used to specific therapeutic ends. Once a water does approach the end of its shelf life, it may yet "prevail against its time" if it is reinfused with fresh herbs and distilled once more.[29]

For the vast majority of Brunschwig's remedies, distillation does not give rise to quintessential incorruptibility, but it has the power to stave off corruption temporarily, by removing excess material dross and unhelpful elemental qualities. It does not produce panaceas, but applied to the whole gamut of plants, animals, and minerals described in the herbal section, it spawns an arsenal of remedies with stable and knowable medicinal properties and shelf lives—a vast improvement on the unpredictable, potentially harmful, and materially unstable nature of undistilled remedies.

Although Brunschwig does not ascribe the ideal virtues of the quintessence of wine to his distilled waters, I contend that he takes more from his reading of Rupescissa than a general notion of distillation's power over matter. Hist distilled waters have an intriguing cognate in Rupescissa's medical cosmology. In that cosmology, as we have seen, all natural things have a quintessence that determines their properties, including their medicinal effects. The quintessence of wine is merely a special case in that its curative power is universal. Rupescissa outlines a complex relationship between these two kinds of quintessence. While both can be extracted by distillation, Rupescissa also describes at great length how ordinary quintessences can be extracted and absorbed, by means of a simple infusion, into the quintessence of wine, which has "such power that it may extract from whichever medicinal thing all powers, properties, natures and effects."[30] Likewise, heating metals and quenching them in distilled wine releases their properties into the liquid. In a delightful extension of his cosmological analogy, Rupescissa refers to this process as "capturing stars to adorn our heaven."[31] As the fifth element of the heavens mediates the effects of celestial bodies, the quintessence of wine reinforces and delivers the medicinal powers of their counterparts on earth.

Brunschwig borrows both the concept and the metaphor: "for as heaven is full of stars, so one may add all other medicine" to the quintessence of wine, which "draws to itself all of their power and virtue, and releases it in its action, as a celestial body does through its movement."[32] Taking Rupescissa's cosmic image one step further, Brunschwig extends the role of stars in this analogy not only to medicinal virtues extracted into quintessence of wine by infusion, but also to his simple distilled waters. "Every natural being must be governed by the movement of the heavens, [but] without the corpus of the heavens; in the same manner the human body [is governed] by a knowledgeable practitioner without the corpus of medicines, but rather with the waters that are separated from the dross."[33] The comparison is apt. As we have seen, distilled waters have a specific effect just as likely to harm as heal, much like astral influences. But they also resemble the stars in that they exert their power with clockwork regularity, much more akin to the perfect motions of the heavens than the often-haphazard permutations of earthly matter. As an astrological practitioner might give advice on the influence of the stars on health and disease, Brunschwig posits the distilling practitioner as an expert in administering the man-made analogs of astral virtues.

Reading Brunschwig alongside Rupescissa shows that he engaged seriously with learned matter theory, and framed distillation as a solution to a fundamental cosmological problem. Like all natural matter, human bodies and medicinal substances were made of corruptible matter that was subject to the passing of time, changing seasons, and the excesses of the four elements. With help from Rupescissa, Brunschwig articulates distillation as the saving technology in humans' struggle with matter. By separating out the corruptible material dross, it channels the stable and predictable materiality of the heavens onto sublunar matter. Although no Rupescissan quintessences, Brunschwig's distilled waters were remarkable material entities, stable or at least predictable both in their shelf life and their therapeutic effect. In his pragmatic reading of medical alchemical texts, Brunschwig carves out a place for his remedies as sublunar healing stars in the medical cosmology of his day.

Brunschwig's commitment to learned understandings of matter and cosmos was serious, but it was oriented towards practical ends. He was no Rupescissa—a religious visionary seeking an alchemical panacea to preserve his band of faith warriors in the impending final battle. Brunschwig was a medical artisan on a mission to provide a broad readership with manageable techniques for making medicines in the workshop or the home. In his writings, he breaks down the quintessence ideal into practical instructions that reveal his own hands-on familiarity with the technicalities of the process. Even when copying Rupescissa's instructions for seven consecutive distillations, Brunschwig adds that the heat should be lowered by a specific degree each time, because the material "is made more subtle and thinner by the preceding distillation, therefore the matter becomes lighter, and from the subtle is separated that which is subtler still, by a lesser fire"—a rationale that is absent in the Latin and Middle German versions of Rupescissa's text.[34] Brunschwig's how-to texts thus offer a unique view of how alchemical matter theory comes to life in the early modern workshop.

Distillation in the Workshop

For all his interest in cosmological rationales for distillation's preservative powers, Brunschwig also wrote extensively about the practicalities of making and storing remedies. Distilled waters were mostly clear liquids kept in identical flasks or earthenware jars, which could make it difficult to tell them apart if they were mixed up. Furthermore, as we have seen, their medicinal virtues gradually changed and faded over time. Distillation made that process regular and predictable, but this was of little use unless one knew how long a distilled water had been sitting on the shelf. Therefore, Brunschwig writes, "it is necessary at all times to write on it the date or year of its decoction or distillation."[35] As is still the case in a modern laboratory, the best sample is only as good as its label. Of course, these inscriptions had to be readable months or years down the line, so paper labels were no good since they were liable to be "nibbled off by mice."[36] Brunschwig speaks from painful experience: "oh how often my labels have rotted away, been gnawed at, or torn off, and I suffered great damage because of it."[37] Instead, labels should be made from sheets of lead that could be inscribed by engraving or pouncing. Protected by individual wicker baskets, the flasks should then be arrayed "according to the letters (a, b, c) to quickly find whichever you desire."[38] Brunschwig's neat shelves of fastidiously labeled flasks in alphabetical order are a charming reminder that, in the everyday lives of artisans, making medicines was just as much a question of prosaic logistics as alchemical matter theories.

The expertise of making and storing medicines was most directly linked with the trade of the apothecary who, as Brunschwig explains, had to distinguish fresh and genuine wares from second-rate or fraudulent ones. With medicinal simples, it can be particularly difficult to tell "whether they are good, fresh, well-selected and not too old, and whether the soul and spirit of its power has not gone," especially if they had been tampered with by an unscrupulous supplier.[39] Brunschwig never speaks of himself as an "apothecary," although others sometimes attached this label to him (see chapter 4). In any case, Brunschwig's experience with hungry mice, proper labeling, and efficient storage speak to a knowledge of materia medica that he shared with other local practitioners. The duties of the assistant apothecary (*apothekerknecht*) in the city's so-called Major Hospital, for example, might almost be copied from Brunschwig's book: "He shall also take care of the distilled waters, and keep them in clean vessels, which should be marked with labels so one may know which is which, and how old it is. For like other medicines in the pharmacy shop, one water may last longer than another."[40] Within the city at large, too, municipal authorities regulated the quality of pharmaceuticals. As we saw in chapter 1, the council ordained in 1500 that every pharmacy shop should be inspected twice a year, by a panel of two physicians, one senior apothecary, and a fourth trustworthy citizen. Just as surgeons inspected injuries or candidates for municipal charity, as we saw in chapter 3, artisanal experts of medicinal substances played their part in the city's apparatus of medical governance.

For purveyors of medicines like Brunschwig, distillation was in large part business. It worked well in economies of taste and preference, as we saw in the previous chapter, but it also had to function in the workshop's economies of material and labor. Especially when describing complicated and laborious processes, Brunschwig often provides alternative techniques and nifty shortcuts. In a recipe for potable gold, Brunschwig suggests improving the efficiency of the process by inserting an oil-soaked piece of sponge in the neck of the alembic: "Then only the spirits distill through the bath sponge, and the moisture, that is the water or phlegm, may not pass through it because of the tree oil; and thus you distill more in one distillation than you otherwise would in three."[41] Making the quintessence of wine was especially lengthy and laborious, as we have seen, but Brunschwig provides an alternative, "with less effort and work, and without great cost," for those unable to face the "frustration or cost" or indeed "neglect of their other business."[42] The "true and just way of experience" is very well for "princes and lords who may achieve it thanks to their affluence," but a shortcut may be necessary for "the workman who does this with great labor, for the pleasure of princes and lords, and for his own sustenance; for in another way he might perhaps have much heavy labor and work."[43] Some of his readers, Brunschwig was well aware, practiced distillation to earn a living.

In addition to apothecaries or surgeons such as himself, Brunschwig may have been thinking of the pharmaceutical assistant at the local hospital, the distilling nuns he mentions in his books, the cottage industry of *wasserbrenner* producing simple medicinal waters, or the local producers of good Alsatian brandy from whom he learned about water-cooled distillation apparatus for strong alcohol (figure 4.7). Drinking spirits were familiar commodities in the German lands from the Middle Ages onwards, and *weinbrenner* (distillers of wine) were a recognized trade. In Berlin, distilled grain alcohol became so cheap and ubiquitous by the end of the sixteenth century that the city consumed more *schnaps* than beer. In Brunschwig's native Alsace, the wine cellar of the empire, the harvest was sometimes so plentiful that vintners ran out of barrels before they ran out of wine, and using the surplus to distill brandy was an economical solution. In the later sixteenth century, the *weinbrenner* of Strasbourg alone exported thousands of liters of prized Rhenish brandy each year.[44]

Brunschwig's reference to "the pleasure of princes and lords" points to noble patronage as one of the most important ways of making a living for practitioners who marketed themselves as alchemical experts, offering secrets of medicine or metallic transmutation in exchange for court positions, materials, and salaries.[45] While the better known among these figures often managed to negotiate long-term relationships with important potentates, there was also a more workaday class of practitioner traveling the German lands, aptly characterized by Tara Nummedal as "entrepreneurial alchemists." The career of Philipp Sömmering illustrates the type; originally a pastor, he lost his position at a mere twenty years of age in 1555, and embarked upon years of travel, stopping here and there to acquire alchemical knowledge. He bought books, but also learned directly from expert practitioners—he paid a woodcarver from Erfurt five thaler to teach him distillation. Together with a like-minded alchemist, Sömmering projected to produce the philosophers' stone for Duke Johann Friedrich of Sachsen-Gotha, and signed a contract that specified material outlay and promised yield as in any other business transaction.[46] Itinerant, business-minded, and with a nose for lucrative opportunities, practitioners like Sömmering viewed their alchemical expertise as a commodity that could be bartered like any other.

Economies of labor and materials continued to play an important part in other kinds of technical writing that involved distillation, such as the increasing number of works on mining, metalwork, and other "arts of the fire." Georg Agricola's *De re metallica*, for example, contains several chapters on distilling *aqua fortis*, a strong mineral acid that was used, among other things, to separate gold from silver and other lesser metals. Making this *scheidewasser* (separating water) involved a lengthy distillation process, and once the separation was accomplished, a further series of distillations could recover not only the silver from the solution, but also much of the acid, which could be reused. These materials were costly and had to be economized, but so did the artisans' time and labor. Weighing these considerations, Agricola writes, "when making strong *scheidewasser*, one has to stay awake at night, and apply exceptional diligence, labor and work." Much like Brunschwig, he offers an alternative that is "not very costly and not very laborious."[47] Artisans' diligent labor and careful attention served pragmatic and economic ends of their work, but it was also their chief way of knowing.

Embodied Knowledge, Disobedient Matter, and the Apprenticeship of Nature

Reading between the lines of Brunschwig's instructions, we can discern the remarkably complex embodied expertise required for dealing with the subtle materialities of distilled healing virtues. As we shall see in the following chapter, Brunschwig goes to unprecedented lengths to codify in print the kinds of sensory cues that signaled the correct temperature of a still or the moment at which a particular substance begins to condense in the receptacle. As Brunschwig notes in his instructions for labeling and storing distilled waters, these were elusive material entities. Faced with row upon row of clear liquids, it could be difficult to tell them apart, and it was "difficult to know when a water is so corrupt that one should pour it out" especially "if one does not know the time of its distillation."[48]

Artisans always find ways of attuning their senses to their materials' elusive properties, and Brunschwig wrote down some of them. A turn in color or a murky deposit at the bottom spelled the water's decay. In addition to visual inspection, the conscientious purveyor of distilled remedies should assess whether their "taste and scent have diminished noticeably or entirely disappeared," in which case they should be discarded.[49] The fluids' viscosity could also provide clues to their remaining medicinal power. This was not a readily visible property, and required specific tests to become available to the practitioner's senses. It could be visualized by "pouring it out from a height; if it runs like a thread . . . it has become corrupt and should be discarded."[50] One could also simply place a drop of liquid on a thumb nail, and "if it does not run off soon, it has become corrupt."[51] The knowledge of making distilled remedies, and of judging their quality, was to a large extent located in the practitioner's body and well-trained senses.[52]

Such direct physical engagement made for an intimate understanding of natural matter and its properties. With many raw materials for distillation, "the matter is initially coarse, and because of its coarseness is not easily obedient to distillation, due to the rawness and impurity of its substance."[53] When it comes to a particularly "hard, that is a coarse or obstinate substance," distillation alone may not be enough.[54] In these cases, the material "must be digested and putrefied, so that it becomes more obedient to distillation."[55] The process of digestion or putrefaction, as we have seen, meant a prolonged exposure of material in a sealed flask to the gentle heat of fermenting horse dung, often for several weeks. To implement this process successfully, Brunschwig writes, "it is necessary to note with diligence the nature of every thing or substance, whether it is of a hard, coarse or soft substance."[56] Distillation thus throws into relief the properties and challenges of natural matter that artisans experienced through their daily work.

This hands-on experience placed Brunschwig in an intriguing relationship with his materials. They are described not only as "hard" and "tough," but also as *widerwertig*, which could be translated as "repulsive," but also "obstinate" or "contrary."[57] He experienced materials not as passive, inanimate things that yielded to human artifice in a straightforward manner, but as complex entities that had to be prized from the natural world and might actively resist manipulation. This was a view of nature shared by many early modern craftsmen, artists, and alchemists who encountered matter in similarly intimate, embodied ways.[58] But Brunschwig was not only one of the earliest writers to articulate that relationship, especially in print; he also did so in strikingly explicit terms that imbue materials with agency and almost a sense of personality. The more stubborn kinds of matter, he suggests, actively resisted the distiller's efforts, and had to be "digested" to render them "more obedient." Artisanal processes of material transformation such as digestion and distillation, in other words, were tools for disciplining disobedient matter so that it might be understood and manipulated.

Brunschwig's writings suggest that distillation played a unique role in shaping artisans' material imaginaries—their view of the material world, its transformations, and humans' ability to imitate and manipulate them. Such early modern understandings of materials were often structured along binaries of qualities experienced by the body—metals could be sweet or sour, sulfur was naturally brittle but could be made malleable, sand used for casting molds could be coarse or subtle, and all kinds of substances could be awkward to work with or easy to handle.[59] Brunschwig articulates distillation as a key technology for pushing materials along the spectrum from coarse to subtle, but also from stubborn resistance to disciplined obedience.

Even with obstinate materials, however, this hands-on encounter was not necessarily or exclusively an antagonistic one. The disciplining process of digestion required intimate knowledge of materials' hardness, subtlety, and so on, and it should not be rushed: "This is to be done sweetly and gently, and not by quick means, since nature does not abide rapid permutation or change. Neither, therefore, should digestion be performed to separate quickly, but in such a way that the artist or worker of nature follows nature in like manner, as a pupil."[60] As its name suggests, the technical process of digestion or putrefaction emulates natural processes that happen in their own good time. Artisans, therefore, should follow suit. "Nature does not like sudden or quick change. Therefore each master craftsman should operate his work by following nature as his master craftsman, for she rejects constraints, or being coerced by force."[61] To master nature, one first had to become her patient apprentice, and breaking her resistance was a matter of gentle coaxing rather than brute force. In Brunschwig's account, material interventions such as distillation and digestion were key tools for negotiating a relationship between artisan and nature—a relationship that acknowledges and absorbs nature's wisdom as well as manipulating and disciplining her more recalcitrant materials.

Imitating nature and making objects that played across the boundaries of the natural and the artificial became a central way of knowing in the century after Brunschwig, expressed most clearly in the *Kunstkammern* of princes and powerful merchants. But where these collections of marvels generally expressed mastery over territories and nature herself, Brunschwig invokes an image of the practitioner as a pupil or apprentice who follows nature by imitation, reflecting the kind of education Brunschwig himself would have received in one of Strasbourg's guilds. As I show in chapters 1 and 6, he absorbed from this environment a sense of manual skill and good craftsmanship honed by endless repetition of the same gestures, by wrestling time and time again with materials and processes, using the entire body and its senses to learn from and manipulate technical apparatus, medicines, and the human body.

In addition to the medical and cosmological challenges of working with matter, Brunschwig's understanding of materials and of distillation as a transformative process was also profoundly shaped by his everyday experience as a craftsman. A distiller and purveyor of remedies had to think within the material economies of the workshop, the marketplace, and city regulations. His expertise of materials and techniques was shaped by close physical and sensory engagement, and thus known in the body as much as from alchemical texts. With his detailed discussion of sensory cues, material tests, and disobedient matter, Brunschwig articulates early modern artisans' ways of knowing with striking lucidity, and remarkably early in the history of the printed book.[62] Over the course of the sixteenth century, perhaps only Paracelsus wrote more explicitly and more insistently about the body and bodily experience as the chief sources of natural knowledge (see chapter 7).[63] While Paracelsus's debt to Brunschwig remains to be elucidated, it is likely no coincidence that this manifesto of artisanal knowledge making came from someone whose own way of knowing nature and the human body was so closely tied to the process of distillation.

Bodies and Matter in Print

In Brunschwig's books, distillation was a powerful process of material transformation. As such, it was the ideal tool for overcoming the threefold challenge a medical craftsman faced when working with problematic matter, be it harmful, elusive, or outright disobedient. We saw in the previous chapter that the first of these challenges was medical. The coarse, unyielding material corpus of materia medica threatened to wreak havoc with the inner workings of a body understood in terms of unobstructed flow. Only distillation, Brunschwig argues, offers an adequate solution by separating subtle medicinal matter from its attendant dross. From a process that mediated the material exchange between the body and outside natural matter, distillation grew into a model for understanding both—fostering, I suggest, a culture of the "distilling body" that remains to be fully explored.

A second challenge arose from the cosmology of matter itself, as I have shown in this chapter. All earthly things were imbued with ill-balanced elemental qualities and unpredictable permutations that, over time and in due or undue season, brought about their decay. In this uncertain world, Brunschwig draws on Rupescissa's concept of the quintessence to explain how even a relatively straightforward distillation process could channel some of the incorruptible, predictable nature of the heavens onto mundane medicines.

Hands-on engagement with materials in the workshop presented a third challenge. Many distillers had to earn a living by their labor, and had to think carefully about material economies of the workshop, the marketplace, noble patronage, or the household. They also faced the epistemic challenge of knowing their materials in artisanal, embodied ways that only came with personal experience and were difficult to glean from a book, even one written by an expert as lucid and aware as Brunschwig. Furthermore, some materials might stubbornly resist any such efforts. Here, too, I argue, Brunschwig presents distillation and its preparatory process of digestion as a technique uniquely suited to disciplining, knowing, and manipulating even the most problematic stuff of nature—if one is willing to become her patient apprentice.

While Brunschwig's account of distillation can be traced to the medical, alchemical, and artisanal traditions outlined in this chapter and the preceding one, I argue that he brings them together as a new project for a new audience. As well as positioning distilled remedies with respect to learned views of the body and the cosmos, Brunschwig places them firmly in the workshop, the marketplace, and charitable or family care. He articulates his own, artisanal version of the quintessence, without the grandeur of Rupescissa's ambitions of perfecting evangelical bodies for the coming apocalypse. Since we must all die one day, Brunschwig writes, "it would be foolish to seek in this life an art which may preserve our body forever."[64] His project is to "seek such art as will extend our life to the limit set by God."[65] In the pursuit of that art, labeling one's flasks was more important than chasing an elusive panacea.

Brunschwig's intended readers and distillers were not warriors of faith at the end of time, but ordinary men and women seeking health and well-being within the limits of their means and status. They were medical practitioners of all stripes, including learned physicians, but also tradespeople distilling for a living, and common folk with little or no medical training. Nevertheless, Brunschwig's distillation books shaped important new agencies for such lay folk. First, distillation enables and legitimates medical self-sufficiency. The subtle material nature of distilled remedies means that they harmlessly penetrate the body, and find the place where they are most needed—without, it is implied, the need for a physician. This argument is strengthened by the ideal of the quintessence as a true panacea that cures any disease—even though the ideal is rarely achieved in Brunschwig's remedies, it informs an understanding of distillation as a process that confers onto earthly remedies some of the heavens' material stability. Second, Brunschwig underscores the agency of artisan-distillers as serious experts in their own right who know nature through the body and senses, as apprentices to nature. The following chapter explores Brunschwig's innovative use of the print medium to codify this embodied expertise in his books.

PART IV

Crafting Medicine

EMBODIED KNOWLEDGE, EXPERIENCE, AND PRINT

6

"The Habit of Those Who Have Done It Often"

FROM EMBODIED EXPERIENCE TO THE VIRTUAL APPRENTICESHIP OF PRINT

The craft of healing dealt with an elusive body of knowledge, and putting it into written words was a challenge. For all his detailed instructions, Brunschwig freely admits when it is time to capitulate to the impossibility of the task. For example, he describes how to set up the apparatus for distilling oil from nutmeg and certain woods, but he has no words to tell the reader how to recognize the crucial point at which the oil has been extracted: "you do not know for sure whether it is distilled out or not, except by good chance and the habit of those who have done it often."[1] The "habit" of sensory expertise had to be formed inside the artisan's body, through repeated practice. Like artisan-writers before and after him, Brunschwig recognizes that key aspects of his craft could only be learned by doing, not reading. As Paracelsus would later put it, "Who could be taught the knowledge of experience from paper?"[2] This chapter explores how the artisans in Brunschwig's local community articulated the acquisition of embodied habit, and how Brunschwig and his publisher rose to the challenge of conveying practical, experiential, and sensory forms of knowledge in the printed book.

In the distillation books, we have begun to see what Brunschwig means by this "habit" of embodied and sensory knowledge (see chapter 5). The distiller had to become nature's apprentice, patiently learning her ways—only thus could he learn how to discipline her more obstinate materials and fine-tune his senses to the elusive properties of distilled waters. Paracelsus later described this way of learning as "sounding out" or "overhearing" nature's secrets, and the French artisan-philosopher Palissy called it a state of being "wide awake" to sensory perception.[3] Pamela Smith has suggested that we can understand this way of knowing in relation to mystics' experience of the ineffable divine, felt or "tasted" in the body rather than known in the mind. Anthropological and sociological approaches to modern-day craft-making provide another perspective. Much like Paracelsus, Tim Ingold describes humans' skilled interaction with the material environment as a kind of listening.[4] Smith concludes that acquiring artisanal expertise is not so much a matter of learning specific rules or information, but rather habits of sensory attention. With experience, these in turn form habits of intuitive response to the contingencies of working with matter.[5] In Brunschwig's account, the "habit" (*gewonheit*) of recognizing the stages of distilling nutmeg oil is just such an intuitive, inexpressible skill.

How does this fundamental difficulty in writing down craft knowledge square with the steady trickle of practical or "how-to" texts, which began long before Brunschwig and became a veritable flood in the sixteenth and seventeenth centuries? Part of the answer is that most of these texts were not primarily aimed at conveying technical information. Much of the practical literature before 1450 was written by scholars and other nonpractitioners, as an exercise in connoisseurship or theorizing about making. After this point, artists and artisans increasingly took pen in hand themselves, but their writings were often aimed not at makers, but at consumers and patrons, to inform their appreciation of skilled craft production.[6] Artisans did not need writing to make things or pass on knowledge, but it was an apt medium for negotiating the status of their craft, and for challenging traditional hierarchies of knowledge that considered the *techne* of the mechanical arts inferior to the certainty and abstraction of *episteme*.[7] Brunschwig, too, was motivated by such concerns. As we have seen, his *Cirurgia* argued that surgery was an honorable and even learned art, and his distillation books advertise the safety, efficacy, and pleasant taste of distilled remedies. As I show in the next chapter, he makes a case for the epistemic importance of particulars and *techne*, and even sets them above the physicians' rational *episteme*.

At the same time, however, Brunschwig clearly did intend his books to convey craft knowledge to the reader. They are addressed to "all those who want to learn surgery" and "those who wish to learn the art of distillation," and their tables of contents list what each chapter is going to "teach" the reader.[8] The writing sometimes takes the form of a didactic dialogue in which the author, in the guise of a teacher, answers questions that the reader might ask. This was common enough in learned texts and in some surgical writings, but in Brunschwig's books, the teacher's voice is not that of an academic, but that of a master artisan instructing an apprentice.[9] The *Cirurgia* could not replace a surgical apprenticeship, but it complemented practitioners' prior training, particularly "in towns, markets, and villages, and especially in castles far away from the big cities."[10] The distillation books reiterate this geography of intended readers, including "common people who cannot obtain or afford medicine."[11] Even the "layman" without any prior knowledge could follow the instructions to make medicines for themselves and their family, independently of learned physicians—a deliberate and controversial use of printed books, as we shall see in chapter 8.

This chapter argues that Brunschwig's books on surgery and distillation should be taken seriously as a project of communicating practical instructions and embodied expertise in print. I begin by exploring what the Strasbourg archives tell us about artisans' notions of expertise as a "habit of those who have done it often," and how it is acquired by apprenticeship in the local guild system. Turning to Brunschwig's books, I argue that they make remarkably innovative use of text and image to provide something of a virtual apprenticeship in print, using three key strategies. First, Brunschwig's cautionary anecdotes aim to provide vicarious experience by anticipating common mistakes. Second, the books push contemporary boundaries in their use of the printed image and its relationship to the text, while at the same time providing careful reflection on its epistemic limits. Finally, I show how Brunschwig's instructions guide not only readers' actions, but also their own sensory experience, and the formation of workmanlike bodily habits. Tracing this move from apprenticeship to print highlights the epistemic possibilities and limitations of the written word and the early printed book, but it also offers a clearer picture of Brunschwig's own understanding of the relationship between craft and medicine.

The Apprenticeship of "Constant Practice"

Brunschwig's books, as we have seen in chapter 1, were shaped by the local culture of artisans, guilds, and their standards of *werschafft*. This expertise of materials and gestures was acquired through an education that began with an apprenticeship of varying duration, depending on the trade and the city.[12] In Strasbourg, most apprenticeships lasted two to three years, but the details were subject to negotiation.[13] In the weavers' guild, for example, the cost and duration varied according to the number of techniques the master agreed to teach.[14] Apprentices learned by observation and imitation, working closely with the master, often living in his household.[15] As journeymen, they usually spent many more years traveling to hone their skills.[16]

The question of skill and training surfaces in an archival document recording a dispute between sculptors and painters around the turn of the sixteenth century.[17] The sculptors' complaint is this: having been inundated with work, they had in the past allowed the painters to do some minor sculpting work. The painters, however, had gone on producing sculpture as a matter of course. Asking the council to put a stop to this, the sculptors cite "ancient custom,"[18] but also cast doubt on the painters' skill for sculpting: they "have never done this nor can they do it; they have never earned it [*erdient*], practiced it [*eriebt*], or learned it."[19] The sculptors, on the other hand, have "earned it and practiced it at great cost."[20] The word *eriebet* comes from the verb *üben* (to practice), and in this somewhat unusual combination with the prefix *er-* it connotes something like "attained through practice." The other key term, *erdienet*, is a similar construction with the verb *dienen* (to serve). Close to the more common *verdienen* (to deserve), it signals the sculptors' view that they had acquired their skill and the right to exercise it through years of service to their master and their guild.

This language of artisanal training resonates in Brunschwig's account of medical knowledge making. He repeatedly emphasizes that much of his own expertise stems from "diligent practice, use, and *practica* of the handiwork."[21] Like the sculptors, he articulates skill in terms of practice (*übung*), associating it with two other key terms, "use" (*bruch*) and *practica*. The German *bruch* (or *brauch*) means "use," "repeated use," or even "custom," underscoring the sense of extensive hands-on engagement. It is a term frequently found in later artisanal writings—according to Albrecht Dürer, for example, *brauch* was the only way for trainee draftsmen to develop their skill.[22] With the Latin *practica*, Brunschwig referenced a technical term used in medicine to distinguish its practical, contingent work from rational *theorica*.[23] Throughout his works, he associates medical *practica* with artisanal *übung*—a good surgeon is one who has "learned through constant practice,"[24] and a good physician is a *practicus*—one who has "practiced [*geübet*] and learned from his *practica*."[25] His books, he claims, are full of things that "I have done and practiced with my hands, through the art of *practica*, . . . not a little, but many times over."[26] This artisanal notion of gaining expertise by practicing the same gestures over and over was a key part of Brunschwig's larger commitment to experience as a legitimate source of knowledge, as we shall see in the next chapter. In his books, rarely articulated forms of expertise and training surface from the obscurity of the body into the light of surviving text. This was by no means a straightforward undertaking, and it required an imaginative use of the print medium.

"What Happened to Me Will Not Happen to You": Pitfalls in Print

Learning by doing involves trial and error. Brunschwig displaces some of this learning process to written text so that readers (and their patients) may be spared the most common and consequential mistakes. Speaking from personal experience, he cautions practitioners not to overlook small lesions: "I have often seen grave afflictions from such small wounds which one entirely neglects; this has happened to myself and many others, so you are required to regard these wounds with great diligence and care."[27] Individual case histories can serve as more specific cautionary examples, such as the following anecdote from the *Cirurgia*: "In a laudable city I have seen barbers, untrained [*ungeübet*] in the art, who were confronted with a wound in the arm that had ulcerated and was full of pus. They went about squeezing out the pus every day . . . until nearly the entire limb was spoiled."[28] These barbers were sadly lacking in the gold standard of craft training: they were *ungeübet*, literally "unpracticed." Brunschwig's didactic purpose in sharing the episode is made explicit: "To this end I have written this for you, that you may know how to treat these cankerous wounds in such a way that the person does not perish."[29] The anecdote's vicarious experience, he hopes, will allow surgical adepts to make better choices when they encounter similar cases.

The art of distillation, too, was riddled with accidents waiting to happen. Furnaces and glassware were apt to "tear or break in the fire" unless they were protected with a "lute"—a sort of putty whose production was a whole art in itself—and cooled down slowly after use.[30] Brunschwig freely shares instructive accounts of his own workshop disasters. While distilling oil of turpentine, he relates,

> I was called away to a patient, so the oil went into the water, and when I came back, a layer of oil was sitting on top of the water. . . . I did not have the sense or wit simply to decant off the oil, . . . so I poured the lot into a new flask and thought I would just extract the water by distillation. . . . But I was called away again, and in the meantime the water evaporated from the oil, and some of it condensed on the side of the flask and dripped back into the oil, which rose inside the flask with a great tumult, and fumes erupted from the flask, blowing off the alembic.[31]

Things got worse when Brunschwig returned late at night, telling his servant to bring a light to investigate the accident.

> When the light arrived, the fumes touched it, and fire burst forth all around, and within a blink of an eye went out again, nevertheless burning off mine and my servant's hair, clothes, and eyebrows. We fell to the ground and did not know where we were, but before long we got up again and fetched a closed lantern so the same thing would not happen again, and threw ashes in the furnace to smother the fire.[32]

He freely admits to a momentary lack of "sense or wit" and does not mind appearing in a rather ridiculous tableau—gazing upon the wreckage of his workshop with smoldering eyebrows—if this provides an opportunity to educate his readers, "in order that what happened to me will not happen to you."[33] Learning from experience could be dangerous and expensive, and Brunschwig aims to supply vicarious experience in writing; rather than break precious glassware or cause major explosions, readers could train their "sense and wit" by reading his books.

Just as an attentive master might prevent his apprentices' worst mistakes as he peered over their shoulder, Brunschwig provides cautionary tales as a kind of shortcut to flatten the learning curve. Although histories of science often emphasize the importance of trial and error, especially in practical and artisanal modes of learning, scholars of the early modern period are only beginning to explore strategies of dealing with failure or writing down mistakes—what Sven Dupré has called the "codification of error"—as a way of textualizing craft knowledge.[34] This was especially relevant for surgeons, whose work inevitably involved messy, unpredictable outcomes and mishaps. As Heidi Hausse has shown, vernacular surgeons of the later sixteenth and seventeenth centuries developed cultures of error that included social and didactic uses of narratives about mistakes and accidents.[35] These codifications of error often foregrounded the mistakes of others, especially in surgery were mishaps could be fatal to the patient and to the already ambivalent reputation of body workers (see chapter 3). Brunschwig, too, focuses on the errors of other surgeons in the *Cirurgia*, whereas he relates his misadventures in distillation with surprising frankness, perhaps because his broken alembic and singed eyebrows were less detrimental to professional sensibilities than botched surgeries.

Images

Brunschwig's books use images as well as text to provide experiential knowledge. In this respect especially, Brunschwig's collaboration with his publisher, Johann Grüninger, was a fruitful one. The Grüninger press is well known for its lavishly illustrated editions of classical drama, poetry, and chivalric romance, but it was also especially forward-looking in its use of images in volumes on natural knowledge. One of Grüninger's masterpieces was a 1485 edition of a popular vernacular herbal, the *Gart der Gesundheit* (Garden of Health), for which he commissioned new, highly detailed woodcuts. His design of a new two-column layout made the volume more economical to produce, and, crucially, allowed the images to be seamlessly integrated into the text column. This successful layout was

FIGURE 6.1. Male and female mandrake. Brunschwig, *Small Book*, fol. 19v. Credit: The Huntington Library, San Marino, California, 85769.

reused in many places in Brunschwig's works, as were the woodcuts of plants in the herbal section of the *Small Book*.[36]

Although these images were the cutting edge of botanical illustration, Brunschwig was remarkably diffident about their potential to convey knowledge. Readers should pay close attention to the written description of "the appearance and species of every herb" because this "cannot be shown in an image to someone who has not known it before."[37] The images are "nothing but a feast for the eyes, and to give an indication to those who cannot read."[38] To be sure, some of Grüninger's images (e.g., of the mandrake, figure 6.1) show a plant's cultural and cosmological meanings rather than its appearance, in the emblematic tradition of medieval herbals.[39] Such meanings were a key part of natural history, but Brunschwig's focus is on identifying the correct medicinal plant, and he admonishes his readers to "pay attention not just to the figures, but to the text," and to practice "identification by sight" of real plants.[40]

There are few anatomical images in Brunschwig's works. The first edition of the *Cirurgia* contained none at all, and the woodcut of a skeleton in later versions was copied from an earlier image, and added mainly to upstage an unauthorized edition of the book printed in Augsburg.[41] As discussed in chapter 2, Brunschwig cautions that the image shows only the rough topography of bones, whereas other structures could only be known through the "inspection [*gesicht*] of dead bodies" during dissection.[42] He insists that an image cannot replace *gesicht* (sight or inspection) of the real thing, much less the hands-on experience of stripping away the body's layers. As early as 1500, Brunschwig voices the kinds of concerns about the potential and the limitations of printed images in conveying knowledge of organic structures that were later formalized by learned botanists and anatomists.[43] Decades before well-known scholars such as Leonhart Fuchs and Andreas Vesalius codified the importance of *autopsia* and producing images "from life," Brunschwig problematizes the relationship between images and knowledge from an artisanal perspective and insists on seeing and doing for oneself.

Brunschwig expresses no such reservations about the illustrations of technical equipment in his books. The *Cirurgia*, as we have seen, presents a full-page illustration of intricately worked surgical forceps, scissors, probes and knives arrayed in a cabinet (see figure 2.2), with smaller pictures of individual instruments interspersed in the text. In Grüninger's two-column layout, they are embedded in exactly the right place in their textual description, which often concludes "and this is its form" (figure 6.2).[44] The reader's gaze moves from text to image and back, without any apparent distinction in value between verbal and visual description. Such confident use of images had precedents in medieval surgical manuscripts, for example, John of Arderne's work on anal fistulae with its detailed drawings of instruments and operations.[45] Indeed, the small inset woodcuts depicting drills and trepans in Brunschwig's work can be traced to Guy de Chauliac's *Chirurgia magna*, and ultimately to the Arabic works of Al-Razi.[46] While the figures in the *Cirurgia* of 1497 are, to my knowledge, the earliest images of surgical instruments to be printed, they reproduced medieval models without adding much that was new. They were rather upstaged by Gersdorff's *Feldtbuoch* with its full-page illustrations of elaborate tools, trusses and prostheses (see figure 2.4 and 2.5), and these images were incorporated into later editions of Brunschwig's surgery by savvy editors and translators, as we shall see in chapter 9.

The images of furnaces, glassware, and other apparatus in the distillation books are a different matter. Here, too, there was medieval precedent of ink drawings, and Brunschwig's were the first printed images.[47] But the visual repertoire of the distillation books is far more extensive, detailed,

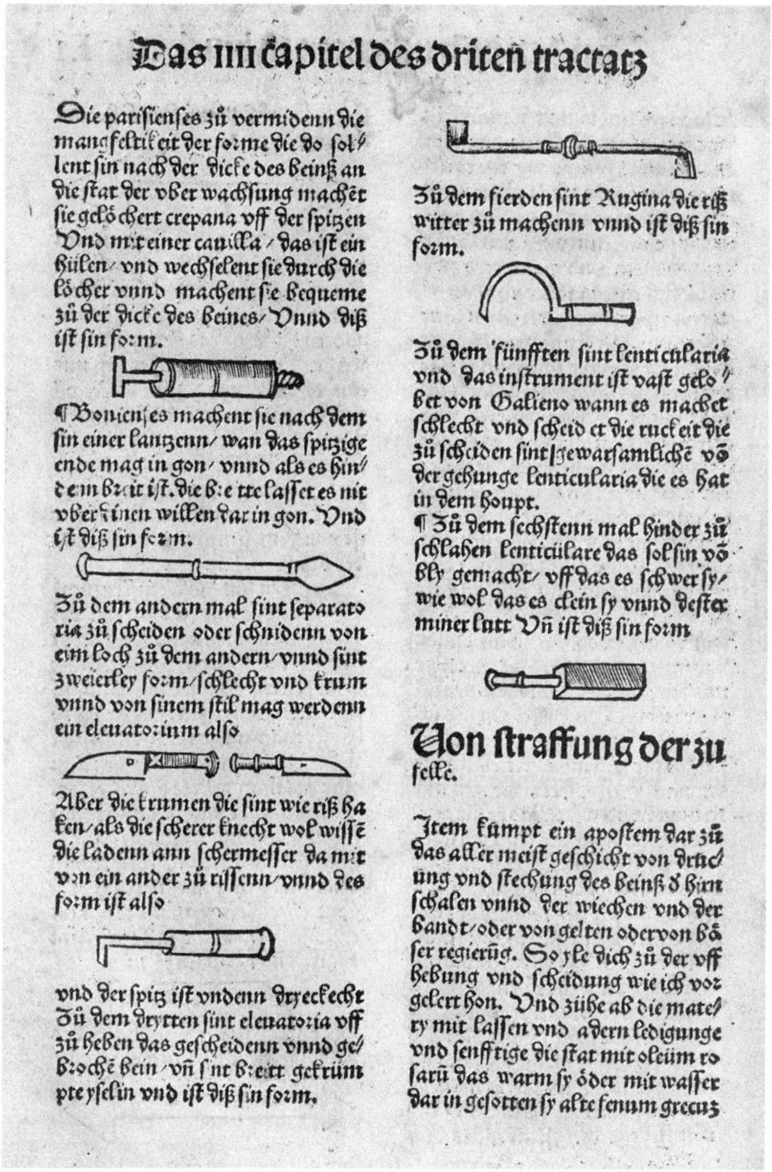

Das IIII capitel des driten tractatz

Die parisienses zů vermidenn die mangfeltikeit der forme die do sollent sin nach der dicke des beinß an die stat der vber wachsung machēt sie gelöchert crepana vff der spitzen Vnd mit einer canilla / das ist ein hülen / vnd wechselent sie durch die löcher vnnd machent sie bequeme zů der dicke des beines / Vnnd diß ist sin form.

¶ Bonienses machent sie nach dem sin einer lantzenn / wan das spitzige ende mag in gon / vnnd als es hin dem breit ist. die breitte lasset es nit vber sinen willen dar in gon. Vnd ist diß sin form.

Zů dem andern mal sint separatoria zů scheiden oder schnidenn von eim loch zů dem andern / vnnd sint zweierley form / schlecht vnd krum vnnd von sinem stil mag werdenn ein eleuatorium also

Aber die krumen die sint wie riß haken / als die scherer knecht wol wissē die ladenn ann schermesser da mit von ein ander zů rissenn / vnnd des form ist also

vnd der spitz ist vndenn dryeckecht Zů dem drytten sint eleuatoria vff zů heben das gescheidenn vnnd gebrochē bein / vñ sint breitt gekrümpte yselin vnd ist diß sin form.

Zů dem fierden sint Rugina die riß witter zů machenn vnnd ist diß sin form.

Zů dem fünfften sint lenticularia vnd das instrument ist vast gelobet von Galieno wann es machet schlecht vnd scheid et die ruckeit die zů scheiden sint gewarsamlichē võ der gehunge lenticularia die es hat in dem houpt.

¶ Zů dem sechstenn mal hinder zů schlahen lenticulare das sol sin võ bly gemacht / vff das es schwer sy / wie wol das es clein sy vnnd dester miner lutt Vñ ist diß sin form

Von straffung der zu felle.

Jtem kumpt ein apostem dar zů das aller meist geschicht von druckung vnd stechung des beinß d hirnschalen vnnd der wiechen vnd der bandt / oder von gelten oder von böser regierūg. So yle dich zů der vffhebung vnd scheidung wie ich vor gelert hon. Vnd zühe ab die matery mit lassen vnd adern ledigunge vnd senfftige die stat mit oleūm rosarū das warm sy oder mit wasser dar in gesotten sy alte fenum greeuz

FIGURE 6.2. Illustrations of surgical instruments embedded in the text that describes them. Brunschwig, *Cirurgia*, fol. 51v. Credit: The Huntington Library, San Marino, California, 85770.

FIGURE 6.3. A workman tending to a large furnace with several alembics. Brunschwig, *Small Book*, fol. 4v. Credit: The Huntington Library, San Marino, California, 85769.

and original than that of the *Cirurgia*. Distilling flasks and cucurbits are depicted in a three-dimensional manner, with some attempts at rendering the effects of light on glass. They are real objects placed in decorative interior spaces, rather than flat diagrams floating on the page (see figure 4.8). In addition to this veritable catalog of vessels, pots and tools, Brunschwig's manuals include images showing how they are put together in a working still and how they should be handled by the practitioner (figure 6.3). The *Large Book* includes more elaborate setups, for example an enormous two-story water bath still, depicted through a combination of three-dimensional and cross-sectional views to show its inner workings (figure 4.6).

One of the most striking uses of a woodcut is the illustration of the brick mold (figure 6.4). As we saw in chapter 1, Brunschwig imagined his readers using the wooden mold to make curved bricks for building round furnaces, or taking the mold to a local brickmaker. To facilitate the transfer of the correct shape from the image to the mold and ultimately the bricks, the figure is printed to scale, across two pages in folio, such that it can be used as a direct template for the wooden frame. Conveying a measure of absolute size as well as relative proportions was an unusual feat in the

FIGURE 6.4. Life-size template for a brick mold, printed across a double-page spread. Brunschwig, *Small Book*, fol. 2v–3r. Credit: The Huntington Library, San Marino, California, 85769.

early history of print.[48] Compared to the *Cirurgia*, the distillation books are much more forward-looking in their use of printed images. Showing an unprecedented inventory of material culture, including fanciful contemporary stills, they push the boundaries of the medium to convey technical knowledge in visual form.[49] In the printed literature on distillation, there was no equivalent of Gersdorff's book to steal the show. On the contrary, many later *Destillierbücher* show a clear debt to Brunschwig's visual material.

Overall, the use of printed images to convey practical knowledge in Brunschwig's books is marked by careful deliberation as well as remarkable originality. Illustrations of technical equipment—the first of their kind in print—are confidently integrated into textual descriptions and convey key information through inventive solutions such as the brick mold template. Where the paraphernalia of human art were concerned, Brunschwig and Grüninger were enthusiastic in their embrace of the printed image. Depictions of God's creation, on the other hand, were an occasion to reflect on the limitations of botanical and anatomical images, and Brunschwig insists that they cannot replace firsthand observation, much less the artisan's hands-on engagement.

Printing the Material Sensorium

Brunschwig's works provide an unusually rich account of this embodied, sensory way of knowing one's craft and one's materials, and a remarkably early effort to convey such knowledge in print. Sensory expertise was especially important to artisans of healing, as it played a key part in recognizing the myriad signs of the body's inner workings and disturbances. The learned medical tradition had developed the study of semiology to rationalize the contingent and often ambiguous nature of bodily signs.[50] While Brunschwig is keen to show himself conversant with Hippocrates's and Al-Razi's semiological reasoning, his own approach to reading the body is distinctly sensory and embodied, spelling out what I have elsewhere characterized as a "material sensorium" of craft knowledge.[51]

When it comes to assessing head wounds, Brunschwig maintains that "nothing makes for better testing and probing [to tell] whether the skull is fractured . . . than gently feeling with the finger to see whether the bone is inwardly loose."[52] Other signs are less directly accessible to the senses. To determine the depth of a skull fracture, Brunschwig advises, "you shall recognize it by this sign. Let the person firmly close their mouth and nose and cough violently. If you then see something exuding from the crack, this means it has gone all the way through the broken bone, and the brain pan is broken."[53] Brunschwig provides two more ways of testing skull fractures, taken from Guy de Chauliac and Lanfranco. If biting on the node of a rye stalk causes the patient so much pain that he cannot manage to bite all the way through, this indicates a skull fissure or fracture. The same localized pain may be revealed by strumming a taut piece of string held between the patient's teeth. In Brunschwig's surgical semiology, reading the signs of wounds and lesions involved the patient's bodily sensations as well as the practitioner's probing inspection and gentle touch.

Although distillation was not part of the same learned semiotic tradition, its complex procedures came with their own set of intricate signs relating to natural materials and their transformations. This, too, involved an expertise situated in the practitioner's body and its senses. As in surgery, the sense of touch is important—the temperature of a water bath, for example, should not rise beyond the point where one can "bear to keep a finger in it."[54] Visual cues also had their place. In the intricate process of distilling quintessence, it is vital not to miss the point at which the alcoholic "spirits" have been fully extracted: "This you may recognize when the streaks in the helmet or alembic begin to branch, and large drops form as if the helmet was sweating on the inside; then the *spiritus* have gone out."[55] To double check, the strength of the distillate can be assessed by taste: "Try it on the tongue, and you will sense whether it is watery or contains more power and spirit."[56] Similarly, one can taste the increasing strength of distilled vinegar: "This is determined by repeated tasting in the mouth and on the tongue: when it starts to become acidic, to tickle the tongue, or to become strong, then the receptacle should be removed, and another put in its place."[57] Once bottled, distilled remedies required a finely tuned sensory expertise to monitor their shelf life, as we saw in chapter 5. As recent work on taste and smell has begun to show, later artisans and chemists shared Brunschwig's view that these senses were a vital part of skilled work with natural materials.[58]

In distillation as in surgery, materials and their properties could be elusive unless one knew how to make them accessible to the senses. How does one know, for example, that the clear liquid obtained in the process has the appropriate medicinal qualities? To assess the quality of distilled quintessence of wine, Brunschwig provides the following test: "If you let a drop of tree oil fall into it, it will sink to the bottom and it will not rise up again, however much you shake the vessel with the distilled wine."[59] A distilled *aqua fortis* to use against fistulae can be assayed in a similar manner: One drop "in still water will go to the bottom . . . and if you throw a needle in it, it will float up. Furthermore, it causes milk to curdle."[60] As we have seen in chapter 5, a distilled water's viscosity—a good indicator of its state of freshness—could be rendered observable by pouring from a height, or by testing how quickly a drop runs off a fingernail.

Human bodies and natural materials were difficult things to know. Particular, contingent, and often inscrutable to the naked eye or the intellect, they had to be touched, smelled, or tasted, and their hidden properties made sensible through skilled gestures. Brunschwig's books discuss this material sensorium of the artisan-healer in unusual detail. He describes materials, gestures, and sensory cues, and where these elude the grasp of the written word—as with the oil of nutmeg—he guides his readers' attention as they develop their own workmanlike sensory habits. While a book cannot replace experience, it can speak to readers in the way a master craftsman might, pointing out pitfalls and showing his apprentices how to attune their senses and actions to the making of embodied knowledge.

Conclusion

It is a commonplace in scholarship on early modern how-to literature that we should be wary of assuming that such texts were necessarily read for practical information, or that they were even written with that intention.[61] Many authors and readers had entirely different aims—the artisan making a statement about their own expertise without necessarily disclosing it, or the connoisseur looking to understand the work that went into the commodities he collected, without ever doing that work himself. Such motives for writing and reading are entirely plausible for Brunschwig's books. But as I have argued in this chapter, they also reflect with unusual eloquence a local artisanal culture of forming bodily skill and habit through repeated practice, and they show a genuine commitment to codifying this expertise in a way that could inform readers' own practice. They are early and rare examples of how-to books that really did intend to teach readers how to do things, and went above and beyond in pursuit of this aim.

Even though Brunschwig was acutely aware that words often failed the task of conveying knowledge that was formed through the constant practice of bodily gestures and sensory experience, he developed remarkable strategies for providing something close to an apprenticeship in print. He wrote cautionary tales of accidents and mistakes, including his own, to provide vicarious experience for virtual apprentices in the early stages of training. In so doing, he anticipated early modern cultures of dealing with error that historians have only recently begun to explore. The innovative use of printed images must be largely credited to the fruitful collaboration with the publisher Johann Grüninger, but here, too, Brunschwig shows acute awareness and careful reflection on the epistemic implications of representing natural and man-made things. Visual, tactile, and olfactory expertise—what I have called the material sensorium—were as crucial to working with bodies and materials as they were difficult to pin down in words. Brunschwig rises to the challenge by articulating something of an artisanal semiology. He describes techniques for locating skull fractures and determining the viscosity of distilled waters—in short, for making sensible the insensible properties of bodies, processes, and substances. Realizing that the sensory ultimately eludes words, he provides instructions that aim to guide rather than replace readers' own experience. The following chapter delves deeper into the relationship between experience and knowledge in Brunschwig's project of crafting medicine.

✷ 7 ✷

Gathering Breadcrumbs

EXPERIENCE AND ERFARUNG IN BRUNSCHWIG'S VERNACULAR EMPIRICISM

In 1537, the Strasbourg city council recorded a case about property ownership brought forward by one Catharina, "the daughter of the experienced [*erfarnen*] Hieronymus Brunschwig, our late citizen."[1] The choice of "experienced" is conspicuous as an honorific. Brunschwig is not remembered as a generally "esteemed" or "honorable" citizen, or as an "erudite" or "learned" author of printed books that had had some success in the print market by the time the document was written. Instead, Brunschwig's legacy is one of experience.

We do not have to look far in Brunschwig's writing to surmise that he would have been pleased to be thus remembered. The verb *erfaren* (to experience), its noun, *erfarung* (experience), and their derivatives are central to Brunschwig's language of knowing. The *Small Book of Distillation*, he claims, contained knowledge that he had "compiled, read, seen, and *erfaren* with much labor over thirty years,"[2] through "his own daily handwork" or from other "experienced masters of medicine."[3] Similarly, the *Large Book of Distillation* was based on "studying and reading," but also things that he had "seen through [his] own and others' experience."[4] While *erfarung* sometimes denoted "knowledge" in a more general sense in early modern German, I argue in this chapter that Brunschwig's use of the term and its derivatives speaks to a commitment to knowledge derived from experience as opposed to rational demonstration.[5]

Questions about the place of experience versus reason in producing knowledge have been central to histories of science and medicine, with a general narrative arc from theory-obsessed antiquity to empirical modernity. In its basic assumptions, ancient philosophy had little room for experience. For Plato, knowing meant grasping abstract "forms" with the intellect alone; anything that could be experienced with the senses was but a poor shadowy reflection. Aristotelian epistemology proceeded from axioms that were confirmed by sensory experience, but it only allowed commonly accepted experience of nature's regular course—extraordinary events or contrived trials distorted the natural order and could thus not give insight into final causes. This marked out two distinct types of experience—general experience of the normal course of things, called *experientia* or *experimentum* in the Latin West, and the contrived experience of a making a deliberate trial (*periculum facere* in Latin). As Peter Dear has suggested, early modern students of nature continued to grapple with these distinct experiences and their epistemic limitations as defined by the prevailing framework of scholastic natural philosophy. Their responses map out key stages in the history of the so-called Scientific Revolution, culminating in the acceptance of contrived experience—increasingly called "experiment"—into philosophically sound investigation of nature.[6]

While it is true that premodern proponents of experiential knowledge often had to work against learned prejudice, the binary conflict between reason and experience is easily overstated in the service of big-picture narratives. Historians of ancient philosophy have shown that claims about the dominance of antiempirical views were often exaggerated, even in their own time. Aristotle was more comfortable dealing with the particulars of experience as well as the universals of theory than has often been made out, and ancient rational medicine, for all its opposition to the "irrational" practices of temple healers, nonetheless shared their terminology, aims, and practices.[7] While Galen thought that proponents of empiricist medicine were wrong to jump from symptom to treatment without ratiocination, he did not deny that they could be good doctors, and even advocated that theories should be tested by experience.[8]

In the medieval West, the reception of Aristotelian writings brought a clear hierarchy of rational knowledge or *episteme* over practical knowledge or *techne*, but it also brought the idea that experience could serve as proof in knowledge making. Scholars such as Robert Grosseteste (1168–1253) increasingly took *experientia* seriously as a source of knowledge.[9] Albert the Great (1200–1280), known as the "experienced scholar" (*doctor expertus*), thought that any rational conclusions that disagreed with experience should be treated with suspicion.[10] Known by the equally splendid nickname *doctor mirabilis*, Roger Bacon (1220–1292) went one step further in claiming that experience was not restricted to producing individual sensory facts, but, as part of what he called an "experimental science" (*scientia experimentalis*), could yield knowledge of underlying causes. Although Bacon's role as a forefather of modern empirical science should not be overstated, he illustrates the extent to which medieval scholars were willing to engage with experience as a source of knowledge.[11] In the writings of Nicolaus Cusanus, the untutored experience of the layman becomes the model not only for knowing nature, but also for knowing God.[12]

Medicine plays a special part in this history because it was always understood as a practical art as well as a theoretical science, so there was always room for experiential insight as well as text-based learning. Precisely how much room was a matter of ongoing debate. Galen—the ultimate ancient authority for most medieval and early modern physicians—distinguishes three medical "sects" that he calls rationalists, empiricists, and methodists. The methodist school is not well understood, and in any case it is largely disregarded from Galen onwards.[13] The real contest, as far as premodern medical writers were concerned, was between rationalism and empiricism. Galen clearly identifies as a rationalist, explaining bodily phenomena through deductive logic, starting from universal principles such as the four humors. The empiricists instead drew their conclusions from the experience of their senses alone. For the rationalist, a good treatment was one that made sense within a theoretical framework that explained internal bodily processes; for the empiricist, a good treatment was one that had reliably been shown to work in cases with similar external symptoms.[14] In Brunschwig's day, this discussion was kept alive in debates about medical authority. As Johann Widmann's letter to the Strasbourg council reminds us, in the rhetoric of learned physicians, the term "empiric" became a commonplace to denigrate nonacademic practitioners and claim sole authority over the body (see chapter 2).

At the same time, the early modern period saw a rise in experience-based—or "empirical"—understandings of the body and its treatment, and they continue to generate lively conversations in the history of medicine. As Michael Stolberg has shown, some physicians who preached a rational stance embraced empirical remedies in practice.[15] And even orthodox medical writing increasingly relied on experiential knowledge and empirically verified cures, recorded as individual *experimenta*, or collated into stand-alone works called *practicae*.[16] Katharine Park has highlighted the importance of *experimentum* and *observatio* in the writings of medieval physicians and the distinct modes of inquiry they signaled.[17] In the Renaissance, experiential categories such as *observatio* and *historia* grew into new medical genres that collated case histories and treatment outcomes, usually called *observationes* or *curationes*, as Gianna Pomata has shown.[18] In recent studies of early modern medical trials, notebooks, and recipe collections, Elaine Leong and Alisha Rankin in particular have uncovered thriving cultures of trying, testing, and recording medical treatments that are rife with the language of experience.[19] We should be wary, however, of reading appeals to *experientia* or indeed *erfarung* as expressions of any sort of unified medical "empiricism." They cover a range of approaches that are best understood, as Alisha Rankin has pointed out, as a plurality of "diverse, overlapping 'empiricisms'" in the early modern period.[20]

Our understanding of these empirical cultures is still skewed towards the epistemologies and *observationes* of learned physicians and the experimental practices of learned natural philosophers. In the history of science, we are beginning to understand artisans' ways of knowing through embodied sensory engagement and alchemical accounts of matter, but here less attention has thus far been paid to medical knowledge.[21] Brunschwig's artisanal perspective thus provides a unique opportunity to gain a more fine-grained historical understanding of the role of experience in early modern science and medicine. In particular, he allows us to escape the gravitational pull of Paracelsus, whose championing of artisans and experience still looms large as a key turning point of early modern epistemology.[22] This is not undeserved, and in fact there is probably more still to say about Paracelsian understandings of the body and how it "knew" through experience. But the focus on Paracelsus has tended to eclipse other, earlier voices such as Brunschwig's, who made similar claims about artisanal experience and chemical practice, decades earlier but with less rhetorical fanfare. On the one hand, he did not have Paracelsus's ambition of revolutionary change; on the other, unlike Paracelsus, he was himself a practicing artisan to whom experiential ways of knowing were self-evident. This means that his underlying conception of knowledge is not always spelled out, but I aim to show here that it is well worth teasing out among his practical instructions and chatty anecdotes. This provides a much-needed deconstruction of Paracelsus-dominated narratives, by characterizing an "empiricism" at the intersection of a range of traditions and genres, learned and vernacular, medical and artisanal, theoretical and practical.

The chapter traces Brunschwig's stance on experience and his distinct experiential personae in his surgical and pharmaceutical writings. I begin by exploring the place of experience in the *Cirurgia*, where Brunschwig is most explicit in addressing its relation to learned medical theory. While this results in a relatively cautious endorsement of experience in surgical matters, I show that the distillation books, addressed to a wider audience, make considerably more room for experience-based validation of medical knowledge. Here Brunschwig engages with different traditions of experience: medical case histories, occult and secret knowledge, and vernacular conversations about *erfarung* as traveling the world. Finally, I analyze Brunschwig's strategies for establishing credible testimony to support his and others' experiential claims. Overall, this chapter shows that Brunschwig was part of a number of early modern conversations about experience, and that he drew from them a "vernacular empiricism" of his own that did not go unnoticed by the learned medical tradition.

"New Experience" and Ancient Learning in Surgery

In the *Cirurgia*, as we have seen in the introduction, Brunschwig draws on the writings of the rational surgeons in defense of his discipline. He aligns himself with a tradition of surgical learning going back to Galen, while also emphasizing the well-practiced manual skill and constant technical improvement that sets surgeons apart from physicians. Along with this double-edged rhetoric, Brunschwig inherited an epistemic tension. He was well aware that his surgery relied, on the one hand, on text-based medical knowledge from "the works of the ancients," and on the other, the "new experience" of individual practitioners.[23] By exploring Brunschwig's treatment of this tension, we can discern what role he ascribed to experience in making surgical knowledge.

Already in the medieval tradition, practical experience served to refine remedies and techniques. In his discussion of a surgical ointment, Brunschwig notes that one of his learned predecessors, Lanfranco of Milan, "has taken this ointment from the book of Rhazes, but he has improved it."[24] Brunschwig does not hesitate to make improvements in his own daily practice. To staunch heavy bleeding, for example, he recommends St. Jacob's stone, based on "the virtue and power which I have seen in the . . . effect of the stone."[25]

Occasionally, Brunschwig brings his personal observations to bear on disagreements between surgical authorities. Galen claimed to have cured a child whose eye had been punctured and lost much of its substance, while the twelfth-century physician Maimonides maintained that such a cure was impossible. The passage is in fact borrowed from Guy de Chauliac's *Chirurgia magna*—with an intriguing addition. While Chauliac sides with Galen because "the spermatic parts in children regenerate daily, as you see," Brunschwig's text adds, "In truth I have seen it myself as it is written above, so I give testimony in Galen's favor."[26] Since the precise manuscript source of Brunschwig's borrowing has not been traced, we do not know for certain who made this addition. Nevertheless, whether it was some earlier copyist or Brunschwig himself, he clearly thought that personal observation was a legitimate way of settling this learned dispute.

As well as determining the efficacy of existing cures, "new experience" could give rise to new treatments in emergencies. The medieval masters liked to display their cunning interventions in heroic anecdotes about apparently hopeless cases.[27] Brunschwig retells some of these—for example how Guglielmo da Saliceto devised a successful on-the-spot treatment for a near-fatal abdominal wound—but he also tells stories about himself in the same mold. In one case, he was called to the aid of a young child whose broken leg had been tended to by "a good and famous barber and surgeon who had another, lesser master with him."[28] This junior colleague splinted the leg incorrectly, leaving it a few inches shorter than the other. When asked for help, Brunschwig "made the bandage in a different way which is much more useful and hurts people much less."[29] In fact, he devised a full-body truss that could be tightened by the kind of ratchet "crossbowmen use to cock their bows."[30] Like Guglielmo, Brunschwig stepped in where others had given up and developed a new treatment that proved to be more efficacious and—if we take his word for it—less painful. As discussed in the previous chapter, this "codification of error"—in this case someone else's—was one key strategy for textualizing practical knowledge in Brunschwig's works.

It is perhaps telling that in this very account of a successful treatment born of "new experience," Brunschwig goes on to caution against rash extrapolation: "If the fracture is old and hardened, it should be left alone; as the ancient philosophers write: to the wise it would seem better that he should live with a limp rather than seek to be straightened and die in the process."[31] Even if new treatments worked, they should not be taken to overturn the accumulated wisdom of tradition. Although Brunschwig praises the fifteenth-century invention of artificial balsam, he criticizes local barbers and surgeons for adopting it rather too enthusiastically and indiscriminately—they behave, he complains, as though they have "reinvented the art," failing to realize that "the ancients knew much better than we do, since their philosophers received the cure through the word of almighty God, and they knew the virtues and complexion of balsam—why it does or does not heal, where to use it, in which wounds or lands, or how to use it . . .—which is very necessary for the surgeon to know."[32] Practitioners should master the ancients' revealed and learned understanding of remedies' properties and their effect on the four humors, depending on the *contingentia* formulated by the rational surgeons. Without a working knowledge of the "natural arts of philosophy,"[33] surgeons were likely to "destroy the entire complexion and nature" of the patient.[34]

Quite apart from endangering one's patients, relying on experience alone was also, in Brunschwig's view, an unsound way of making surgical knowledge, sadly all too common among some of his contemporaries: "If you ask such a surgeon why the balsam or plaster heals the wound or not, he can give no answer other than saying 'because it is good and

I have seen other masters use it also,' or 'I do not know why it did not cure him, maybe it is not good for this wound, it has never failed me.' O what a foolish answer you receive!"[35] In his condemnation of an exclusively empirical surgery, Brunschwig again points to his medieval antecedents. Guy de Chauliac, he writes, "in his youth followed the young [masters]. But when in old age he regarded the works of the ancients, he avoided the new experience [*erfarung*], for he grew suspicious of it."[36] Guy thought it foolish to "reject the art of the great and the good because of the teaching of one person, or even three or more."[37] This applies especially in surgical interventions that are already risky enough to begin with. In his chapter on skull fractures, Brunschwig once more borrows Guy's words: "by learning from experience [*erfarung*], medicine causes damage, so one should choose the most common and best way. Therefore . . . do not stray from the way of Galen."[38] Treating depressed skull fractures is difficult and dangerous at the best of times—this is not, Brunschwig insists, a good time to deviate from established surgical procedure for the sake of "new experience."

There is a place for individual experience in Brunschwig's surgery. Observations of successful cures were a legitimate indication of the efficacy of treatments, and they could help settle disagreements among learned medical authorities. The pragmatic demands of surgical practice might call for new and improvised cures; these could even enrich the surgeon's expertise and make him a better practitioner. But an indiscriminate reliance on experience could be dangerous, especially in the hands of those who failed to weigh it against a surgical rationale based on ancient learning and the medieval canon of rational surgery. In the dialogue between "new experience" and established learning, the safety of tradition generally outweighed the lure of the new.

Medical Experience for the "Common Man"

In his distillation books, Brunschwig offers a different take on experience. It seems unlikely that this reflects a change in Brunschwig's outlook over time. He was already a highly experienced practitioner in his late forties when the *Cirurgia* was published in 1497, and the *Small Book of Distillation* appeared only three years later, with the *Large Book* already in the works. More likely, the differences represent a conscious shift in register, corresponding to a wider audience that included the "common man" as well as formally trained practitioners.

In the preface to the *Small Book*, Brunschwig states that it contains things that he had "learned by means of *experimenta* from many people, learned and lay, men and women."[39] This was knowledge that came from all kinds of people—you did not have to be a learned man, or even a man at all. Although Brunschwig recommends "an experienced and learned healer," he then pivots to the "common people" who might not have access to professionals: "If they can have neither healer nor medicine, much help is to be had through the *experimenta* and experience [*erfarung*] which are proven through the effect of the waters."[40] Writing for laypeople who could not be expected to grasp ancient medical theory, Brunschwig offers *erfarung* that promised practical benefits independently of learned expertise.

Brunschwig's gathering of experience encompassed practice as well as texts. In the *Large Book*, he writes that he found insights by "studying, reading, and seeing, through my experience and that of others; I have done not a few but many *experimenta* and compiled them together." Reflecting on the merits of his earlier *Small Book*, he refers to its "praiseworthy *experimenta* for the common or layman," and again to the fact that he experienced them "not a few times, but many." As well as emphasizing the immediacy of his sensory experience, Brunschwig makes a point of stressing the number of *experimenta* and repeated experiences on which his knowledge was based. In surgery, Brunschwig agrees with Guy that it would be foolish to prioritize a new cure over an established one on the word of two or three people; in the distillation books, multiple experiences carried epistemic weight.

The project of making knowledge from experience did not stop with the author, but implicated readers, too. With the help of his books, Brunschwig suggests, medical knowledge is "easy to experience [*erfaren*] through *experimenta* . . . with the distilled waters." Once an aspiring distiller—a "learner" (*lerenden*) and "experiencer" (*erfarenden*)—had understood a principle such as the purifying of quintessence, "he should believe that it works the same way in other things, as he has learned and experienced from the *experimenta*." This kind of extrapolation from experience is precisely what Brunschwig warned against in the *Cirurgia*. Here, on the other hand, *erfarung* and *experimenta* become the lynchpin of his and his readers' knowledge making.

These were loaded terms, and to explore what Brunschwig meant by them, we turn to the "Thesaurus pauperum," the collection of recipes that forms the final part of the *Large Book*. True to its name, this "treasure of the poor" largely contains simple and inexpensive treatments such as poultices, ointments, and herbal baths. As we have seen, this part of the book enjoyed huge success as a stand-alone publication. Brunschwig describes the collection as "breadcrumbs fallen from the books of medicine, and gathered by me as *experimenta*" and, elsewhere, "breadcrumbs fallen from my *experimenta* and those of others, which I have done with my hand, and also seen from other truly famous masters."[41] Loath to let go of a good metaphor, he dubs the collection a *micarium*—a "crumbery" perhaps if we were to attempt an English coinage. The message is clear: *experimenta* were independent nuggets of experiential insight, gathered incidentally in his reading, his practice, and that of others, brought together in a printed collection that promised practical utility. Brunschwig's "crumbs," as we shall see, come in all shapes and sizes, and from different traditions of experiential knowledge: medical recipes, accounts of cures, and secrets. In what follows, I explore how Brunschwig engages with each of these traditions to gather his breadcrumbs and legitimate them as valid knowledge.

Recipes

Many of Brunschwig's *experimenta* take the form of recipes. As "texts of action," recipes are a key genre for the codification of practical, material, and processual knowledge, although by the same token they are also challenging historical sources that confront us with unfamiliar materials and gaps of tacit knowledge.[42] As one of the oldest and most recognizable written forms, they lent themselves to satire and humorous subversion.[43] From Assyrian glassmaking recipes from the seventh century BCE to modern laboratory science, recipes have been key vehicles for different sorts of practical knowledge.[44] They were a staple of the tradition of secrets, encoding alchemical processes or feats of natural magic, and formed the focus of attention for early modern "academies of secrets" devoted to a quest for hidden natural knowledge.[45] Artisans, too, used recipes to exchange, test, and modify their experiential insights.[46] Recipes were (and still are) central to pharmacy, and arguably represent the foundation of written medical traditions in the vernacular.[47]

For such an iconic form of writing, their structure is remarkably fluid. In surveys of early modern recipes, literary scholars have found that they usually include a title, a list of ingredients, instructions for preparation and application, and perhaps notes on storage or efficacy—but these can appear in any order, and the ingredients appear to be the only strictly necessary feature. This textual fluidity extends to collections of recipes as a whole; rather than a continuous text, they present independent units that can be removed, altered, or shuffled around without changing the work's overall meaning (although it may affect its usability). While they often have no single identified author, they are held together by an overall framework of utility and purpose, forming what the linguist Michael

Hoey calls a "discourse colony."[48] Owing to this modularity and versatility, recipes and recipe collections became one of the most fruitful sites of knowledge making in the early modern period. Elaine Leong has shown how recipes circulated within and between printed books, manuscript collections, and personal correspondence to generate practical knowledge through regimes of collecting, modifying, and testing, often within the gendered context of household and medical care.[49]

The epistemic status of a recipe depends on its trajectory within these networks of circulation and testing. It might be read as an open suggestion or invitation to trial and error—perhaps an alternative procedure scribbled in the margin, later to be transcribed into a fair copy if it worked, or crossed out if it did not. A recipe in a published book or a manuscript collection passed down within the family might present itself with the authority of a tried-and-tested piece of knowledge, perhaps signaled by endorsements such as "good" or *probatum est*. Locating Brunschwig's recipes on the spectrum from the "open" to the prescriptive within a thriving recipe culture helps us understand a key part of his attitude towards experience as a source of knowledge.

The *Cirurgia* devotes an entire section to the plasters, ointments, and other remedies required in surgery. Brunschwig likes to point out that these are canonical recipes, from authorities such as Mesue or the *Antidotarium Nicolai*, with well-known effects.[50] Where a more recently fashionable remedy is introduced, such as the "good green wound plaster which the new surgeons use," Brunschwig does not deny that it "heals any fresh wound." But just as he criticized indiscriminate users of artificial balsam, here he chides "unlearned barbers" whose only rationale is that the green plaster "heals wounds and that they have seen others use it."[51] This purely empirical explanation was not good enough: Brunschwig insists that "every surgeon should know the virtues and powers of the simples that go into the plaster or ointment."[52] These rational deliberations were necessary to ensure that the plaster would perform the right surgical *intentiones*, such as cleansing, adhesion, or skin growth.

Some recipes are not meant to be attempted by readers at all. The *Cirurgia* contains recipes in Latin, for example an "ointment to be made at the apothecary's shop" that lists medicinal ingredients and tersely concludes, "make the ointment according to the art."[53] Turning this recipe into a remedy required the apothecary's knowledge of specialist jargon and the tacit skill of making ointments. Brunschwig includes similar recipes in Latin in his plague treatise, where he explains his reasons for doing so to his readership of common citizens: "For the most part these recipes and ingredients which go in it cannot be succinctly expressed in German, and are commonly found in pharmacy shops. I have therefore left them in Latin . . . like the physicians usually write them, such that anyone who needs such a recipe may copy it, send it to the apothecary, and have it made."[54] These recipes were meant to be used without being understood or even read; instead, they functioned as coded messages that could be copied on a piece of paper, to be decoded and translated into remedies by an expert. As instructions for making, such recipes are effectively closed to most readers, tapping instead into established networks between learned physic, artisanal standards, and specialized skill. In the *Liber pestilentialis* and the "Thesaurus," closed recipes are simply a pragmatic alternative for those who live near a pharmacy shop, and they are presented alongside simpler recipes in plain German. In the *Cirurgia,* they appear as expressions of a recipe apparatus that reduces practitioners' agency and prioritizes canonical ingredients and procedures.

Outside his surgical writings, Brunschwig's recipes take on a more dynamic epistemic role. If his plague treatise presents "closed" recipes alongside "open" ones, his distillation books even tell his readers how to open up "closed" prescriptions. Tucked away at the back of the 1509 edition of the *Small Book* there is a "short introduction to pharmacy," which explains the meaning of such abbreviations as "Rx" for "recipe," and the most common weight units.[55] The *Large Book* contains a more extensive discussion of simples and units of measurement, as well as the shorthand used for different weights and mixing techniques. Thus, Brunschwig tells his readers, "If a doctor writes or prescribes something for you [to take] to the pharmacy shop, you may understand whether that which he has written for you is good or bad,"[56] explicitly inviting them to crack the code of a physician's prescription, and even to judge its quality.

In addition to opening up the black box of prescriptions, Brunschwig provides recipes that are more open to readers' interpretation, intervention, and experiential testing. He suggests a range of alternative ingredients, including seasonal ones—"in the summer I have added chopped leaves of white horehound"—and alternative modes of application—"sometimes I have used it dry."[57] Such alternatives are common in the "Thesaurus." Green ginger is good for a phlegmatic stomach, but "if you are poor and do not have ginger, make a pap of young nettles."[58] Besides specific substitutions, Brunschwig sometimes offers an even wider range of choice. For a draft to promote sweating, he provides a list of herbs that looks like a prescriptive recipe at first glance, but in fact states that one may use "any or all of these."[59] These recipes are open to experimentation, but also to a range of readers with varying levels of access to ingredients and professional expertise.

Brunschwig's breadcrumb metaphor suggests that his *experimenta* are validated by his own experience ("which I have done with my hand"), but it also implies that such crumb-gathering was both cumulative, building on the collections of other "famous masters," and open-ended, since no such "crumbery" could ever be considered in any sense complete or definitive.[60] Over the course of the sixteenth and seventeenth centuries, the potential for infinite accretion and modification placed recipes and recipe collections at the center of artisanal and household cultures of making knowledge through iterative trial and error.[61] If trying and testing can be said to be integral to the recipe genre itself, Brunschwig further encourages his readers to "experience through *experimenta*" by providing a range of alternative recipes for the same disease, and suggesting modifications and substitutions within individual recipes.[62] Annotations in surviving copies of Brunschwig's books (discussed in chapter 10) suggest that readers often followed his invitation. Using crosses or other marks in the margin, they singled out recipes that they perhaps intended to try out or had already found to be efficacious. Sometimes their approval is more explicit, noting "this is a good piece," or "est certe probatum."[63] Some also scribbled new recipes into margins and on flyleaves, extending the collection with new breadcrumbs of their own. Although Brunschwig's recipes were the product of his experience, they were not an endpoint, but a starting point for readers' own experiential knowing.

Brunschwig's use of recipes reflects the experiential personae he adopts in different genres of writing. His presentation of canonical recipes that have stood the test of time, or have at least been approved by a number of medical experts, is an expression of his conservative approach to surgical procedures and his caution regarding "new experience." In the distillation manuals, on the other hand, a more empirically minded Brunschwig collates incidental crumbs of experience and implicates the reader in the ongoing process of testing recipes.

Experiential Narratives Between Case History and Anecdote

In addition to recipes, Brunschwig's *experimenta* appear in the form of personal narratives about particular cases. In the *Cirurgia*, as we have seen, they served to legitimate both the practitioner and the treatment, although any "new experience" ought to remain in dialogue with established surgical rationale. In the distillation books, anecdotes about new cures go further in sanctioning experienced healers and remedies. For example, Brunschwig tells us that, in 1499, he was "called upon by a famous barber" to help a patient whose grossly swollen uvula threatened to suffocate him.[64] He proposed phlebotomy to draw the blood away from the swelling, but was persuaded by the patient's friends and family to come up with a less invasive treatment and "bound him tightly in four places with tourniquets," repeatedly constricting and releasing to divert the blood flow.[65] The cure was no less successful for being improvised, and Brunschwig does not hesitate to adopt it into his repertoire: "From this experience [*erfarniß*] I have since performed many other treatments that are necessary for those who have little or nothing, or are far from doctors or pharmacies."[66] Indeed, the treatment crops up elsewhere in Brunschwig's books, in descriptions of individual cases and as a treatment for a range of ailments, for readers to try for themselves if no other help was available.

The contrast to his caveats about new experience in surgery is all the more poignant considering that the swollen uvula episode starts out as a surgical case—Brunschwig was brought to it by a "famous barber," and thought that it called for bloodletting. When this proved impossible, the improvised, noninvasive tourniquet treatment became an *erfarniß*, a nugget of experience that could be mobilized from the context of his own practice into the text of his book, and used outside the bounds of learned or formally trained medical practice. As such, it was more a "secret" than a case history, more *erfarung* than learned *observatio*.

Anecdotes like these make claims about practitioners as well as remedies. As we saw in chapter 6, Brunschwig commends the skill of a *practicus* whose artisanal education embodies within him the requisite manual expertise. In addition to this gestural knowledge, the *practicus*'s years of training make him more "experienced" (*erfaren*)—over the years, he comes to understand "certain treatments that he has practiced in his *practica*, and has learned how the disease ended."[67] In his ideal of a medical practitioner, Brunschwig thus blends the experience of artisanal training with the experiential insights of medical *practica*.

The opposite end of the spectrum is "a highly learned but unexperienced [*unerfarnen*] doctor." He should be "avoided at all times" because he lacks practice in diagnosis and treatment, and it is too late to start learning on the job once the patient is already ill—"by the time he has learned to recognize [the disease], the sick person is dead."[68] As a general rule, therefore, "an old healer is always to be preferred over a younger one; even if he is not very learned, he is well practiced, and more experienced than a younger man."[69] Brunschwig implies that one could do without medical learning in a pinch, whereas the experience of the *practicus* was indispensable. He was not alone in this view: as Sabine Schlegelmilch has shown, city authorities looking to hire a new municipal physician only accepted a candidate's learned credentials if his application letter also demonstrated sufficient *experientia*.[70] Writing towards the end of his career, Brunschwig makes much of his own credentials as an experienced *practicus*, in the double sense of the well-trained artisan and the medical practitioner's accumulated *practica*.

Here it is worth returning to a close reading of what we might call Brunschwig's tourniquet episodes for a richer story about experiential knowledge and what it meant to be a *practicus*. In his original account, Brunschwig is remarkably open about the difficulties of being called to attend a critical case by another practitioner, admitting that "the more he hurried, the more I delayed, saying to myself I would rather [the patient] died before we arrive than have him die when we are present."[71] When he reasoned that the blood causing the swelling could be diverted by phlebotomy, he was challenged: "There were many women who said 'What do you think you are doing? Do you not see that he is close to death? And when he dies many will say that you killed him.'"[72] Although convinced that his proposed treatment would work, Brunschwig respected their sensibilities and devised a new cure within the social constraints of the encounter.

In another tourniquet episode, Brunschwig faced an even more complicated situation. The case was once more a swollen uvula; the patient was a butcher who had seen his sister suffocate from a similar complication. Rather than suffer the same gruesome death, he seemed resolved to kill himself: "If he had got hold of a knife, he would have slit his own throat."[73] To make matters worse, he struggled so much that it proved impossible to give him the last sacrament. Since suicide was a cardinal sin, the spiritual stakes of the case were perhaps even more crucial than the patient's physical health. In naming the patient as a butcher groping for one of his knives, Brunschwig illustrates how artisanal expertise could complicate medical encounters; here was a strong man skilled in severing muscle and tendon, and in his desperation, it took eight men to prevent him from forfeiting his soul by cleaving his own flesh.

Although similar to the anecdotes in the *Cirurgia*, these vivid accounts place a different emphasis on the experience of the *practicus* and the contingencies in play. These could be practical—for instance, Brunschwig tells us that in the case of the butcher "there was no time to wait for the tourniquets," so he improvised by using "four strings used to tie grain sacks."[74] But these episodes also highlight the socially and culturally embedded nature of experience. Rather than insisting on the phlebotomy treatment dictated by his surgical reasoning, Brunschwig knew when and how to negotiate his own precarious position and others' sensibilities within the social web of the medical encounter. In the butcher, he saw not just a swollen uvula, but a human in a desperate situation complicated by soteriological and artisanal concerns. Where the *Cirurgia* emphasized the surgeon's need for learning to master the *contingentia* of individual cases within a rational framework, in the distillation books the experience of the *practicus* negotiates the practical, social, and spiritual contingencies of providing medical aid to "common people."

If the ideal practitioner distinguished himself first and foremost by experience, how did Brunschwig feel about the much-maligned medical sect of the "empirics"? At first glance, he appears in agreement with contemporary learned physicians' rhetoric against their rivals, stating that "the *empiricus* is to be avoided at all times."[75] However, he then goes on to state two exceptions to this rule. First, one might look to an empiric for treatment "from the outside," for example a surgeon who "has exercised and learned through constant practice,"[76] thinking perhaps of such itinerant specialists as oculists or "stonecutters." Second, Brunschwig acknowledges that empirics sometimes had privileged access to knowledge: "Such a healer often knows an *experimentum* that is hidden from a hundred learned doctors, or not accessible to them through reason, but only to him, through practice, and his constant use."[77] This perhaps explains why Brunschwig cites among his sources not only such established authorities as Avicenna and Albert the Great, but also "the *Emperici* of our time," whose knowledge was "experienced and found through their *practica*."[78] At a time when most medical writing emphasized the ignorance of "empirics," this was a strong endorsement of their experiential ways of knowing. In his accounts of medical cures, Brunschwig makes considerable room for experience and allows that even a straight-up empiric can discover what is hidden from the learned. Such hidden knowledge was the purview of another tradition that was formative for Brunschwig's engagement with experience.

"Secrets and Hidden Art"

In the preface to the *Large Book*, Brunschwig explains that, after the Fall, mankind lost much of Adam's knowledge of nature and its healing powers. The aim of his work, therefore, is to "bring to light and reveal these secrets and hidden art."[79] Such secrets of nature were at the core of a longstanding tradition of pursuing occult (literally "hidden") knowledge. It concerned any phenomena that could not be explained by "manifest" qualities such as the hotness or coldness, dryness or moistness of the four elements. While Aristotelian metaphysics could account for the heat of a coal fire, it had no ready explanation for the attractive force of a magnet or the shock of the *torpedo*, what we would now call an electric ray. Such phenomena were hidden from humans' declining reason, and therefore had to be dealt with outside of natural philosophy—by "occult" traditions such as alchemy, natural magic, and astrology. While the causes of occult phenomena could not be explained, their effects could be known by experience, and used for practical purposes.[80] They were a chink in the armor of scholastic dogma that was eagerly exploited by medieval proponents of experiential knowledge such as Albert the Great, who distinguished occult *experimenta* from general *experientia*.[81]

The medieval interest in experiential knowledge gave rise to new textual genres. *Experimenta* appeared not only in medical *practicae*, but also in the emerging literature of secrets—a loose term referring to a wide range of collections of practical knowledge, including what were known as "books of secrets" in Italian or "little books of art" (*Kunstbüchlein*) in the German lands. They circulated widely from the Middle Ages onwards, first in manuscript and, from the later fifteenth century, increasingly in print. In addition to occult phenomena, they also contained the techniques and experiential insights of artisans. Much like the secrets of nature, these secrets of craft eluded rational theorization because they dealt with particulars rather than universals. Just as occult properties were hidden within nature's inner workings, craft knowledge was formed in artisans' bodies through repetition, and was thus difficult to write down, as we have seen in chapter 6. Both kinds of secrets, therefore, could only be accessed, replicated, and understood by experience, and they were often found side by side in the same texts.[82] In medical works, the term *experimentum* itself became associated with occult or secret knowledge.[83]

In proposing to uncover "secrets and hidden art," then, Brunschwig drew on a solid intellectual tradition. He applied the label "secret of nature" to John of Rupescissa's quintessence of wine (see chapter 5), and to potable gold as developed by Arnald of Villanova and Marsilio Ficino—in both cases, Brunschwig underscores the idea of secrecy by using both the Latin-derived *secret* and the German *heimlichkeit*.[84] In addition to these alchemical authorities, he also cites other experts of the occult, including (pseudo-)Geber, (pseudo-)Lull, and Albert, the *doctor expertus* himself.

But Brunschwig was equally at home in the world of more miscellaneous "secrets" that were hard to explain but simply worked. As an aside in a chapter on composite distilled waters, Brunschwig notes, "here I must reveal to you a secret about frog spawn, which I have often done myself."[85] Washing one's hands in frog spawn, he claims, confers a marvelous healing touch that will cure other people's hands of worm infestations. Other practices reminiscent of folk magic seem even more outlandish to modern ears. Brunschwig writes that his father could cure the heaviest nosebleed by whispering the words "Max, Nax, Pax" in the patient's left ear.[86] He had "often see him do it," and although he "did not particularly believe in it," he tried it himself. The results were mixed—it "sometimes helped and sometimes not"—but Brunschwig was willing to let experience be the judge, and included his father's magic spell alongside other cures because it had proved successful on occasion.[87]

Brunschwig thus positions his own empirical approach within established traditions of experiential knowledge. Building on occult and artisanal epistemologies as well as the literature of secrets, he collects his own secrets of craft and nature for his medical purpose. Whereas Brunschwig's surgical *esperimenta* were modeled on medical *practicae*, where they functioned as a carefully circumscribed part of a rational endeavor, the traditions and genres of secrets offered a different model, in which the epistemic role of *experimenta* was less fettered by learned concerns.

Fahren, Erfarung, Gefahr: *Experience and the Dangers of Vernacular Travel*

As well as engaging with wider traditions of experiential knowledge—recipe culture, medical case histories, and secrets—Brunschwig was also part of more local conversations about the values and dangers of *erfarung*, driven by Strasbourg's intellectual elite. The German word *erfarung* (or *erfahrung* in modern spelling) is derived from the verb "to travel" (*fahren*), and originally referred specifically to impressions, experiences, or knowledge gathered by moving around in the world and seeing new things.[88] During the Middle Ages, the meaning of *erfarung* broadened to include knowledge and the process of learning things more generally, and we can see this usage in Brunschwig's day—he does not connote literal traveling whenever he tells of something he has *erfaren*. But neither does he use *erfarung* in an entirely generic way. As we have seen, it implies something distinct from book learning and medical theory, a kind of knowledge that emerges from practical engagement with particulars. The vernacular sense of travel resonates in Brunschwig's language and shapes his notion of experiential knowledge.

Brunschwig's use of the vernacular meaning of experience forms part of a larger trend in German literature from the late Middle Ages onwards to recharge the term *erfarung* with its original connotations of travel. Authors of chivalric tales liked to play with the slippage of *erfarung* as travel (*fahren*), but also danger (*gefahr*). Travel literature was lastingly popular, none more so than the tall tales of Sir John Mandeville; Brunschwig might even have seen a German copy printed in Strasbourg in 1499, whose title summarized the knight's travels as an *erfarung* of the Holy Land. The equally popular *Fortunatus* describes the sense of curiosity associated with travel, which lures the pilgrim away from his quest for salvation and into sensory explorations of the material world and its marvels.[89]

This sort of idle *curiositas* was the subject of long-standing criticism by theologians from Augustine onwards, who saw in it a sinful distraction from spiritual contemplation.[90] Among the most vocal critical voices, certainly in the German-speaking area, were two humanist scholars from Strasbourg, introduced in chapter 1—the lawyer and author Sebastian Brant, known for his satirical *Ship of Fools*, and the priest Johann Geiler von Kaysersberg, who published a series of sermons based on Brant's work. Writing in German, Brant and Geiler saw in *erfarung* the vernacular expression of *curiositas* and all that was wrong with it. One of Brant's chapters is dedicated to "*erfarung* of all countries," poking fun at fools who want to know "strange things" and measure the world with their feet but neglect to plumb the depth of their own soul.[91] Geiler unpacks Brant's verse into a series of sermons dealing with different categories of fools, and he dubs *wanckelnarren* (dithering or unsteady fools) those who like to *erfaren* new things, be it through travel or reading useless books.[92]

At face value, this is classic Augustinian denigration of *curiositas*, but at a closer look, Brant's seemingly brash verse reveals a slightly more nuanced picture. Anticipating the gist of the *Fortunatus* story, Brant's fool "always desires to experience [*erfaren*] much, but does not thereby improve himself." Aligning *erfarung* with traveling, Brant specifies that "he is a fool who travels through [*durchfert*] many countries and learns neither art nor virtue."[93] Between the lines, Brant implies that the experience of travel is only a fool's errand if one does not learn from it, and it is suggestive that the knowledge that makes experience worthwhile includes not only spiritual "virtue," but also practical "art."

Geiler, too, deserves a second look. As we saw in chapter 1, he knows his artisanal congregation and speaks their language in his sermons on the ant. He speaks of artisans as the nimble hands of the civic body and praises their dedication in acquiring skill through practice. Far from criticizing the experiential knowledge of the craftsman, he compares it favorably with the unsteady to-and-fro he associates with *erfarung*. The dithering fool, he writes, is like a craftsman trying to learn too many trades—in the words of a popular saying, "twelve crafts, thirteen times unlucky."[94] While grinding a slightly different axe, this time against the flood of books on anything published by anyone, he again compares this with the rigor of artisans, "who must first learn their craft and have permission or join a guild before they may practice it . . . but those who fabricate books are free and not bound by any law; nowadays anyone who wants can produce books without experience [*erfarnis*]."[95] His word for experience or expertise, *erfarnis*, is close to *erfarung*, but while Brunschwig appears to use the two interchangeably, Geiler upholds the value of artisanal expertise while parceling off the *erfarung* of travel as a different and lesser kind of experience. Intriguingly, he specifically condemns it in the medical context. One of Brant's chapters deals with the follies of sick people, especially those who trust in old wives' cures, "fool's wort," or even the devil rather than a doctor. In his sermon on this chapter, Geiler unpacks what Brant had summarized as "superstition"[96] by putting different words in the fool's mouth: "I have experienced (*erfaren*) that this or that helps; I got well because of it."[97] As medical epistemology, this was unsound and outright dangerous: the fool who thinks he is cured by magic may be in the clutches of an evil spirit who is toying with him by temporarily making him feel better. In medicine and on the road, *erfarung* was dangerous and unreliable in Geiler's books.

The local intellectual climate at Strasbourg, then, was on the whole against *erfarung* as a way of knowing the world by moving about in it and seeing new and marvelous things—even though both Brant and Geiler make subtle exceptions for experiential knowledge of art and craft. In part, Brunschwig might have agreed with Brant or even Geiler that experience could be dangerous in medicine (especially in surgery), and that idle curiosity led nowhere. His own curiosity, however, was never idle. As we have seen, *erfarung* was how artisans knew their craft, and how empirics and common people knew medical secrets. As I aim to show, Brunschwig writes against the grain of the Strasbourg humanists' critique because he makes the connotations of travel, of moving in the world, a key part of his project of experiential and artisanal knowledge. He would have avoided being the fool in Brant's book, not by avoiding the *erfarung* of travel altogether, but by making sure that he learned art (and perhaps virtue) on the way. To Geiler, he might simply have replied that one cannot make *erfarnis* without *erfarung*.

Brunschwig does not overtly portray himself as a great traveler. Indeed, when he states in the *Cirurgia* that he has "learned from many *herfarnden* healers," using a participle that tellingly slips between "experienced" and "itinerant" or "traveling," we get the impression of an author at home in his native city who collects the travel experience of others.[98] This is an important part of Brunschwig's project, but there are also passages that speak to his own comings and goings. As discussed in the previous chapter, traveling was built into artisanal training. While Brunschwig does not explicitly write about his days as a journeyman surgeon, we do know that he took part in military campaigns, notable during the Burgundy Wars in the 1470s. Brunschwig's contemporary, the Strasbourg surgeon Hans von Gersdorff, specifically notes that he worked together with his master in battles at Murten and Grancy. As we saw in chapter 2, Brunschwig and Gersdorff write of their own and others' exploits on the Burgundy battlefields—new sites that brought unexpected circumstances, new challenges, and opportunities for improvisation and experiential learning. For all that Brunschwig's experience was focused on the city of Strasbourg, he knew his way around the surrounding area, especially when it came to sourcing ingredients and equipment. His local expertise amounts to something of a map of materials and artisans (as we saw in chapter 1), from the place for the best honey to the small village where potters make good earthenware jugs for keeping distilled waters.

We also see Brunschwig moving around in his chattier accounts of making medicines and treating patients. In the preface to his plague treatise, he describes the horrifying consequences of an epidemic in an unnamed city "where a terrible pestilence reigned."[99] One of his surgical anecdotes is set "in a praiseworthy city," and another case, of a child who had stuck a pea up their nostril, he had "seen in Würzburg." Perhaps on the same trip, Brunschwig observed the bandaging technique of one "Hans Pfarrer of Würzburg." Even in cases where he did not leave the city, Brunschwig was often on the move. As we saw in his accounts of cures, he was frequently called to a patient's bedside, sometimes by another practitioner at their wits' end. He never quite knew what he would find, and often had to improvise.

Being called away to deal with a patient is a key turning point in the dramatic episode of a turpentine distillation gone wrong, quoted at length in chapter 6. Brunschwig is summoned away from his workshop twice, and each time the process takes a disastrous turn, culminating in an explosive conflagration. The tale is a cautionary one, included for the benefit of the reader, "so that what happened to me will not happen to you."[100] As I have shown, this is part of Brunschwig's project to provide an apprenticeship in print; by learning their lesson from a detailed written account, readers can avoid making the painful experience themselves. Brunschwig may only have traveled a short way across town in this episode, but he nonetheless uses an experiential narrative similar to those of later travel writers. In his *Cosmographei* of 1544, Sebastian Münster claimed that he had read, seen, and *erfaren* so much that his writing could serve as surrogate experience for armchair-traveling readers.[101]

Fully conversant with vernacular debates about *erfarung* and its connotations of travel, Brunschwig takes its epistemic potential far beyond the reservations of local humanists. His *erfarung* does not entail narratives of travel to faraway places, but he gathers valuable experience from others' and his own movements between battles, cities, workshops, and patients' homes. Moving through the material world and navigating its particulars and challenges was part of his artisanal way of knowing, and he wrote down his experiences to allow his apprentice-readers to travel along in print. This is a remarkably early statement of a positive attitude towards *erfarung*: in travel writing as in works of natural knowledge, it was not until later in the century that the sinful vagaries of travel were increasingly recoined as the necessary labor of insight.[102] When Paracelsus, who learned much from artisans such as Brunschwig, wrote that one should forget about books and instead read the book of nature with one's feet, this was a radical proposition.[103] It was not until the seventeenth century that traveling became an official part of scientific programs such as Francis Bacon's. Members of the Royal Society embraced the ideal by interviewing travelers and sending out questionnaires to investigate and confirm facts from faraway lands. Even within this highly organized regime, however, there remained a key problem. It was difficult enough to know whether any individual witness was a truthful one, but travelers in particular, as a common proverb went, had a "license to lie."[104] These questions about testimony and credibility were in the background of all early modern efforts of making knowledge from individual experience.

Witnessing and Reporting Experience

The problem of credibility is most visible in the case of personal accounts of travels and deeds, but it ultimately concerns all forms of experiential knowledge that Brunschwig employs in his books. Empirical cures, recipes, secrets, and other *experimenta* were at an epistemic disadvantage in an intellectual climate dominated by scholastic philosophy. Without the certainty of rational demonstration from the principles of medical theory, why should anyone believe that frog spawn conferred a healing touch? Since the *experimentum* could not provide its own proof, the burden fell instead on the person of the witness who reported it. In her study of the emergence of the early modern eyewitness, Andrea Frisch has identified two main modes of legitimating statements of experience. In the Middle Ages, the prevalent ideal was the ethical witness, whose credibility was based not on their testimony as such, but on their character and status, or that of the "oath-helpers" they could mobilize to support their story. Over the course of the sixteenth century, this was only gradually and partially superseded by the epistemic witness, whose credibility was grounded in the quality of their experiential accounts.[105]

Versions of these strategies can be discerned in the context of making natural knowledge. Medieval physicians interested in the virtues of individual healing springs and other particular phenomena sought to buttress their observational claims with the ethical prestige of noble patrons, and by emphasizing their disciplined habits of sensory inquiry.[106] Questions about the social negotiation of truth and credibility have been central to narratives about the so-called Scientific Revolution in the seventeenth century, particularly the shift from philosophical demonstration to a new science based on the observation of particular natural and experimental phenomena. New scientific communities such as the Royal Society were at pains to show how individual testimony of such observations could establish "matters of fact." Robert Boyle and others argued that if the testimony of two gentlemen was enough to send a murderer to the gallows, surely it was enough to generate consensus about particular phenomena. As Simon Schaffer and Steven Shapin have shown, Boyle's revaluation of the experimental witness combined ethical and epistemic strategies. On the one hand, he promoted a culture of genteel discourse and trust that rested on noble birth and social status. On the other hand, the experimental philosophers developed literary techniques to mobilize assent to their testimony. By composing their reports in clear, unadorned language and with a plethora of circumstantial detail, they recruited their readers as "virtual witnesses" to the observed phenomena.[107]

While craftsmen, servants, and other "invisible technicians" rarely appear in the writings of Boyle and his colleagues,[108] historians have begun to unravel the new science's indebtedness to artisans' practices of making and writing knowledge.[109] It is likely that the new modes of witnessing in particular owed a greater debt to artisans than has been realized. Craftsmen and painters were an early exception to the primacy of the ethical witness as posited by Frisch: in their writings and works of art, they articulated a confident stance of epistemic witnessing based on their sensory expertise, autoptic observation, and accurate representation or imitation of nature.[110]

Writing around 1500, Brunschwig offers a rare glimpse of the ways in which an artisan and collector of experiential knowledge dealt with testimony and authority a century and a half before Boyle and others explicitly articulated the issue. While Brunschwig was happy to collect all kinds of *experimenta,* including the secrets of empirics and common folk, he sometimes felt the need to provide ethical credentials. In his observations on an outbreak of urinary tract infections in the early 1500s, Brunschwig names local dignitaries as witnesses to his description of the symptoms:

"as I have seen myself on an honorable alderman of the laudable city of Strasbourg, and on the noble squire Wilhelm von Bach, Hans Ingollt, an excellent merchant, and others besides."[111] The clergy were another pool of respectable witnesses. A recipe for potable gold bears the following testimony: "as I Hieronymus Brunschwig have myself seen a vicar . . . at Strasbourg, around a hundred years of age, who used this water as he was feeble and sick."[112] Some are mentioned by name, such as "Hans, priest at Würzburg"[113] and "Herr Burckhart of the Carthusians";[114] others are merely referred to as "an old priest"[115] or "a man of the clergy."[116] The fact that Brunschwig felt the need to stress the honesty and status of his witnesses does not mean that their knowledge was not valuable as such; on the contrary, Brunschwig's framework of validation underlines how keen he was to draw on a wide range of experience from all walks of life.

Nor does it mean that experience never translated to epistemic testimony. The recipes and cures of the medieval masters were sanctioned by a rational framework and textual authority, but contemporary practitioners, too, could lend authority to treatments such as a powder that Brunschwig uses with confidence because "it was sent by the surgeon of Emperor Friedrich to another good surgeon from whom I have learned of it." Referring perhaps to Hans Seyff, who famously amputated the emperor's leg (see chapter 3), Brunschwig appeals to both medical expertise and noble patronage.

A similar range of ethical and epistemic witnesses appear in local court proceedings. In 1523, the knight Wolf Wilhelm von Andlain wrote to the council to bring charges against a chandler named Conrad, who had promised to cure a wound on Wolf's leg by means of baths and fumigations, for a fee of six gulden. Conrad had promised not to use mercury or cinnabar, but was later found to have secretly added these corrosive substances to his fumigations. Wolf was outraged at having his life thus put at risk and demanded his money back. At the end of his lengthy narrative, he added, "in order that the story and the foundation of its truth may be entirely known [*erfaren*], I ask that the honorable knight Hans Bork, Doctor Michael Meyer the city physician, master Hans of Konstanz, and Jacob the barber near the penny bridge, should be questioned and heard in this matter."[117] To lend credence to his experiential account (as signaled by the verb *erfaren*), Wolf mobilized the ethical credentials of a fellow knight, but also the learned authority of the municipal physician and the expertise of a local barber.

Brunschwig himself acted as forensic witness in his official role as wound inspector, as we saw in chapter 3. He sets down his own expertise with some force, asking those who "have not learned or experienced much" to accept his verdict "as a lesson and let it stand unquestioned" until they have acquired "sound experience" of their own.[118] His meticulous distinctions of different wounds and first-person accounts ("I have seen many . . .") underscore his own "sound experience" of witnessing and assessing injuries.[119] We can imagine Brunschwig and his colleagues appearing in court as highly credible witnesses both in ethical terms, as officially appointed inspectors, and in terms of their expert "insight." In his own writings, however, Brunschwig the wound inspector appears first and foremost as an epistemic witness whose credibility rested on his experience as an artisan of the body.

Beyond the context of legal witnessing, the craftsman's specific way of acquiring such experiential knowledge provides another model for Brunschwig's validation of testimony. In his account of the spectacular surgical extraction of an arrowhead from the king of Hungary's arm, Brunschwig notes that the surgeon performing the operation had told him "in person with his own mouth how he got it out," emphasizing the direct report of a trusted colleague.[120] Reflecting on the *experimenta* in his distillation books, Brunschwig writes "I have practiced [them] through the art of *practica* with my hands . . . also I have seen with my eyes and heard with my ears, not a few times, but many."[121] The emphasis on repetition as well as the immediacy of oral instruction, manual experience, and sensory perception reflects the artisanal mode of learning by apprenticeship. As I have argued in chapter 6, Brunschwig does his best to recreate that immediacy in print, to provide a kind of surrogate apprenticeship. To achieve this, he describes techniques, potential mishaps, and sensory cues in great circumstantial detail. It is tempting to read this as an artisanal anticipation of seventeenth-century experimental philosophers' emphasis on thick description as a way of securing epistemic legitimacy. In presenting his artisanal testimony as an experienced master, Brunschwig recruited his readers as virtual apprentice-witnesses.

Brunschwig's crumb-collecting was subtended by a network of experiences that connected empirical cures, medical secrets, and artisanal expertise. Its strands of testimony were validated through complementary strategies modeled on legal and artisanal contexts. For many secrets, recipes, and case narratives, producing ethical witnesses opened up a broader and more democratic range of contributors. Especially regarding his own testimony, Brunschwig is an early example of an author who articulates the stance of an epistemic witness, specifically that of an artisan of "sound experience." By teaching his readers how to look, listen, and do for themselves, he recruited them as witnesses in their turn.

The Knight and the Ivy Cup: Brunschwig's Vernacular Empiricism and Its Legacy

Written by a well-read and experienced artisan, Brunschwig's books offer a rare view of experiential knowledge making at the intersection of a range of genres and traditions, theoretical and practical, medical, occult, and artisanal. Brunschwig speaks of experience in such learned terms as *practica* and *experimenta*, but also in the vernacular of *erfarung*, rich with connotations that were hotly debated in the intellectual circles of his hometown.

What I have characterized in this chapter as Brunschwig's "vernacular empiricism" was in conversation with many genres and traditions, and so it is not surprising that it was not a monolithic project. In surgery, experience serves to revive the medieval rhetoric of rational surgery as both manual work and learned art. While experiential knowledge forms surgical skill and even legitimates the occasional innovation in treatments, it must ultimately remain in dialogue with learned theory. In his pharmaceutical works, Brunschwig places experience at the center of his efforts to provide medical knowledge for a wide audience of nonspecialists, including common folk. In the form of the recipe, he opens up the process of modifying and testing to his readers. His accounts of cures stress heroic action and improvisation in unusual circumstances as a site for gaining experiential insights. While theoretical learning was good and necessary in a healer, it was less essential than the experience of the *practicus*. In fact, even a mere empiric could know a useful cure. Familiar with traditions of occult knowledge through his alchemical work, Brunschwig naturally accepted that nature held many valuable secrets that eluded rational thought and could only be uncovered by experience. For all his learned references to *secreta*, *experimenta*, and *practica*, Brunschwig was never entirely part of the learned world, and his experiential knowing is solidly grounded in the vernacular. The *erfarung* of navigating the material and social worlds of healing, and the *übung* of artisanal training provide essential modes of producing, writing, and witnessing knowledge.

Brunschwig's strategies for mobilizing authority and assent around experiential knowledge claims highlight a persuasive rhetorical dimension of craft writing that is often overlooked. He deliberately presented ethical and epistemic legitimation of his own experience and others' testimony. It is tempting to read his thick sensory descriptions as attempts to recruit readers as virtual witnesses avant la lettre—at the very least, it shows that strategies of witnessing experiential knowledge were developing in artisanal and medical cultures long before the Royal Society (re)discovered the vernacular of experience as a language of science.

To get a clearer sense of Brunschwig's empiricism, it will be helpful to see it at work in one final telling example, a passage describing a cure for blindness.

> I have seen a knight of the king of Sicily . . . whose eyes were darkened, and he was nearly blind. Every day he drank from a wooden cup made of the wood of ivy, Edera in Latin, which had been procured from an oak tree and shaped by a woodturner between the two feasts of Our Lady; and so his sight returned. . . . Melchior Blum, mayor of Franckfurt, did the same, and made me a gift of fifteen florins in return. In the same way, a lay sister in Strasbourg who could not see the light and drank from such a cup within days could thread a needle, as I have seen with my own eyes, and marveled greatly.[122]

The remedy is a marvel and a secret, working by some combination of an occult property of the ivy wood, the artisanal skill of the woodturner, and perhaps the propitious time of its making. The reported healing effect does not lend itself to causal explanation and is known through experience alone. It is attested by three separate witnesses with strong ethical credentials—a knight, a mayor, and a pious woman. Brunschwig's claim to have seen it with his own eyes provides further testimony and, together with the reference to the patients' regained eyesight, locates the episode firmly in the realm of sensory experience. The fact that the protagonists or their testimony traveled from Sicily and Frankfurt to Strasbourg makes this a true piece of *erfarung* in the full vernacular sense of the word.

Brunschwig's project of experiential knowledge making did not pass unnoticed. The physician Lorenz Fries lived just over a day's ride from Strasbourg, but he shared a publisher with Brunschwig, and knew his books.[123] Fries's own book, the *Spiegel der Artzny* (Mirror of medicine), published in 1518, was an early example of a vernacular work on learned medicine, "the like of which has never before been published in German by a doctor."[124] Much like Brunschwig, Fries presents his book as essentially a work of experience, "thoroughly compiled in part from the teachings of famous ancient masters, in part from my own labor and work, with many good experienced [*erfarnen*], tried-and-tested [*bewerten*] pieces."[125]

Fries's stance on experience becomes clearer in his discussion of the medical sects. What he has to say about the "methodics" is characteristically vitriolic: they are "spider-munchers, stray hangman's bastards, tinkers, and tooth pullers."[126] His account of the empirics, by contrast, is surprisingly gentle. To be sure, as a rational physician, he is bound to point out that both methodics and empirics are "in error and not true servants of nature, for they know *what*, but not *why*, and do not recognize the cause."[127] But Fries also acknowledges empiricism's roots in the writings of ancient authorities: "Dioscorides experienced [*erfaren*] that fennel is good for the eyes, for when the snakes came crawling out of their holes in the spring, Dioscorides saw that they rubbed their eyes against the fennel."[128] And to anyone familiar with Fries' penchant for feisty invective, even his description of contemporary empirics speaks to a kind of gruff respect—they are "barbers and stonecutters who on occasion prescribe potions, plasters, or ointments, and other fellows who know their business," including "many an apothecary."[129]

One practitioner in particular caught Fries's attention. He lived near the fish market, and wrote "two very nice books"[130]—from Fries's gentle mocking of a particular passage in those books, it is clear that he is talking about Brunschwig, as we shall see in the following chapter. Coming from someone who clearly did not pull his rhetorical punches, Fries's account of Brunschwig practically amounts to a glowing reference: "He essentially understands the teachings and Latin. He has done enough inasmuch as he knows the art of medicine, and is well practiced [*geübt*]."[131] Although Fries thought of Brunschwig as an empiric, he commended his experience, accrued over years of practice (*übung*). Similar to Brunschwig's *Cirurgia*, Fries cautions that "it is not good to lend credence to experience without cause and thorough proof [*bewerung*]."[132] If proof meant experiential testing, his caveat about experience is in large part a question of sufficient empirical verification.

In a later work, Fries even had occasion to comment on *erfarung* in the context of travel. In 1525, he provided the explanatory text for Johann Grüninger's edition of the popular Waldseemüller map of the seas. In a chapter on maritime winds, Fries explains why sailors distinguish so many more different kinds of winds than the usual sets of four, eight, or twelve known to the learned tradition. These standard sets may be sufficient on land, Fries writes, but the sea is bigger and flatter, giving rise to many kinds and directions of wind that have implications for sailing a ship, "so the sailors have arrived at such a number of winds from their daily *erfarung*."[133] He mockingly contrasts these expert travelers with the "old masters" on dry land who "in ancient times sat in their little huts to search for the causes of natural things" and came up with a mere four winds to match the four corners of the earth.[134] Far from idle curiosity, the *erfarung* of travelers was a model for knowing the world.

Returning to medicine, the second edition of Fries's *Spiegel* was edited by the Strasbourg humanist Otto Brunfels, well known for his illustrated herbal that combined philological text work with detailed observation and images of plants drawn "from life."[135] In his preface to the *Spiegel*, Brunfels explicitly endorses Fries's emphasis on *erfarung*, noting that medicine should be restricted to "those who are knowledgeable, and also practiced and experienced [*erfaren*] in anatomy and natural things."[136] He even called for a "reformation" in favor of "the learned and experienced [*erfaren*]."[137] Brunfels admits that he was initially skeptical of Fries's vernacular project, but he soon had to concede that "although it is translated into the German language, it nevertheless has everything in it which is found in the best authorities of medicine, which I myself had not previously believed possible."[138] Like many other writers, Brunfels wondered (not without reason) to what extent the German vernacular, with its many dialects and a grammar that lacked the rigor and clarity of Latin, could function as a technical language.[139] The previous year, Brunfels was already coming round to the idea that technical knowledge could be conveyed in the vernacular—as we saw in chapter 2, he translated a medieval surgical text to benefit those of "true experience and knowledge" in their work of upholding the peace and order of the city. Like Brant and Geiler—Strasbourg humanists of the previous generation—Brunfels sees artisans and their labor as central to urban society; but unlike his predecessors, he fully endorses the vernacular project of *erfarung*, in practice and in writing, as the basis of the expertise that held the city together.

Brunschwig's vernacular empiricism and its reception in learned medical culture suggests an alternative story about the development of medical empiricisms in the early modern period. Whereas previous accounts have focused on learned genres, Brunschwig's works represent a rare pre-Paracelsian articulation of a medical empiricism in the vernacular language of *erfarung*, drawing on cultures of knowledge largely outside the bounds of rational medicine. It is an empiricism with a mission to promote the medical agencies of ordinary men and women through a kind of knowledge that they can understand and work with. Brunschwig's mission convinced at least one local physician and his humanist editor: Fries approves of Brunschwig's experience-based medicine and addresses his own vernacular work on medicine to "striped laymen"—as I show in the next chapter, this emerging audience of the literate middling sort was at the center of contemporary debates about the education and agency of the layman, and Brunschwig's books made a bold contribution to these debates.[140]

PART V

Crafting Text

READERS, EDITORS, TRANSLATORS

8

Dreaming of Vernacular Readers

"COMMON MEDICINE" AND THE STRIPED LAYMAN

Towards the end of his final publication, the *Large Book of Distillation*, Brunschwig tells his readers a colorful origin story. As he sat down to write the final section, he fell asleep at his desk, exhausted from the effort of writing the earlier parts of the book. In his dream, he heard the "screaming and shouting" of voices trying to rouse him:

> "How entirely you have neglected us!" I awoke, saying "Who are you?"—"We are the poor."—"Who are the poor?"—"Those who haven't a penny to their name."—"Who else?"—"We are the peasants from the villages."—"Who else?"—"We are those living in castles and small towns. Although we have enough money, we have neither healers nor medicine to help us."[1]

A wonderful woodcut illustration of this dream scene shows the slumbering author as a sort of medical St. Jerome, surrounded by a woman clasping her hands, a peasant in ragged clothes, and a knight whose costly dress suggests that he could afford professional help, if only he did not live in a castle so remote that he had to consult physicians by letter (figure 8.1). This clamoring chorus of imagined readers, Brunschwig claims, prompted him to take up his pen once more and write the final part of his magnum opus, titled "Thesaurus pauperum" or "treasure of the poor," to provide "common medicine" for those who could not obtain the "high medicine" of learned physicians.[2] This origin story, although plainly fictional, vividly illustrates that Brunschwig imagined a new vernacular readership, not just for the straightforward recipes of the "Thesaurus." Elsewhere in the *Large Book*, he writes that "this book and other books that I have made are for the commoner and those who do not have a learned and experienced doctor."[3] In this chapter, I trace Brunschwig's intended audiences across his works, in text and image, to illuminate his project of providing "common medicine" for "common" people.

FIGURE 8.1. Brunschwig asleep at his desk in the dream scene from the "Thesaurus Pauperum." Brunschwig, *Large Book of Distillation*, fol. 283r. Credit: Wellcome Library, London, EPB/D/1114.

Historians of early print agree that vernacular books addressed to the "common man," especially on medical topics, became one of the fastest-growing sections of the market.[4] But we still only have a vague notion of who this new reader was—as Lyndal Roper has pointed out, the term "common man" had no clear definition, but was a matter of local negotiations about who belonged to the community.[5] In Strasbourg and the Upper Rhine, there was an intriguing local twist to debates about common or middling folk with the emergence of the "striped layman" as a controversial figure in humanist satire and visual culture.[6] As I show here, this visual trope of the striped appears as a social label in Brunschwig's books, and more generally in the visual output of his publisher, Johann Grüninger. With their intriguing textual and visual commentary on readers, practitioners, and their agencies, Brunschwig's books provide fresh insights into the emerging vernacular readership embodied by the "common man" and the "striped layman," and into the ensuing debates about what (and how much) they should know.

I begin by tracing Brunschwig's statements about his intended audience. While the emphasis changes across his publications, the "common man" figures as a stable core of socially and intellectually middling readers eager to learn more about medicine. I then explore the trope of the "striped layman" in the local humanist discourse and visual culture to show how it is deployed in Brunschwig's books, transforming from a troubling figure of humanist critique to a shining image of the ideal practitioner, patient, and reader. Finally, I show that Brunschwig's project of providing "common medicine" for the striped layman was endorsed by the local physician Lorenz Fries, in something of a medical Luther moment that contributes to a more nuanced understanding of the politics of knowledge leading up to the Reformation.

Writing for the "Common Man"

In his first published work, the *Cirurgia*, Brunschwig specifically addresses aspiring surgeons to bolster their training with practical instructions and a solid theoretical basis. His plague treatise, the *Liber pestilentialis* of 1500, speaks to a similar audience of "prospective master surgeons and barbers," but it also includes instructions for the "common man in villages and castles."[7] Recommending pomanders to protect oneself from corrupt air, Brunschwig provides a range of recipes for different kinds of readers: "some costly, for lords; some middling, for citizens; some for the common folk in the country who cannot get to an apothecary."[8] From villages and castles to the city, most of these plague remedies offer self-help for individuals and households. This is not primarily a book for "the learned" who, in Brunschwig's words, "have greater and better things to do."[9]

The *Small Book of Distillation*, published the same year, is not addressed to practitioners of any particular trade—unlike surgery, distillation was not a recognized craft with its own guild and formal training program. This is a book for any and all "who desire to learn the manner and art of distillation and to begin this work."[10] Brunschwig included distilled waters "for the common man who has neither doctor nor medicine, and could not pay for it," and saved more complex procedures that were "not for the common man" for later publication in his *Large Book*.[11] The title page illustration underscores the impression that Brunschwig's intended readers were not necessarily full-time distillers; for all its bustling activity, the scene is one of bucolic leisure more than professional routine, more side-hustle than day job (figure 8.2).

In subsequent editions published by Grüninger in 1505 and 1509, the *Small Book* appears as part of a larger volume. It now contains a new section on more complex distilled waters and a collection of medical recipes, both

FIGURE 8.2. Title page of Brunschwig, *Small Book of Distillation* (Strasbourg, 1500). Credit: Wikimedia Commons.

likely by Brunschwig, as well as the first German translation of Marsilio Ficino's *Book of Life*. While the added recipe collection is aimed at readers "in villages and castles who cannot reach a large town with an apothecary," the instructions for composite distilled remedies include such costly and sophisticated procedures as making potable gold or extracting the quintessence of a medicinal herb.[12] Together with the medical writings of so eminent a natural philosopher as Ficino, this gives the volume a distinctly grander and more learned bent. At the same time, even Ficino's text is explicitly adapted to vernacular lay readers. As the introduction explains, only the first two of Ficino's three books on the preservation of life are included, because "the third book is about life which comes from heaven . . . This requires high understanding and is therefore left out here."[13] Like the first edition of the *Small Book*, the 1505 and 1509 volumes address a lay audience, but one with an interest in the pricier and more philosophical aspects of medical distillation, if not the "higher understanding" required to fathom its theological implications. The title page neatly reflects the layers of these composite volumes: the original illustration is cut in half and topped with an image of Ficino instructing a male figure, clad in striped clothes that are clearly a cut above those of the distillers in the lower half of the image (figure 8.3).

Even as he was writing the *Small Book*, Brunschwig was planning to "publish the large book for learned people of this work."[14] Published in 1512, the *Large Book of Distillation* certainly delivered on Brunschwig's promise of learned reading matter. It devotes more than twenty chapters to quintessence and the alchemical theories behind it (see chapter 5 in this book), as well as cures explicitly intended for "internal medicine, which belongs to the physician."[15] Despite its greater scope and ambition to speak to "learned people," the book includes sections such as the "Treasure of Poor Men" with its illustrated dream of common readers.

Across his works, Brunschwig's intended audience develops, from journeyman surgeons enrolled in formal training to common citizens in times of plague or distilling householders with little prior knowledge. In later editions of the *Small Book* and especially in the *Large Book*, Brunschwig expands his remit to include readers with more philosophical interests and ready access to complex equipment and costly materials. But he never loses sight of his core readership.

Who was this common man at the center of Brunschwig's imagined audience? Over the course of the sixteenth century, an increasing number of books, especially in the vernacular and on medical topics, were addressed to the "common man." In the German lands, this emerging readership has been estimated to represent a middling layer of society; the 20 or so

Liber de arte distulandi Simplicia et Composita.

Das nüv büch d̄ rechtē kunst

zū distilliere. Ouch vō Marsilio Ficino vn̄ anderē hochberōmptē Ertzte natürliche vn̄ güte künst / zū behaltē den gesunden leib vn̄ zū vertreibē die kranckheitē mit erlengerūg des lebēs.

FIGURE 8.3. Title page used for the enlarged 1505 and 1509 editions of the *Small Book*. Credit: Wikimedia Commons.

percent of the population who could read the vernacular (although generally not Latin), and who could afford to buy books, pay an apothecary for simple remedies, and consult a surgeon (but usually not a physician).[16] As the household became a crucial unit of society, the "common man" was increasingly identified with the *hausvater*, the father of the house who managed the spiritual and physical well-being of family and servants.[17] If the term "common man" was in frequent use across the German lands, this did not mean that it mapped onto well-defined or consistent social categories. As Lyndal Roper has shown, being "common" signified a person's belonging and status in a community, and was thus a matter of local negotiation in specific contexts.[18]

In Brunschwig's books, too, it is clear that "common" did not simply mean "everyone." By addressing his plague treatise to the "common citizen," he invokes his readers' civic status in the city, usually cemented by membership in a guild.[19] As a respectable member of urban society, the common man could not have been an itinerant day laborer, but neither was he part of the patrician elite. In terms of wealth, too, the "common man in castles and villages" was distinct from "the rich" in Brunschwig's book.[20] He presumes at least a modest disposable income for purchasing medicines, but often indicates cheaper options for readers on a budget. For example, he includes an alternative recipe for incense balls against the plague that are "not as costly as the above," explicitly for "the common man who can get to an apothecary and can afford to pay for it."[21]

In terms of education, Brunschwig often elides "common" and "lay" readers, that is to say those without a formal medical education, much less a university degree. He clearly expects his readers to be literate in the vernacular—although the images might provide "some indication to those who can neither write nor read"—but he does not presume a knowledge of Latin, which he himself likely had not fully mastered.[22] He provides German equivalents for anatomical and pharmaceutical terms, and an extensive errata list so that readers "may be spared what happened to me when I was reading medical books printed in Latin."[23] For the most part, Brunschwig also does not expect more than a basic acquaintance with learned medical theory, passing over the finer points that would be "more harmful than useful for the commoner."[24] As we have seen, he did expect an aptitude for artisanal learning by doing.

Socially, financially, and intellectually, Brunschwig's "common man" was neither at the top nor at the bottom end of the spectrum, but on respectable middle ground. He represents a readership that became the fastest-growing market for printed books over the sixteenth century.

While Brunschwig's words help us map the contours of this emerging audience, in what follows I suggest that the images in his books were even more significant in shaping and advocating for a new kind of readership.

Fashionable Fools and Dangerous Know-It-Alls: The Striped Layman in Humanist Satire

Many of the illustrations in Brunschwig's books show people wearing conspicuous striped clothing, their bold black strokes calculated to stand out against the fine lines and intricate hatching of the scholars' fur-trimmed robes. It is clear that these striped figures represent the core of Brunschwig's intended readers—surgeons, distillers, householders, and other practitioners. This was no coincidence. However, since Brunschwig does not comment on these images or the striped figures, we must look for clues elsewhere. As we have seen in the previous chapter, Lorenz Fries, a physician from Colmar, published a work called *The Mirror of Medicine* (*Spiegel der Artzney*) that endorsed Brunschwig's project. On the title page, Fries intriguingly addresses his vernacular readers as "striped laymen," and he sees no need to explain the phrase—perhaps because it was commonly used and understood, at least in the Upper Rhine area. It was, in fact, coined by Strasbourg humanists responding to what they perceived as a troubling new fashion.

The tight-fitting striped hose and short doublets that we see in Brunschwig's images took on new social meaning as they were brought north of the Alps from the Italian courts by mercenary lansquenets towards the end of the fifteenth century. The look caught on among other social groups who had disposable income and were increasingly keen to spend it on their appearance: merchants, artisans, and even farmers expressed their new-found social confidence in conspicuous clothes. To the learned and political elites, this was no mere harmless fad, but a sure sign that the world was going mad: ordinary people were getting above themselves and forgetting their proper place in the social order.[25]

Learned criticism of this newly confident middling sort was driven by the Strasbourg city scribe and lawyer Sebastian Brant and the local preacher Johann Geiler von Kaysersberg, whom we already met in chapters 1 and 7. In Brant's satirical *Ship of Fools*, overreaching social ambition is a central theme. Even peasants, he complains, were no longer content to wear their customary rough linen, but dressed in fashionable garments, "slashed and lined with all sorts of colors."[26] In his sermons based on Brant's text, Johann Geiler elaborates on the hubris of the peasant who "leaves behind his father's garb" and the craftsman who "dresses himself and his wife as if he were a knight."[27] Moreover, the fashionable composite or "slashed" patterns themselves signaled dishonesty and make-believe. Geiler points out that they were "pieced together from small patches so that they cost more to make than the cloth is worth."[28] If the mantle of Christ, woven in one piece, was the material ideal of moral integrity, anything slashed, striped, checkered, or cobbled together was devious fabrication, calculated to obscure the true value of persons and materials.

Striped clothes signaled intellectual as well as social claims: Brant's "book fool" hoards books without reading them, and yet thinks himself "striped and learned."[29] Geiler calls such fools "half-learned laymen" (*laici semidocti*) and, in the German version of his sermons, "striped laymen" (*gestreifletten leyen*), coining the phrase that Fries later used to describe his audience.[30] Geiler's layman was striped in the sense that he was not of any one color, neither ignorant nor fully educated, but an uneasy, half-learned hybrid. A little learning was a dangerous thing. Especially in theological matters, Geiler warned, it gave rise to "errors and heresies."[31] Thomas Murner, another local humanist, continued the satirical admonition of fools who only knew "four words of Latin" but presumed to ask "sharp questions" about the holy Trinity and the virgin birth.[32]

This was symptomatic of the increasingly heated debates that would eventually lead to the educational and spiritual reforms we now associate with the Reformation. As Verena Schmid Blumer has shown in an illuminating study of a pamphlet called *The Striped Swiss Peasant* (*Der gestryfft Schwitzer Baur*), the visually striking striped layman became a poster boy for the Reformation project of claiming back knowledge and agency from learned authorities.[33] This was not a monolithic project, however, and as we shall see, the striped layman was a marker of local and complex negotiation rather than a straightforward Catholic/Protestant dichotomy in the politics of early print.

As well as dabbling in learned knowledge, the striped layman had a distinct form of knowledge of his own that was just as troubling to the humanists. In a Latin translation of Brant's *Ship of Fools*, the original "striped and learned" (*stryffecht und gelert*) is rendered as "learned and experienced" (*docti tamen atque periti*), glossing "striped" as "experienced."[34] As I have shown in chapter 7, the Strasbourg humanists were on the whole wary of experience or *erfarung* as an unsteady and misguided pursuit of worldly matters. While it had its place in the expertise of artisans, it was not conducive to spiritual and physical welfare. Brunschwig, of course, placed experiential knowledge at the heart of his medical project for the common man. As we shall see, the imagery of stripes in his works powerfully underscores this endorsement of the layman's way of knowing.

Picturing the Striped Layman

The images of striped figures are rich visual sources, but their interpretation comes with caveats. Unlike the images of plants and distilling equipment, they have no close relationship with the text—in fact, Brunschwig does not comment on them at all. Furthermore, they are frequently reused and recombined across Brunschwig's works and indeed other publications of the Grüninger press, which is well known for its illustrations. To interpret the meaning of the striped figures, we therefore need to view them against the larger background of Grüninger's visual practice. As I have shown in detail elsewhere, Grüninger's frequent recycling of visual material was by no means haphazard.[35] In publications before and after Brunschwig, he deliberately uses stripes as a visual label for specific social and moral types and was clearly aware of the striped layman's increasing social significance. Across diverse genres including Plautine comedy, tales of chivalry, and legal works, Grüninger's images use stripes to designate people of liminal status, in between traditional social categories, often with ambitions to move up in the world. As we have seen, Brunschwig's text imagines the "common" reader as just such a middling figure, socially, intellectually, and economically. I argue here that the illustrations are powerful visual statements in support of these nonacademic audiences, turning the often-ambivalent connotations of stripes into a ringing endorsement of "half-learned" practitioners, patients, and readers.

The images in the *Cirurgia* show a clear division of medical labor. Manual activities such as stitching up wounds or setting bones are invariably done by a striped practitioner. We see surgeons in tight-fitting striped clothes at the center of the action, breaking a badly healed fracture or suturing entrails back into place (figure 8.4), while scholars in fur-trimmed robes observe from a distance without getting their hands dirty. These images illustrate the traditional hierarchy of physic over surgery, or perhaps of the seasoned expert over the apprentice, but they also place the manual skill of the striped craftsman at the center of medical practice.

In addition to embodied artisanal expertise, Brunschwig's defense of surgery emphasized learned knowledge, as we have seen in his engagement with the medieval tradition of rational surgery (see chapter 2). This, too, is captured in a large woodcut that first appeared in the *Cirurgia*, but also prominently featured in his other works. It shows a scholarly figure lecturing to a small audience (figure 8.5). The recurring figure of the scholar looks similar to the portrait of the sleeping author in the "Thesaurus" (figure 8.1) and was perhaps intended to represent Brunschwig himself, graduated from the surgeon's stripes of his early training to scholarly robes

FIGURE 8.4. A practitioner dressed in stripes performing abdominal surgery. Hieronymus Brunschwig, *Cirurgia* (Strasbourg, 1494), fol. 78r. Credit: The Huntington Library, San Marino, California, 85770.

through his self-directed study of texts in later life.[36] One student listens demurely with hat in hand while another looks distractedly out of the window. The striped student in the foreground strikes a more active pose, gesticulating as if arguing a point of his own. Although not in command of large scholarly volumes, he holds on to other written forms of medical knowledge, perhaps a single recipe. Literate and attentive, this striped student is confident that he can learn from a scholar and appears to hold his own in a discussion with the learned expert.

These images illustrate the ideal surgeon's expertise and training. Like any craftsman, he hones his skill through hands-on practice, but he also

FIGURE 8.5. A man wearing stripes and three other students listening to a medical lecture. Hieronymus Brunschwig, *Cirurgia* (Strasbourg, 1494), fol. π2r. Credit: The Huntington Library, San Marino, California, 85770.

knows the basics of learned medical theory. Like the striped layman of the humanists' imagination, the striped surgeons depicted in the *Cirurgia* are in-betweeners: literate in the vernacular if not Latin, they are half learned, half artisans. In contrast to the humanists' arrogant upstart, Brunschwig's surgeon represents an ideal revived from the writings of his medieval antecedents: half-learned is exactly what he should be, and the other half is not ignorance, but the manual expertise of the artisan. In a similar way, the images play with and subvert a second stereotype. As we have seen in chapter 3, part of surgery's image problem was its association with the dishonorable body work of the executioner, and Brunschwig goes to great lengths to assert surgery's honor and integrity. In some places, executioners were required to wear conspicuous stripes so that decent citizens might avoid them in public. By associating stripes with dishonor, civic authorities also hoped (usually in vain) to curb the public's taste for the new fashion.[37] This is reflected in several images of the Grüninger press showing executioners wearing stripes.[38] The images in the *Cirurgia* inevitably draw attention to the visual connections between stripes, surgeons, and executioners, but only to explode the stigma of dishonorable body work and replace it with the epistemic stripes of the ideal surgeon.

While the surgeons in the *Cirurgia* plied their handiwork under the beady eye of learned men, the striped distillers of the *Small Book* roam more freely through the landscape of practice depicted on the title page. They are engaged in the practical work of decanting distilled waters, digging up herbs, and tending a still, with no fur-robed scholars in sight (figure 8.2). Brunschwig is cautious not to fall foul of academic authorities, deferring to the "honorable and highly learned doctors of the free arts of medicine, whose humble servant I am at all times, and whose advice one should duly heed," but he slyly adds "if they are available."[39] Throughout the distillation books, Brunschwig addresses readers to whom physicians were not available, such as the crowd of sufferers from his dream. Moreover, as we have seen in chapter 4, he frames his distilled remedies as gentle and self-directing cures that can be safely made and used without consulting a physician.

In the *Liber pestilentialis*, Brunschwig points out that even city dwellers might be left to their own devices during epidemics as doctors fled and official support collapsed. Consequently, he tells surgeons how and where to let blood in cases of plague—something that was usually determined by a physician for each individual case. Here, the striped surgeon is shown bleeding a patient without learned supervision (figure 8.6). Brunschwig also instructs common citizens to protect themselves against the plague by fashioning a pomander that is "neither too hot nor too cold, but can be used at all times," or by lighting fires in their homes to purify the air, as demonstrated by a stripy-trousered householder in the corresponding woodcut (figure 8.7).[40] In text and image, Brunschwig's books negotiate a remarkable degree of independent medical agency for their lay practitioners.

FIGURE 8.6. A surgeon bleeding a patient. Brunschwig, *Liber pestilentialis* (1500), fol. 25v. Credit: The Huntington Library, San Marino, California, 85772.

Some of the images show striped patients as well as practitioners (figure 8.8), and Brunschwig provides instructions for them, too. As I show in chapter 7, he decodes pharmaceutical jargon so that patients can judge their physician's prescriptions. This striking subversion of the physician's learned authority continues in Brunschwig's advice on choosing the right practitioner. "If you can get a doctor," he writes, "I will reveal and explain to you which one to choose and which others to avoid."[41] The one to choose, as we have seen, was the *practicus* who has both learning and experience—the complementary stripes of the ideal practitioner. The striped layman could shop around for medical care as well as fashionable clothes, and Brunschwig reminds them of their power of choice as paying clients.

The two women in the title illustration of the *Small Book* call attention to the highly gendered language of common men and striped laymen (figure 8.2). Indeed, there was no straightforward female equivalent of these phrases. "Common women" were prostitutes, not female citizens.[42] Nor do we hear of "striped laywomen," perhaps because women had their

FIGURE 8.7. Household measures against the plague: purifying the air by lighting fires and incense. Brunschwig, *Liber pestilentialis* (Strasbourg, 1500), fol. 7v. Credit: The Huntington Library, San Marino, California, 85772.

own ways of dressing to impress that did not reference the fashionable stripes of the lansquenet mercenaries. Nevertheless, as we saw in chapters 4 and 5, women were key players in premodern cultures of distillation, from humble householders to professional distillers of brandy or medicinal waters to princesses with their own distilleries. Brunschwig was aware of female experts. He acknowledged his debt to "many learned and lay men and women" and described the distillation techniques of local nuns.[43] Although excluded from the rhetoric around the common man and the visual trope of the striped, women were part of the picture, and their expertise was valued.

Despite its gendered language, the trope of the "common man" encompassed an inclusive sense of community and the common good—a notion that is also reflected in Brunschwig's books.[44] The plague treatise promises free medicines for fellow citizens who cannot afford them, and

FIGURE 8.8. Patient wearing striped clothes. Brunschwig, *Cirurgia* (1494), fol. 20v, 27r. Credit: The Huntington Library, San Marino, California, 85770.

is dedicated to the "gracious dear gentlemen of the imperial city of Strasbourg," who asked him "to write something about it [the plague] for them and the entire community."[45] As we have seen in chapters 1 and 3, the surgical and pharmaceutical books, too, make the case that the healing crafts serve the community by ensuring the integrity of artisans' bodies, livelihoods, and honor. Brunschwig pictures his writings as a service to the city of Strasbourg, "to the pleasure of all her citizens and inhabitants, including the entire community."[46] His books were more than repositories of useful knowledge for new readers—they were themselves intended as social agents for the common good.

The Striped Layman and the Luther of Medicine

While the images in Brunschwig's works provide powerful social commentary about new readers, the identity of the "striped layman" is never spelled out in the text. Perhaps this was unnecessary in a visual culture where stripes were so obviously charged with meaning, and perhaps Brunschwig did not think it wise to align himself so openly with such a contentious figure. Lorenz Fries, a learned authority and inveterate contrarian, had no such compunctions. Writing for a similar audience to Brunschwig's and with a similar didactic thrust, he made explicit reference to "striped laymen" as readers on the title page of his *Mirror of Medicine*. In fact, Fries traces a direct line from Brunschwig's audience claims to his own. He mentions an apothecary "to whose house the poor come running when they go to buy fish, shouting 'dear sir, you help all people, help us, too. We are in the villages and castles, but nowhere else can we find as much fish to buy as near your house.' Thus he has made two nice books for them, in which you will find much pretty instruction."[47] Brunschwig is not mentioned by name, but anyone familiar with his work would recognize him as the author of two books on pharmacy with a shop near the fish market, if they had not already clocked Fries's parody of Brunschwig's dream episode in the "Thesaurus," quoted at the beginning of this chapter.

Fries takes some delight in mocking this vision of poor prospective readers from faraway places who somehow still manage to crowd into Brunschwig's shop on their way to buy fish in the city. Nonetheless, he is sympathetic towards Brunschwig's mission to provide ordinary people with medical knowledge in the vernacular. As we saw in the previous chapter, Fries approved of Brunschwig's expertise, his "nice books," and especially their commitment to experiential knowledge. His own work reads like a continuation of this project to divulge medical experience and learning to new vernacular audiences. Fries's imagined reader asks, "Dear master, you speak of complexions, but not in the way I have heard our priest talk about it; also, I am something of a striped layman, and I have read that there were only four of them."[48] This self-identifying striped layman is already familiar with medical theory through religious sermons and through his own limited reading, and Fries has no scruples about introducing him to the finer distinctions of humoral complexions. In fact, he specifically caters to his readers' limited prior knowledge, noting that "striped laymen need more explanation than the learned."[49]

As Fries no doubt intended, his book stirred the controversy around the striped layman. In later editions of his *Mirror*, he reports that other physicians attacked him for divulging learned medicine in the vernacular.

In the preface to the 1532 edition, he takes them to task for monopolizing useful knowledge instead of using it for the common good, and proposes to "spare them no more than Martin Luther did the pope."[50] In the face of physicians' arrogant elitism, the striped layman was well within his rights to venture into learned territory on his own account.[51] Posturing as something of a Luther of medicine, Fries epitomizes Brunschwig's project of educating the striped layman through vernacular books, to make him a more informed practitioner and customer of healthcare.

In retrospect, it is easy to see the striped layman as a straightforward symbol of the Reformation, and of course he was co-opted as such in Lutheran pamphlets.[52] At a closer look, however, attitudes towards the striped layman do not map onto a simple Catholic/Protestant binary. This is partly a matter of chronology—Brant, Geiler, and Brunschwig all wrote well before the Reformation came to Strasbourg in the 1520s. But even Fries's views were more complex than his self-proclaimed Luther moment suggests. He remained Catholic all his life, which clearly did not stop him from aligning with Lutheran criticism of learned experts' monopoly on knowledge—a discourse, incidentally, that goes back to Brant's satire. Johann Grüninger, too, remained Catholic, and all but ruined his business by continuing to print anti-Protestant pamphlets long after the city had converted.[53] And yet, he marshaled his considerable visual expertise to imbue the illustrations in Brunschwig's works with powerful statements about the agency of striped experts. The striped layman divided contemporary opinion, but not along neat confessional fault lines.

This story also shows that the social and intellectual changes generally subsumed under the term "Reformation" were about much more than questions of theology or who should be allowed to read the Bible and in what language. The early Strasbourg humanists, including Brant and Geiler, were wary of the striped layman's claims to social status and knowledge of scripture, but they were by no means opposed to lay education in other areas. As Brunschwig's books make abundantly clear, medicine was an important and only slightly less controversial part of these debates about educational reform and the agency of vernacular books and non-learned readers.

The striped layman is a historically revealing figure precisely because he refuses to be pinned down. As his striped clothes suggest, he is neither black nor white, neither Catholic nor Protestant, neither rich nor poor, neither ignorant nor learned—he is always resolutely both. Thus he symbolizes a new social and intellectual middle ground that is being carved out in the early decades of the sixteenth century; a space where affluent farmers and mercenaries find the confidence to dress like their betters, and where people of modest education glean useful knowledge from vernacular books to further their spiritual and physical well-being. Brunschwig, himself a distinctly middling figure as a hybrid expert and artisanal author of medical books, was seminal in opening up this middle ground in the sphere of medical knowledge and practice. In text and image, his books packaged significant political as well as epistemic claims. In the chapters that follow, we will see how his project was translated to other vernacular audiences, and we move from dreamed-up readers to real readers, exploring different modes of "book use" among the annotations and other material traces left in surviving copies.

9

Cutting Words, Distilling Experience

THE VERNACULAR TEXT WORK OF TRANSLATION

As the preceding chapters have shown, the vernacular was central to Brunschwig's project, as an epistemology and as an intervention in the politics of knowledge. Rooted in practice and experience, Brunschwig's medicine had a distinct local flavor. In spreading far and wide as it did, it went through numerous transformations as it was translated from one vernacular to another. This included presentations in different kinds of German. As Walther Ryff points out in his own edition of the *Small Book*, the native Alsatian of Brunschwig's original was "mixed with the Swiss language, so not everyone can easily understand it."[1] The Low German of the North was even further distant in linguistic terms, and warranted its own translation of the *Cirurgia* in 1518.[2] One surviving copy has annotations from a reader who wrote and thought in the same local idiom, and perhaps no other.[3] But Brunschwig's books also reached beyond the German-speaking lands. In addition to Low German, the *Cirurgia* was translated into Dutch and thence into English in 1525. The *Small Book* made the same linguistic journey with a couple of years' delay. The *Large Book* was never translated in full, but the physician Philip Ulstad borrowed entire chapters on quintessence, complete with images, for his Latin work *Coelum philosophorum* of 1526 and its German translation the following year. The recipe collection titled "Thesaurus pauperum" that forms the final part of the *Large Book* was published on its own in numerous editions, and it, too, was translated into English, albeit decades later, in 1561.

Apart from the brief excursion into Ulstad's Latin, the transmission of Brunschwig's books was, true to his original project, a vernacular story. While historical studies of translation have grown into a rich field, we still know less about the transmission of knowledge through and between vernaculars compared to the learned languages.[4] Brunschwig's books trace what Elaine Leong has called "epistemic itineraries" through several languages, raising key questions about vernacular translation: How is Brunschwig's expertise—practical, particular, rooted in local experience—transmitted across linguistic, cultural, and material contexts?[5] Who were the people with the expertise to undertake this new kind of translation? And how did Brunschwig's controversial empowerment of vernacular lay readers work within different cultures of knowledge?

Every act of translation speaks to the reach of a text and tells its own story, but in this chapter, I focus on the itineraries of the *Cirurgia* and the *Small Book*, which were translated as a whole, including many of their images, across multiple languages, and within a transnational network that we can at least begin to reconstruct. The English editions of both works, *The Noble Experyence of the Vertuous Handy Warke of Surgery* (1525) and the *Vertuose Boke of the Distyllacyon of All Maner of Waters* (1527), were translated not from Brunschwig's original German, but from Dutch intermediary editions associated with a circle of Antwerp printers including Jan van Doesborch, who knew the English market well. In fact, Doesborch employed as a translator one Laurence Andrew, who went on to translate and publish the *Vertuose Boke* (the English version of the *Small Book*). As we shall see, Andrew also collaborated with the printer of *The Noble Experyence* (the English *Cirurgia*) and was likely the translator of both Brunschwig books. As something of a professional translator, Andrew is an unusual figure who has received little attention. The same is true of his treatments of Brunschwig: translations of how-to texts, and of vernacular texts more generally, were rare at this early stage in the English book trade.[6] In terms of their contents, too, they were the first English-language books on their respective topic, and remained the only ones for decades. Thus, they are not only early examples of practical medical knowledge in translation, but also speak to distinctly vernacular cultures of text work and early modern print.

Translation has been a central issue for historians of science from the beginnings of the discipline.[7] It has turned out to be far more than a means of rendering a text in another language. As Bruno Latour put it, there is no such thing as diffusion of knowledge—whenever knowledge moves between people, places, languages, or media, this involves translation of one sort or another.[8] To make knowledge in the first place, as James Secord has suggested, there has to be communication, co-construction, and translation.[9] Some theoreticians, including Kuhn, Feyerabend, and Foucault, view languages and cultures of knowledge as incommensurable, casting doubt on the very possibility of translation. Yet, throughout history, people evidently did translate, and historians of science (and especially early modernists) have made productive studies of the ways in which they got on with the job, despite its challenges and limitations.[10]

Translation is closely tied to politics of knowledge and language. Already in the fourteenth and fifteenth centuries, as Peter Jones has shown, English friars translated medical texts as part of their mission to improve the physical and spiritual health of laypeople, writing against the grain of university scholasticism.[11] As we have seen, Brunschwig's vernacular books and their audience of striped laymen speak to debates that eventually built up to the fever pitch of Reformation controversy around language and education. In the seventeenth century, as Sietske Fransen points out, learned scholars such as Galileo, Descartes, and Van Helmont still made a point of publishing their works in the vernacular first, so as to reach a growing nonacademic readership as yet uncorrupted by establishment dogma.[12] Latin and other cosmopolitan languages remained dominant idioms of learning for much of the premodern period, while vernaculars were more often the realm of local, practical, and particular forms of knowledge.[13] These were by no means hermetically separate worlds, but their exchanges and translations were often hierarchical. In her study of glossaries appended to medieval works of learned Arabic medicine, Shireen Hamza has shown that these acts of translation make visible the vernacular knowledge of local farmers and foragers, but they also silence the knowers.[14] Examining the complex publication history of Nicolás Monardes's sixteenth-century natural histories of the New World, Alisha Rankin has argued that the underlying "archive" of Indigenous practices is all but effaced in some learned editions.[15] Translation privileged or silenced specific forms knowledge and language, often invisibly, even as it mediated between them.

The textual labor that went into making books was often hidden from view, in what Anthony Grafton has called the "literary underworld" of early modern printing houses.[16] The labor of translation was no different, even in scientific institutions where it mattered most. As part of its commitment to particular, experimental knowledge and the mechanical arts, the Royal Society generally conducted its business in the vernacular, "preferring the language of Artizans, Countrymen, and Merchants, before that, of Wits, or Scholars."[17] Letters, experimental reports, and even published works from international correspondents were regularly translated into English to be read aloud in meetings or perused by fellows. Some of the more prominent text workers are known to us Robert Hooke translated a number of travel accounts and coordinated various other translation projects. However, as Felicity Henderson has shown, much of this work was entangled, collaborative, and tucked away in drafts and archival scraps. In the same way that much of the society's experimental work was done by "invisible technicians," much of its translation work, too, happened behind the scenes.[18]

As these examples show, translation by turns reveals and obscures local expertise, vernacular knowledge, and textual labor. Brunschwig's books and their translations offer a rare opportunity to trace itineraries of practical knowledge across multiple vernacular languages. They allow us to recover the labor of translators and their strategies for rendering technical, experiential, and embodied forms of knowledge. This chapter aims to shine a light on what I call the "vernacular text work" of artisans and other non-learned translators.

Historians have studied translation from a variety of methodological angles. Literary scholars in particular have studied translator's paratexts and theories of translation to characterize strategies and regimes of translation.[19] Others trace the fortuna of individual texts as they are adapted, reconfigured, and often substantially transformed.[20] Recent studies have productively emphasized translation's continuity with other textual and epistemic practices, from humanist techniques to household recipe collections.[21] While we have rich accounts of these macro-level transformations of translation, less attention has been paid to close comparative reading of original and translation at the sentence and word level, in part because many early modern translators treated their sources rather freely. As I show in this chapter, the translations of Brunschwig's works reward attention at both levels, from the compilation of text and image to the word-for-word correspondence of technical processes and terminology.

Comparisons of translations against source texts often raise questions about "fidelity"—how they follow the original, to what extent they introduce their own interpretation or agenda, and what makes a translation "faithful" in the first place. Needless to say, our modern notions of accuracy, fidelity, and even authorship would have been alien to early modern translators.[22] They had their own ideas of fidelity, and historians have often charted these in terms of changing attitudes towards "word-for-word" versus "free" translations. John Florio, the English translator of Boccaccio and Montaigne, famously quipped (with the casual misogyny of his day) that translations, like women, can be either beautiful or faithful, but not both. While some Renaissance humanists leaned towards one or the other extreme, many aimed for a middle path.[23] The general narrative proposes that medieval translators favored the *ad verbum* or word-for-word approach, which could result in unfamiliar or "foreignizing" formulations, whereas the early modern period saw a rise of *ad sensum* translation, where the sense of the original could be rendered in more familiar, "domesticating" language. Humboldt and the Romantics later rejected this approach in favor of preserving "a certain tint of foreignness" in their translations, for example by retaining Greek sentence structure in German.[24] While questions of "free" versus "word-for-word" translation were asked by some important historical actors, as historians we should be wary of reducing our questions to this binary, which is itself historically contingent. For example, as Rita Copeland has shown, what we might see as similarly "free" translations in Roman antiquity and the Middle Ages in fact signified completely different practices and frameworks.[25] As we will see, the translations of Brunschwig's works combine significant macro-level changes with remarkably close word-for-word correspondence, and I argue that both strategies and the way they deal with different kinds of information can be seen as "faithful" to Brunschwig's original project of providing workable, sellable how-to texts for vernacular readers.

While this chapter aims to characterize shared cultures of vernacular text work that shaped the medical crafts, for clarity's sake the complex itineraries of the *Cirurgia* and the *Small Book* are discussed separately. In tracing these itineraries from German to Dutch to English, I ask how successive text workers dealt with technical detail, embodied experience, and intended audiences. In the second part of the chapter, I explore how these strategies influenced the translation practices of subsequent generations of surgical and "chymical" practitioner-translators in Elizabethan London.

Vernacular Labors in Surgery

The English version of Brunschwig's *Cirurgia*, titled *The Noble Experyence of the Vertuous Handy Warke of Surgeri*, was printed in London by Petrus Treveris in 1525. The prologue states that "this boke is translated out of duche in to englisshe," and the colophon elaborates that it was "translated out of the speche of hye Almayne into lowe Duche. And afterwarde in to our moderstonge of Englysshe."[26] Although one could not ask for a clearer statement that the unnamed translator of the English *Cirurgia* worked from a Dutch translation rather than the German original, the Dutch intermediary text proves somewhat elusive. We do not know of any manuscript translations of Brunschwig's works, and while there is a printed Dutch edition of the *Cirurgia*, it was not published until 1535, ten years after the English version. Book historian Piet Franssen has suggested that there was an earlier Dutch edition that served as the source for *The Noble Experyence*. According to him, the surviving Dutch edition, which was printed by Jan Berntsz in Utrecht in 1535, was in fact a reprint of an earlier, now-lost edition by the Antwerp printer Jan van Doesborch. Doesborch left Antwerp around 1530 and moved to Utrecht, where he struck up a collaboration with Berntsz, and for a time, Berntsz printed little else except reprints of previous Doesborch publications. It is thus entirely plausible that the 1535 Dutch *Cirurgia* was just such a reproduction. Furthermore, the English edition reproduces a woodcut—a generic image of a scholar holding a book—that does not appear in Berntsz's edition, but does appear in some surviving Doesborch publications, providing a firm link between the English *Cirurgia* and the Doesborch press. Since we know that Berntsz habitually reprinted Doesborch's books without making any changes to the text, we can use the 1535 *Cirurgia* as a stand-in for the lost Doesborch edition with some confidence.[27]

The unnamed translator of the surviving Dutch edition gave the work a significant structural overhaul. Brunschwig's division of the work into seven "treatises" was abandoned, and the chapters numbered consecutively throughout. After a slightly longer prologue, the translation includes a section on anatomy that was present in all but the earliest German editions. Treatises 1 and 4, the shortest ones with three chapters each, are absent in the Dutch edition. Some of the material, on the proper behavior of the surgeon and the signs of death or recovery, was absorbed into the prologue, but the chapter on the inspection of wounds and bruises (discussed in chapter 3) was cut, as were three chapters on specific treatments for injuries caused by blunt force, drowning, and torture. After the anatomy section, the translation picks up in what used to be Brunschwig's second treatise on wound treatment, skipping the first two chapters, which provide short introductions to causes of injuries and surgical instruments. From a chapter on small wounds onwards, the Dutch *Cirurgia* follows Brunschwig's chapters in order to the end of the book, with few exceptions, which I will discuss in detail. The first five chapters of what used to be treatise 3, which deal with wounds to the head and skull, are expanded into ten chapters in the Dutch text, by moving passages around and adding material from other sources, as we shall see.

The images, too, speak to significant differences. The original title illustration of the "wound man," a medieval icon of surgery, is replaced with a more recent image, showing a device for stretching a broken arm, first published (and perhaps invented) by Brunschwig's younger local colleague, Hans von Gersdorff, in his *Feldtbuoch der Wundarznei* of 1517 (see chapter 2, figure 2.5). This central image is surrounded by four more generic images illustrating aspects of medical treatment. Many of Gersdorff's images are reproduced throughout the book, often replacing the illustrations in the German original. Where Brunschwig's anatomy section only provided an image of a skeleton, the Dutch edition reproduces the rather more impressive woodcuts from the anatomical fugitive sheets by Hans Waechtlin that were associated with Gersdorff's book (figure 2.7 and 2.8).

Brunschwig's presentation of surgical paraphernalia in a single image of a well-equipped cabinet is replaced with Gersdorff's more detailed images of individual instruments, inserted into the translated text in the appropriate place. Where readers of the German original had to flip back many pages to identify "the aforementioned scissors" in a crowded image, the Dutch translation refers to an image of just this instrument on the same page, "as is pictured above."[28] The makers of the Dutch edition could not resist including more of Gersdorff's impressive images of surgical contraptions, including some that bear little or no relation to Brunschwig's text. For example, the striking woodcuts showing the use of stabilized drills to recover depressed skull fractures are inserted in between two of Brunschwig's chapters on head wounds, with a translation of the instructions included in the original images themselves, but without Gersdorff's brief preamble explaining in what cases they should be used. In one of the copied images, the original tripod structure supporting the drill is missing a leg, suggesting perhaps that the images were included for prestige and aesthetic reasons more than technical information.

Like the original, the Dutch version includes a series of smaller images showing different trepans, but upon closer inspection it turns out that these were not copied from the German *Cirurgia*, but from a 1507 Dutch translation of Guy de Chauliac's *Chirurgia magna*.[29] Perhaps aware that Brunschwig's images and descriptions of these instruments were themselves drawn from the manuscript tradition of Chauliac's work, the translator decided to go back to a printed version of this source instead of adopting Brunschwig's treatment. Thus, the image of Brunschwig's favored drill—one described not by Chauliac but by another medieval surgeon, Lanfranco of Milan—is not included. The Dutch edition expands on Brunschwig's visual presentation and makes it stronger, adding instruments and expertise from medieval and contemporary sources.

The interventions in the text are in a similar vein. While much of the Dutch edition follows Brunschwig's text quite closely, there are significant changes, particularly in the chapters on wounds to the head and skull. The original five chapters are expanded into ten chapters in the Dutch text, with some passages reshuffled, removed, and inserted from other sources. Many of these additions, too, can be traced to the Dutch Chauliac.[30] Even when working from Brunschwig's original, the translator seems to have kept an eye on parallel passages in Chauliac. Where Brunschwig refers to head wounds with contusions as *zerknoerschung*, a somewhat idiosyncratic coinage from a regional verb that signifies something like "crunching-up," the translation uses the equivalent terms from the Chauliac edition, *conquassacie oft contusie,* closer to the original Latin *conquassatio* and *contusio.*[31] This suggests that the translator read Brunschwig's text alongside the Dutch Chauliac, to borrow passages that he deemed a worthy addition, and to check Brunschwig's renditions of medieval surgical terminology against an existing Dutch translation. More than simply translating, this text worker engaged in a vernacular philology of sorts, conscious of shaping a nascent technical language of surgery in Dutch.

The translator also referred to the text of Gersdorff's *Feldtbuoch.* The original *Cirurgia* contained a chapter on fractures that had healed with the bones at the wrong angle. If baths and splints were not enough to straighten the kink, it had to be forcefully broken and reset to heal once more—a procedure vividly illustrated in a woodcut (figure 9.1). This chapter is left out in the translation, and replaced with Gersdorff's chapter on the same topic. His approach seems more gentle, using softening baths and adjustable splints with a ratchet mechanism that straightened the crooked limb gradually rather than fracturing an already healed join (as in figure 2.5). But the translator's choice could just as easily have been swayed by the striking images of Gersdorff's contraptions, showcasing new solutions that combined surgical procedures with inventive uses of the armorer's craft.[32]

The translator also condensed and streamlined some of Brunschwig's more rambling passages. In a chapter on head injuries, Brunschwig cites authorities including Galen and Avicenna and their views on suturing wounds, as well as weighing the pros and cons of applying rose oil. The translator makes an executive decision in favor of the rose oil, and condenses this passage into a single, coherent set of instructions. In another chapter, Brunschwig writes that the crushed tissue in a wound should be "resolved with honey, or various other things as indicated in the chapter on contusion."[33] Perhaps to save readers the trouble of looking up a whole raft of alternatives in a different chapter, the translator picked one of them and inserted it into the instructions, which now simply read "resolved with salted wine and honey."[34] While the translator does not contradict Brunschwig's instructions, he is not afraid to excerpt and condense, sometimes decisively choosing among alternative explanations and treatments.

While this is not a "faithful" translation from a modern point of view of authorship and the integrity of the work, I suggest that it is remarkably faithful to the spirit of Brunschwig's project, which was itself a compila-

FIGURE 9.1. A surgeon refracturing a badly healed leg. Brunschwig, *Cirurgia* (1497), fol. 98v. Credit: The Huntington Library, San Marino, California, 85770.

tion of surgical texts in vernacular translation, combined with artisanal experience. The Dutch translator in many ways produced a stronger presentation of surgical expertise by streamlining some of Brunschwig's more rambling instructions and incorporating more recent translations of his main authority, Guy de Chauliac, complemented by the latest images and technical know-how from Gersdorff's work. Brunschwig might have disagreed with some of the cuts—especially his favorite trepanning drill—but he could hardly have failed to appreciate the continuation of his project to codify an authoritative vernacular surgery from learned texts and the experience of artisans.

Noble Experience and Virtuous Handiwork: Englishing the Cirurgia

The English translation of the *Cirurgia* was published in 1525 as *The Noble Experyence of the Vertuous Handy Warke of Surgeri*. It follows the Dutch edition of 1535 very closely, which suggests a close resemblance between the latter and the putative earlier Dutch version from which the English translator likely worked. The revised structure of the Dutch *Cirurgia* is replicated precisely in the English, with corresponding chapter numbers, even those that are incorrect in the original.

The title page of *The Noble Experyence* reproduces the two woodcuts showing the raising of depressed skull fractures from Gersdorff, no doubt the most striking of the surgical images in these works. Apart from this, the placement of images follows the Dutch example. The illustrations of shears, forceps, and other surgical instruments are clearly copies of woodcuts similar or identical to those used in Berntsz's 1535 edition, but some other images were apparently copied from the earlier German editions of Brunschwig and Gersdorff.

The illustrations of Gersdorff's trepans on the title page are more cropped than those in the Dutch edition, and the disembodied hands manipulating their screws have disappeared (figure 9.2). The missing third leg of the tripod, on the other hand, is restored, perhaps with reference to Gersdorff's original. Another Gersdorff image that was not included in the Dutch edition makes a comeback in the English translation: the full-page woodcut of a surgeon extracting an arrow in the middle of the battlefield (see figure 2.1). When it comes to the anatomical illustrations, however, the English edition resorts to older images. Where Berntsz's edition reproduced the strikingly detailed skeleton and dissected organs from Waechtlin's fugitive sheets associated with Gersdorff's *Feldtbuoch*, the English skeleton was modeled on the image in the original Brunschwig editions, while the depictions of internal organs and the brain were copied from smaller images originally published in the Strasbourg edition of Gregor Reisch's *Margarita philosophica*. However, the text of at least one of the labels in the skeleton image, keyed to what we would now call the sacrum and coccyx, is taken word for word from the Waechtlin woodcut, making the illustration a composite of both images.

It is difficult to know whether these differences between the English and surviving Dutch versions were introduced in the English translation itself or in its now-lost source, the putative Doesborch edition. Since Berntsz usually reproduced Doesborch's publications more or less unchanged, it seems likely that at least some of these decisions were made in England. In either case, the choice of simpler anatomical woodcuts when more

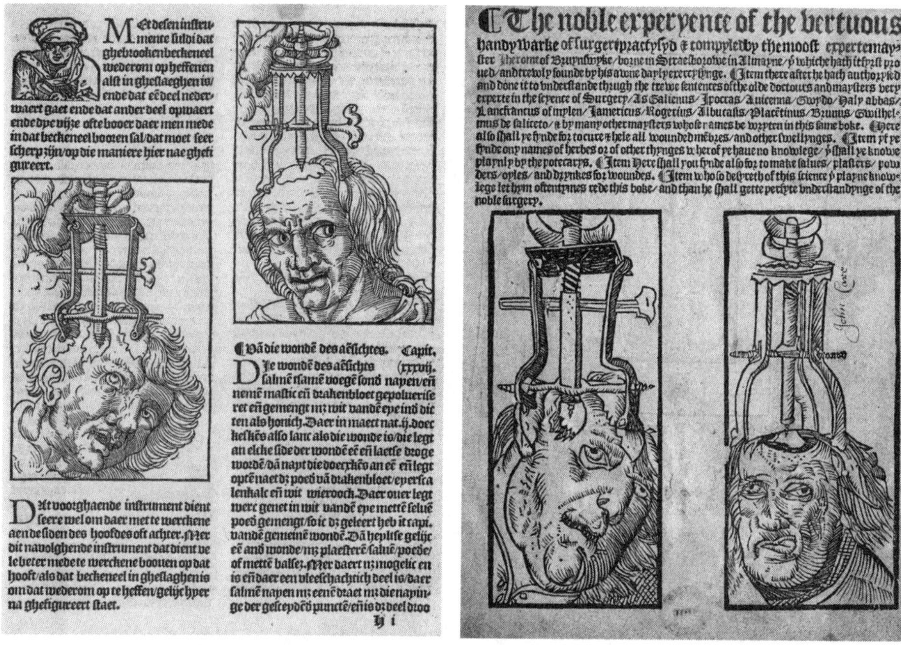

FIGURE 9.2. Tools for recovering depressed skull fractures, from the Dutch and English translations of Brunschwig's Cirurgia. Left: Brunschwig, *Dits dat hantwerck der cirurgien* (Utrecht, 1535), fol. H1r. Credit: Rare Book and Special Collections Division of the Library of Congress, Washington, DC, RD30.B69 1535. Right: *The Noble Experyence of the Vertuous Handy Warke of Surgeri* (London, 1525), title page. Credit: Wellcome Collection, London, EPB/D/1117.

impressive alternatives were clearly available might have been a question of finding—and paying—a skilled woodcutter who could reproduce the original's intricate detail. While this is achieved rather well in the arrow extraction scene, most other woodcuts are noticeably cruder in their execution, including some of the simpler line drawings of instruments. In one illustration of a sickle-shaped knife, the conventions of the printed woodcut are abandoned altogether: instead of working the lines in relief by cutting away the white space, they were simply incised into the woodblock, resulting in a crude and visually disruptive "negative" image of white lines on black background (figure 9.3). Of course, this required much less skill and labor—we can even imagine a hapless apprentice knocking out such a woodblock at the last minute. If adequate woodcutting skills were as hard to come by as some of these images suggest, then the anatomical images from the Brunschwig and Reisch editions were certainly less of a challenge to copy.

The text for the most part follows the Dutch version closely, and mi-

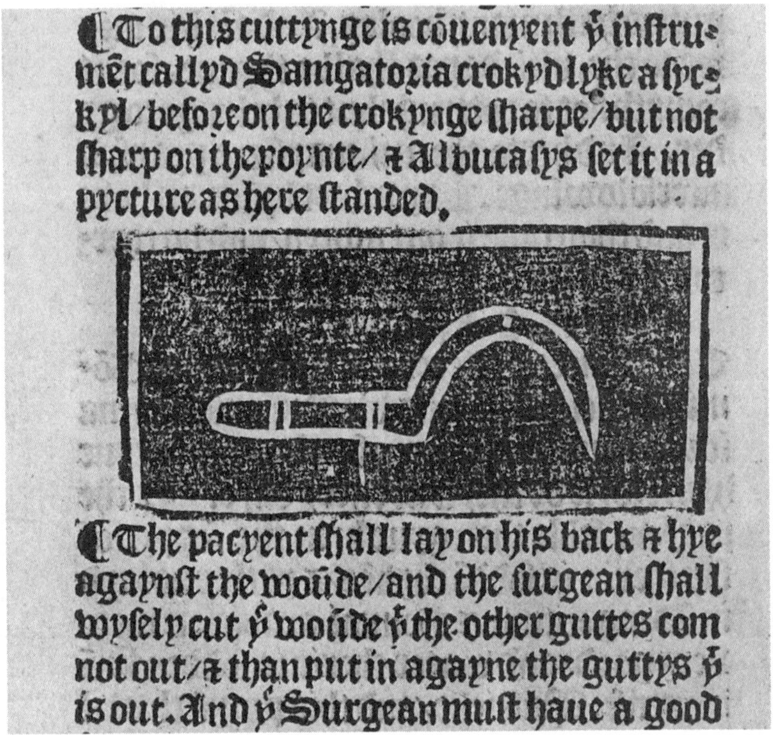
To this cuttynge is couenyent ỹ instru-
mēt callyd Samgatoria crokyd lyke a syc-
kyl/ before on the crokynge sharpe/ but not
sharp on the poynte/ & Albucasys set it in a
pycture as here standed.

The pacyent shall lay on his back & hye
agaynst the woūde/ and the surgean shall
wysely cut ỹ woūde ỹ the other guttes com
not out/ & than put in agayne the guttys ỹ
is out. And ỹ Surgean must haue a good

FIGURE 9.3. “Inverted” woodcut image of a surgical knife. Brunschwig, *The Noble Experyence of the Vertuous Handy Warke of Surgeri* (London, 1525), fol. M2v. Credit: Wellcome Collection, London, EPB/D/1117.

nor deviations from a literal rendition—a plaster “dressed” instead of “boiled” with wine—are relatively few.[35] Several discussions of different opinions and treatments, however, were shortened in translation, including some of the passages from Chauliac that were newly inserted into the Dutch *Cirurgia*. The English edition also omits two of Brunschwig’s seven tests for detecting skull fractures: hitting the patient’s head with a rod to hear what sound it makes, or thrumming a taut string held between the patient’s teeth to see if the vibrations cause pain in the skull (see chapter 6). The Dutch edition does describe these tests, but it leaves out an earlier passage stating that these two were in fact “the surest signs, more so than all the others.”[36] Without this information, the English translator unwittingly cut the tests that Brunschwig deemed most reliable. There are a few small additions in the English compared to the Dutch, such as the rare Latin gloss “*ex provincia Helvetiarum*” to explicate the phrase “out of Zwytselonde.”[37] However, the English translator never makes the kinds of incisive and knowledgeable interventions we saw in the Dutch text, emending technical terms or adding passages from other works. The itinerary of Brunschwig's surgery thus illustrates the remarkable labors of vernacular text workers, but also the inadvertent losses that may occur in transit across multiple languages, despite their meticulous efforts.

Translating the "Moost Experte Mayster"

The most significant changes concern the framing of the book and of Brunschwig's authority. The title page of the original German edition simply reads "This is the book of *Cirurgia*, the handwork of surgery, by Hieronymus Brunschwig," followed by the image of a wound man illustrating the range of injuries that can befall the human body.[38] In the Dutch translation, Brunschwig's name is removed from the title page and only mentioned in the prologue. Instead, the title lists at great length the different injuries and treatments covered in the book, as if it were spelling out all the ailments of the wound man, whose image is replaced with one of Gersdorff's traction instruments. The English title page, in contrast, foregrounds the author and his expertise, naming Brunschwig as "the moost experte mayster" who "practysyd & compyled" the work.[39] As we have seen, the emphasis on practical experience as well as reading and compilation goes back to Brunschwig's original edition. On the English title page, however, this is highlighted up front, and with a significant shift in emphasis. Brunschwig's expertise was "*fyrst* proved / and trewly founde by his ane dayly exercysynge," and only "*there after* . . . authorysed . . . through the trewe sentences of the olde doctours and maysters very experte in the scyence of surgery" (my emphasis). As the title states, what the book promises above all is the "noble experyence" of surgery.[40]

As I have shown in chapter 7, experiential knowledge played a key part in Brunschwig's project, but in his surgery, it remained subservient to established knowledge and procedures. Some of the passages in which Brunschwig is at his most cautious about experience-based cures are left out in the translations—for example, his criticism of local surgeons who use empiric remedies and can offer no theoretical justification. His reference to Guy de Chauliac's warning about the dangers of "new experience" is included in the Dutch edition, but left out by the English translator.

As well as dampening Brunschwig's reservations about experience in surgery, the translations tone down some of his witness statements. In translating the description of Ulrich von Baden's extraction of a bullet, the Dutch translation reproduces Brunschwig's assurance, "I have heard it from a truthful man," but the English text simply states "or els do as hans ulrik of baden dyde," recasting the episode as an established set of instructions.[41] In another anecdote of surgical heroism, Hans von Dockenburg's spectacular extraction of an arrow, even the Dutch translator perhaps thought that Brunschwig's statement "he told me himself with his own mouth" was laying it on a little thick, and shortened it to "as he told me himself," whereas the English translation cuts the statement altogether.[42] Similarly, some of the more specific endorsements of individual remedies are stripped away in the translations, such as the chain of transmission by which Brunschwig obtained a powder "sent by Emperor Friedrich's surgeon to another good surgeon who taught it to me."[43] A recipe for a plaster from Gersdorff, who attributed it to the Tübingen physician Johann Widmann, simply becomes "an expert plaster" in the Dutch and English translations.[44] In part, these omissions no doubt reflect a geographical and cultural distance that makes local events and people less significant. But they also suggest that the translators deliberately removed Brunschwig's self-reflective caveats about experience, as well as instances where he tried a little too hard to establish the credentials of witnesses and experiential reports. It is precisely by doing less of this sort of heavy lifting that the translations, especially *The Noble Experyence*, present Brunschwig's insights as established and authoritative.

Brunschwig's own voice as an experienced practitioner is largely unaffected by these changes, as the translations reproduce his assertions such as "I iheronymus brounswyk born in straesborow have sene, . . ." or "I have sene & herde . . ."[45] This first-person voice, we must remember, silently incorporates that of another craftsman, since the passages from Gersdorff that were added to the Dutch edition are not marked as such. Readers would not know that the "I" in statements such as "that I here wryte I have my selfe occupyed" originally represented a different author.[46] Such an amalgamation of authors' voices was not unusual for the compilation practices of the time—as we have seen, Brunschwig himself on occasion borrowed Guy de Chauliac's first-person narrative (see chapter 2). What is significant here is that the translations fashion an authoritative voice that can be attached to a strong authorial persona. While the identity of that voice does not make it to the title page in the Dutch edition, The *Noble Experyence* makes a point of crafting Brunschwig's identity as an "experte mayster" who no longer needs to justify every experiential statement. Where the original Brunschwig used experience to complement and cautiously build upon canonical texts, the English Brunschwig's experience is placed front and center, legitimated or "authorysed" by canonical texts.

Finally, the English text, unlike the Dutch, mentions the fact that it was a translation, and offers suggestive glimpses of what that might mean for the text and its audience. The colophon, as we have seen, describes the text's itinerary from German into Dutch and then English, adding that this transformation is "moche necessary & profytable for surgyans / as wel for them that have conynge as for them that be lerners."[47] The act of translation itself is bound up with the text's "profytable" project of providing knowledge for surgeons at different stages of training. As we shall see, this intertwining of translation, profit, experience, and disciplinary training came to characterize the nascent surgical literature of early modern England.

Distilling the Vernacular

Brunschwig's *Small Book of Distillation* appeared in English as the *Vertuose Boke of Distyllacyon* in 1527, two years after the translation of the *Cirurgia*, and in similar circumstances. It, too, was a translation from a Dutch intermediary, and its translator and publisher was Laurence Andrew, a close colleague of Peter Treveris, who had printed the English *Cirurgia*. In 1526, Treveris published a *Grete Herball* that he billed as something of a sequel to the *Cirurgia*, in collaboration with Andrew.[48] As we shall see, Andrew's expertise, connections, and translation style suggest that he was likely the translator of Brunschwig's English surgery as well as his distillation book.

What is known about Laurence Andrew marks him out as an intriguing and unusual figure. Born in Calais, he worked as a translator for the Antwerp printer Jan van Doesborch (the publisher of the putative original edition of the Dutch *Cirurgia*) before establishing his own press in London. Antwerp was the only continental city that regularly produced English-language books, and Doesborch was one of a few printers who maintained direct connections with the English market and traveled to London for extended stays. It is perhaps through Doesborch's network that Andrew was able to secure the support of Treveris and a loan of money and type from another printer, John Rastell, which would have been a great help in setting up his own shop in Fleet Street.[49] Through his work for Doesborch, Andrew also knew that books on the wonders of nature, travel, and practical knowledge sold well in England. His own translations for the Antwerp printer included *The Valuacyon of Gold and Silver*, a pocket-sized octavo volume with woodcut images of current coins and their exchange value, and *The Noble Lyfe*, a translation of the animal section of a popular herbal in the *Hortus sanitatis* tradition.[50]

Andrew was unusual both in the extent of his translation work and in his specialization in vernacular works of natural and practical knowledge. Translations of such texts were still relatively few, and they were usually incidental projects by practitioners or subject experts who rarely produced more than one translation in their lifetime.[51] As something of a professional translator, Andrew was an outlier. In the *Vertuose Boke,* he emphasizes his role as translator on the title page and elaborates on it in a "prologue of Laurence Andrew the Translatour." He modestly mentions his previous "sondry small volumes & tryfles of myrthe & pastaunce / Som newly composed / som newly translated," and justifies his undertaking to translate such a technical text: "For me thynke where the masters in all Science expert wyll take no suche paynes / it is nat dyscomendable for a man of more base lernyng to put to his helpynge hande."[52] In this show of appropriate humility about his knowledge of distillation, Andrew nevertheless reassures his readers of his expertise as an experienced vernacular text worker. Before exploring his "helpyinge hande" at work, we must first turn to the Dutch text from which he worked.

Condensed and Clarified: Brunschwig's Distillation Book in Dutch

Die distellacien ende virtuyten der wateren was published in Brussels by Thomas van der Noot in 1517. While there is some confusion in previous scholarship as to which of Brunschwig's works served as the source for this translation, a comparison of text and images clearly points to the *Small Book,* and more specifically to Grüninger's 1509 or 1515 editions with their improved index.[53] In these editions, Brunschwig's distillation manual was paired with excerpts from Ficino's *Book of Life,* but these were not included in the translation. The overall structure closely follows the German model. The depictions of distillation equipment in the technical section were carefully reproduced, and the image of a furnace with four helmets was used on the title page. In her recent study on Netherlandish woodcuts, Andrea van Leerdam notes that van der Noot's images show what she calls a "reduction" strategy of visual copying; the landscapes or decorated interiors in the background of the original images are reduced or removed altogether, focusing on the instruments rather than the activity and spaces of distillation.[54] The images of plants in the herbal section are not included in the Dutch text. This could be explained if van der Noot's edition was, as has been suggested, a reprint of a now-lost Doesborch publication that was intended as a companion volume to an illustrated herbal.[55]

The translated text follows the original German closely, almost word for word. The only lapse is a mistranslation in the amount of time distilled waters should be rectified in the sun, where the original "fourteen days or longer" is rendered as "forty days or longer"—an easy mistake to make, and perhaps not too consequential, since Brunschwig indicated that the interval could be prolonged.[56] The wordy prose of the original is slightly condensed here and there, but there are few larger cuts. The translator leaves out a small number of somewhat extraneous passages, such as references to the *Large Book* (which was not available in Dutch) and personal anecdotes. The only omission of technical information is at the end of the chapter describing the necessary equipment, where the description of individual instruments, complete with images, moves into a jumbled list of basic paraphernalia such as bellows and coals of different hardness and size. Of these, the translation only retains the jars and jugs for storing distilled waters. Given the translator's attention to detail, this was likely a deliberate move, perhaps reflecting an intended readership that was already a little more conversant with the technical basics than Brunschwig's entry-level readers, at least to the extent that they knew how to light and maintain different kinds of coal fires. In concert with the "reductive" translation of the images, the translation of the text subtly tightens the work's technical focus.

Otherwise, the translation cleaves to the original with remarkable attention to technical detail, and occasionally even achieves clearer phrasing than Brunschwig. For distilling flowers in the sun, two glass vessels are connected mouth to mouth, separated by a wooden cross to prevent the flowers from falling into the bottom flask. Where Brunschwig describes, somewhat vaguely, "a cross of pieces of wood, such that it is also locked with a small wooden cross," the translation reads "make a cross of pieces of wood, and lock the same with a small wooden cross in the mouth of the glass," making it slightly clearer that there were two wooden crosses, one locked in place by the other.[57] (As we shall see, the English translation explains this even more clearly). The text of the herbal section was translated more freely, with more significant omissions, as we shall see in comparison with the English translation.

Laurence Andrew and the Vertuose Boke

Laurence Andrew's English translation, the *Vertuose Boke*, was published in London in 1527 in two different versions, and again in 1528. There are few major changes with respect to the Dutch source text, the most obvious one being the omission of Brunschwig's preface to the book. Some of its key statements appear elsewhere in the book—his claim to "thirtye yeres studye and Labour," for example, moved to the title page. In place of the author's preface, as we have seen, Andrew included his own translator's prologue. The individual chapters in the body of the text are the same and in the same order as in the Dutch edition, except for the somewhat

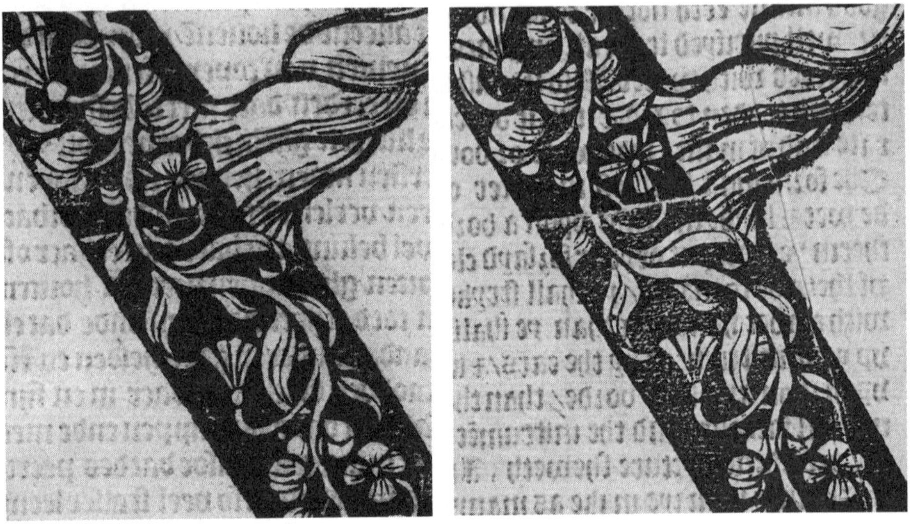

FIGURE 9.4. Woodcut with small crack in *Die distellacien en virtuyten der wateren*, fol. b2r (left). In the English translation (fol. a4r), the same crack has widened over time (right). Credit: Rare Book and Special Collections Division of the Library of Congress, Washington, DC, RS81.B673; Wellcome Collection, London, EPB/C/65968.

puzzling move to incorporate the description of distillation in a sand bath into a previous chapter on the description of different furnaces.

The images of distillation apparatus are more clearly foregrounded in Andrew's English edition. The first sentence of the title advertises the "fygures of the stillatoryes" to be found in the book, and the title page of the first edition shows the same furnace with four helmets as the Dutch edition. In the second 1527 printing and the 1528 version, Andrew replaced this single image with four smaller depictions of individual vessels, showcasing the wealth of technical illustrations found within. A close look at these images suggests that many of them were printed from the same woodblocks as those used in van der Noot's Dutch edition. Comparing the illustrations of the brick mold (discussed in chapter 6), for example, one can see a small gap in the van der Noot image resulting from a crack in the woodblock, and a corresponding gap in Andrew's image, slightly larger as the crack had widened over time (figure 9.4). Somehow, then, these blocks traveled from Brussels to London in the decade between the two editions, although it is unclear how Andrew obtained them. If Doesborch printed an earlier version of the book, as has been suggested, perhaps these were his woodblocks to begin with; he might then have loaned them to van der Noot and arranged for his former collaborator Andrew to take them with him to London. For the few blocks that did not make it to London,

Andrew commissioned new ones, some as poorly executed as the worst of the woodcuts in the English *Cirurgia*.

The most obvious change with respect to the Dutch edition is the reappearance of images in the herbal section. These, too, were printed from woodblocks borrowed among Andrew's circle—those used to illustrate Treveris's *Grete herball*. Andrew contributed to the text of this herbal, and a later edition bears his printer's mark. Like the images of distilling apparatus, these herbal woodcuts have a connection with the Low Countries. They were close copies of illustrations in a popular Dutch herbal with a similar title, *Den groten herbarius*, printed in Antwerp by Claes de Grave in several editions between 1514 and 1547. Having served for Treveris's and Andrew's projects, these copied woodblocks in turn must have made their way to the Low Countries, where they were used in 1532 by Andrew's associate Jan van Doesborch for another edition of the *Groten herbarius*.[58]

The translations of both *Cirurgia* and the *Small Book*, then, occasioned a lively exchange of images and physical woodblocks between London and the Low Countries. While their complex and circuitous routes are only glimpsed here and there in the surviving editions, they show that images of technical and natural knowledge could be highly mobile across national and linguistic boundaries. More specifically, they speak to an important Dutch influence in the visual culture of early English print that is often overlooked.[59] We may not know precisely what part Andrew played in this story, but it is clear that he was not just an unusually active translator, but also a broker of visual knowledge. If decent woodcuts were as hard to come by in England as the cruder images in the *Cirurgia* and *Small Book* suggest, and if Andrew brought along a stock of decent woodblocks as well as his translation skills, this would have been an additional incentive for printers such as Treveris and Rastell to help this well-connected newcomer to set up shop in London.

In text as well as image, the section on different distillations shows a clear technical focus and a similar attention to word-for-word precision as the Dutch edition. Andrew rarely omits even small details, the only notable instances being the indication that the labels for distilled remedies should be made of lead, and Brunschwig's recommendation of earthenware jugs from Siburg or Hagenau. While specialist equipment such as Hessian crucibles sometimes did make it across the channel, Andrew perhaps decided that English readers were unlikely to have these relatively ordinary jugs shipped from the southern reaches of the Holy Roman Empire, and simply listed them as "stone cruke" and "earthen canne."[60]

Most of Andrew's micro-level interventions in the text serve to clarify or explain, sometimes on the basis of his own expertise. Brunschwig's measurements in hands, cubits, and fingers are rendered in "ynches," with an eye for consistent conversion already evident in Andrew's previous translation of *The Valuacyon of Gold and Sylver*. Where the German and Dutch editions recommend "white triangular felt" for filtering out impurities, Andrew specifies that the material should be "lyke fyltes of hattes."[61] In translating Brunschwig's instructions for securing a wooden cross in between two flasks, Andrew evidently judged that they were too sparse to be workable. The Dutch translator, as we have seen, had already made small changes to the passage, and Andrew adds another level of practical interpretation: "Make a small crosse of woden styckes and laye that upon the mouth of the glasse than make an other crosse of smalle styckes somwhat longer than the other that mai be bent crossewyse over the other streyght within the glasse to make them both sure."[62] To provide such additional detail, he must have thought very carefully about how the two interlocking crosses could be made secure, seen someone else do it, or perhaps even tried it himself.

Some of Andrew's interventions speak to a general familiarity with artisanal practices. Where Brunschwig describes the correct temperature for certain operations with reference to commonly available heat sources, such as "a baker's oven when it is warm through and through," the translator narrows this down to a more specific point in the heating cycle, "whan the brede is drawen."[63] Occasionally, Andrew is tempted to use a more technical formulation than the original. In the chapter on laying bricks for a distilling furnace, where Brunschwig says to lay the second row of bricks "such that no seam come to lie on top of another," Andrew replaces these literal instructions with the much more businesslike "cowched mason wyse."[64] While Andrew strives for widely intelligible instructions, his use of somewhat specialist technical terms suggests that he embraced the turn, already begun in the Dutch translation, towards a slightly more technically literate audience.

In the herbal section of the *Small Book*, Andrew translates more freely. He adds the Latin name for lily of the valley and a lengthy physical description of water lilies that was not included in the Dutch text. In a section dedicated to hot and dry herbs, his selection of plants diverges from Brunschwig's. Both the Dutch and English translations occasionally make more extensive changes. In the chapter on sorrel, for example, some of the details concerning the herb's medical application are truncated. Brunschwig recommends sorrel water against jaundice, to be taken for "six or eight days, morning, noon, and night, three or four ounces each time."[65] The Dutch translation indicates the times of day and the number of days, but fails to mention the dosage. Andrew's translation states no more than that the water should be "dronke .vi. or .viii. dayes."[66] The loss of such medically pertinent information stands in contrast to the attention to detail in the technical section. Perhaps the translator(s), under some pressure to economize, decided to make cuts where it mattered less. After all, readers could readily find details of plants' medical virtues and applications in contemporary publications, such as the Dutch *Groten herbarius* or its English counterpart, Treveris's *Grete herball*, whereas the distillation section was one of a kind.

Overall, Andrew's focus was on a precise translation of distillation techniques, but this does not mean that he lost sight of the book's medical and alchemical dimensions. The herbal, and in particular the register of diseases and their corresponding cures, served readers as a customizable medical reference work, as we shall see in the following chapter. Furthermore, the translation faithfully renders Brunschwig's medical and alchemical rationales for distilled waters as safe and efficient remedies for common people, to the extent that they were spelled out in the original *Small Book*. Distillation separates "the subtyll from the grosse" so that "the corruptyble shal be made incorroptyble," and the resulting "quycke spyryte" penetrates more easily to work its "helthfull operacyon in the body of man."[67] While these arguments are folded into general discussions of distillation in Brunschwig's original, Andrew displays one of his central claims on the title page: "Avycenna testifyeth in his fourth Canon that the waters be farre better than the Herbes in theyr Operacyons."[68]

English Readers of Craft and "Scyence"

While Andrew's micro-interventions in the technical part suggests an audience with some practical expertise, his overall framing of the text marks a shift to a slightly more learned register. On the title page, he names the original author as a "most famous and expert master of Phisyke" even though Brunschwig clearly self-identified as a surgeon and not a physician. The subject of distillation itself is elevated from an art or craft (*kunst*) to a "scyence," and the intended audience includes "Surgyens / Phisycyens / and Pothecaryes," as well as "all maner of people." Brunschwig, as we have seen, addressed the original *Small Book* to "common" readers, especially those who had no access to physicians, and widened his audience appeal in later works and editions. Whether or not Andrew was aware of this, he pitched his translation at a readership resembling that of Brunschwig's later works, which included learned practitioners.

This was not an unusual move in vernacular medical works. Even if they described treatments and aspects of theory that most students of medicine should already know, they could still serve as nosological reference works, glossaries, or aide-mémoire to surgeons, apothecaries, and even physicians.[69] The *Vertuose Boke,* moreover, was the first printed work in English on a topic that had long been of interest to physicians, as illustrated by the twelve signatories of a 1456 petition to Henry IV who asked for a license to distill "quintessences" and other alchemical remedies.[70] In his study of the libraries of Cambridge physicians, Peter Jones found that they frequently read vernacular books, including Brunschwig's *Vertuose Boke,* which appears on a list of useful readings handed down from a retiring fellow to his successor.[71]

Learned or not, Andrew had good reason to speculate that a translation of Brunschwig's how-to book on distillation would find an audience. Through his work for Doesborch, he had a shrewd idea that books on trade, travel, nature, and medicine sold well on the English market.[72] Collections of remedies were among the most popular medical publications on medicine, especially cheaper quarto formats such as the anonymous *Treasure of Pore Men* (1526).[73] In publishing the *Vertuose Boke,* Andrew was adding to a small but growing body of accessible vernacular works with medical applications. What made it stand out was its detailed, step-by-step introduction to distillation techniques, which Andrew spared no effort to improve and emphasize, through his meticulous translation as well as the reproduction of the "fygures of styllatories" advertised on the title page. Readers would have been hard pressed to find a better how-to guide. Other popular medical works included recipes for aqua vitae and other distilled medicines, but their instructions rarely got more specific than those in the *Treasure of Pore Men*: "put [the ingredients] in a styllatory and sprynge above them good whyte wyne and dystyll it."[74] Andrew's translation was the first book on the English market to describe distillation (or indeed any "chymical" preparation) in technical detail, and remained the only one for decades. It was not until the 1550s that other chymical books began to appear, growing into a small industry of translations by practicing distillers in the 1570s, as we will see.

Practitioner-Translators and Medical Craft in the Early English Print Market

Looking at the two translations of Brunschwig's works side by side, there are considerable grounds for speculating that Andrew, who signed as translator and publisher of the *Small Book,* was also responsible for the translation of the *Cirurgia.* It was published two years previously, in 1525, by Petrus Treveris, with whom Andrew collaborated from at least 1526 on the *Grete Herball*. Both Brunschwig works were translated from Dutch editions connected with the Antwerp printer Jan van Doesborch, who had employed Andrew as a translator. They continued to exchange images and woodblocks that appear in both Brunschwig publications. It takes much less speculation to suspect Andrew's involvement in the *Cirurgia* than it does to contrive the alternative: that Treveris, rather than asking the Dutch translator he was already working with, instead found another equally well placed in terms of expertise and access to images.

Whether or not Andrew was responsible for both translations, they share a careful approach to the originals' practical instructions. The text of *The Noble Experyence* closely follows the Dutch edition's overhauled structure. While it cuts and condenses here and there, it never makes such incisive changes as the Dutch translator's inclusion of passages and entire chapters from other sources. The translation of Brunschwig's distillation techniques is even more meticulous. Andrew clearly understood both the value of this how-to text and the importance of rendering its practical details. Such small changes as he made show that he was generally conversant with the practicalities of artisanal work. As a vernacular text worker focused on practice, Andrew was in his own way no less meticulous than the most philologically scrupulous humanist. The illustrations in both translations were chosen with deliberation, drawing from a range of German and Dutch sources. Since we know that Andrew procured images and even woodblocks for the distillation book, it is likely that he did so for the surgery as well.

Drawing on an international network in the printing business, he was as much a broker of visual knowledge as of how-to texts. Andrew knew something of the English print market, and pitched his translations accordingly. While the image of surgery as a science as well as an art is maintained, the very title of *The Noble Experyence* marks a shift in emphasis towards skill and manual expertise. Treveris or Andrew may well have guessed that this would speak to London surgeons in their struggle for identity and recognition, much as Brunschwig had in his own context. As we shall see, surgeons in later decades did indeed assert their mastery and control over experiential knowledge and continued to use translations as literary vehicles for doing so. The *Vertuose Boke of Distyllacyon* was even more distinctly reframed by Andrew to appeal to more learned, but also more technically literate readers. This prefigures an audience of more specialized chymical practitioners who eventually also turned their hands to translation and publication, much as the surgeons did. In fact, there was considerable overlap between these practitioner-translators.

The translations of Brunschwig's *Cirurgia* and *Small Book* were the first books of their kind on the English market, and remained the only ones for decades. Subsequent publications began to appear in the 1540s, and they, too, were often translations. One of the first surgical works after Brunschwig, *The Questyonary of Cyrurgyens*, published in 1542, contained excerpts from Galen and Guy de Chauliac translated from French by the clergyman Robert Copland. As to works on distillation, the bookseller William Cooper's *Catalogue of Chymicall Books* of 1688 shows an even wider gap, from Brunschwig's *Vertuose Boke* in 1527 to 1550s publications such as the well-known books of secrets attributed to the nobleman Alessio Piemontese, which included "distillations, parfumes, confitures."[75] It, too was translated from a French version into English, by the Cambridge physician William Ward. The first book after Brunschwig to provide a more focused treatment was *A New Booke of Destillatyon of Waters*, a 1565 translation by the clergyman Peter Morwen of Conrad Gessner's compendious *Thesaurus Euonymi*, the epitome of the German *Destillierbuch* tradition (see chapter 4).

These early translations by scholars were swamped in the 1570s by a new wave of surgical and "chymical" publications by translators who were themselves expert practitioners. The core of this movement was a group of surgeons including Thomas Hill, George Baker, and William Clowes. They sought to build a corpus of surgical works in English, but also shared a keen interest in chymical techniques and remedies with other artisans such as the "Practicioner in the arte of Distillation," John Hester, who translated surgical as well as chymical works.[76] Upon his death, Thomas Hill left behind two unfinished translations, which were passed on to his friends. Hill's translation of Gessner's work was taken over by George Baker, who published the finished version in 1576 as *The Newe Iewell of Health*. In in his preface, Baker pointed his London audience towards local experts in distillation, including Thomas Hill and "Iohn Hester dwelling on Powles wharfe."[77] Hester's reputation as a distiller was such that Gabriel Harvey, one of England's finest scholars, dubbed him "the great alchemist of London."[78] In addition to selling remedies from his shop on Paul's Wharf, Hester published a number of chymical works. He was the recipient of Hill's other unfinished translation, a plague treatise by the unorthodox Italian surgeon and "spagiric" practitioner Leonardo Fioravanti. Fascinated by this author, Hester made it his mission to translate and publish his other works, and became the most prolific of the Elizabethan practitioner-translators.[79] Among this apparently close-knit group, vernacular translation became intertwined with the identities and expertise of medical artisans.

London surgeons were striving to consolidate their status as reputable medical practitioners with learned as well as practical expertise—much as Brunschwig did in his *Cirurgia*. The formation of the Barber-Surgeons' Company in 1540 was a crucial step towards corporate recognition, but the surgeons kept aloof from the barbers to form an elite group.[80] They policed their members' moral and honorable conduct to protect the company's reputation, and devised a program of surgical education that included anatomy classes as well as new surgical texts in English, written by prominent members.[81] Thomas Gale, an early contributor, highlighted translation as a key part of this project. He deplored that aspiring surgeons had so few texts to draw on because most authors either "wryte in the tonge whiche the most of you vnderstand not," or else they use "corrupt and barbarous names of sicknes and medicines" and are "ful of errours."[82] Gale had clearly read "Brunswicke," and mentions him by name (albeit as a bad example) in a work on the treatment of gunshot wounds, informed by his own experience in Henry VIII's army.[83] Other texts, however, were to be supplied by translation. Other nations, Gale admonishes, "have bin more diligent, and careful to further their common welthes . . . by meanes wherof they have brought manye noble artes into their owne tounges, which we have not so largely set forth in our Englishe tounge."[84]

By the time Gale issued his call for new translations in the 1560s, the impact of new humanist approaches to translation could be felt in surgery as in other fields of learning, albeit with a small delay. Scholars across Europe had been hard at work recovering ancient texts from their original Greek sources, including previously unknown surgical writings from Galen's *Methodus medendi*. For the first time since the decline of the medieval rational surgeons, learned scholars once more took a serious interest in surgery. Some physicians called for a united medicine encompassing both surgery and physic, as Galen's rediscovered writings suggested. Substantive efforts towards such a union were relatively few in practice, but they hinged upon translation. Two physicians from Lyon, Jean Canappe and Pierre Tolet, translated relevant parts from the new Galen into French to improve the theoretical education of skilled local surgeons who could not read Latin. Their translations soon issued from Lyon's flourishing printing houses as a series of affordable small-format volumes.[85]

In England, the translation and compilation of new surgeries was largely a vernacular affair. As we have seen, even the early translations by scholars were often based on French editions. When Gale and his fellow surgeons later produced English versions of newly available Galenic texts, they translated from Canappe and Tolet's French versions rather than going back to the humanist standard editions in Latin or even Greek.[86] This did not mean, however, that they paid any less attention to language and terminology, to avoid "corrupt and barbarous names," as Gale put it.[87] George Baker spelled out these linguistic labors in his 1579 publication *Guydos Questyons*, which was in fact a revised version of Copland's 1542 translation from Guy de Chauliac. Baker complained that he found in it "so many faults, so evill Orthographie, so ill poynted, ill distinct, and in many places whole lynes left out."[88] We can imagine that he had quite a job checking the text line by line, correcting typos and filling in the blanks—perhaps with reference to an English manuscript copy of Chauliac such as the one owned by his colleague William Clowes.[89] Much like the translator of the Dutch Brunschwig, who checked his terminology against an existing Dutch edition of Chauliac, Baker and his colleagues did intricate vernacular text work to shape up a corpus of English surgical literature, including "the mending of the Orthographie" and "the correcting of many wordes of Arte."[90]

English imports of chymical literature, too, were largely vernacular products, no doubt emboldened by their continental sources' apologiae of the vernacular as a language of knowledge. While the surgeons based their status in translations of Galen or Chauliac, distillers such as John Hester could model their nascent professional identity by translating Paracelsian works and above all Fioravanti, who knew better than anyone how to market idiosyncratic but fashionable cures and create a medical reputation in print.[91]

"Making Trial" and Distilling Experience

As Brunschwig's books show, experiential knowledge was at stake in both surgical and chymical translations. Neither Andrew nor the Dutch translators were practitioners, as far as we know. They translated Brunschwig's experience and technical skill as precisely as possible, without much intervention on their own account. Especially in the aptly named *Noble Experyence*, Brunschwig's experiential credentials are pushed to the fore. Later translators kept experience in their sights. Defending the use of the vernacular, Gale writes "the tounges maketh not an artist but reasonable knowledge being ioyned with experience maketh the artist."[92] In translating the hybrid expertise of surgery, experience was a part of the equation. Gale even suggests that experiential knowledge can itself be translated. He tells his readers to "take counsaile of other auncient and grave men that have had good experience . . . for in seinge of other cunynge mens workes, you shall learne and see those thinges, whiche you did never knowe nor see before."[93] In the same way Brunschwig envisioned his readers learning from his written experiences, Gale offers the translated experience of the ancients as a sort of virtual apprenticeship. William Clowes's *Short and Profitable Treatise* reiterates that "reason with experience joined, a perfect artist make," but he, much like Brunschwig and Gersdorff, offers his own battlefield exploits as a source of experiential training as well as heroic identity.[94]

Where experience remained untranslated and untranslatable, it offered a means for the London surgeons to control access to their elite circle. While their books—once approved by the master and governors—were out in the open, they also emphasized the need for practical training, in particular anatomy lessons, and such in-person experience could be tightly controlled against potential intruders from other trades, not to mention itinerant practitioners with dubious licenses and worse cures.[95] In the preface to his edition of a Spanish work on head wounds, the surgeon John Read demarcates surgeons even from other artisans of the body with whom they shared a company: "the Barbors crafte ought to be a distinct mistery from chirurgery" because "Chirurgians ought to be seene in phisicke."[96] Gale, too, is very clear that he means "onely to instruct those, that be brought up in the arte of Surgeri under some experimenters," and fervently wishes "that politike lawes might be made, to constraine every man to folowe that art, in which he hath bine well instructed."[97] If the law did not oblige, experiential knowledge could be controlled through books and lecture halls.

In chymical translations, experience was not just a matter of techne, but increasingly played into claims about remedies' efficacy. Brunschwig's works, as we have seen, explained how distillation produced medicinal virtues so subtle that they could safely and efficiently penetrate the body, and Andrew's translation prominently displayed these claims on the title page. Baker discusses the point at greater length in the preface to his translation of Gessner, explaining that "the vertues of medicines by Chimicall distillation, are made more vailable, better, and of more efficacie" than traditional remedies. He invites readers to "make tryall betweene the one and the other, and you shall see that the decoctions, Iuices, Syrupes, or such lyke, shall neuer come neare to the dystilled waters" or other chymical medicines. "Three droppes of the oyle of Antimonium," he claims, "doth more than one pound of those decoctions not dystilled."[98] Hester, too, makes a case for chymical remedies, focusing on the material process of their preparation: "For we haue seene and proued diuers tymes, that the first vapour or smoke of any Hearbe or Spice is the beste that is therein, and yet our learned Phisitions commaunde it to be boyled vntill halfe be consumed, &c. Then note if the beste flye awaie in the boyling, what strength can the medicine haue."[99] Where Brunschwig grounded the efficiency of his remedies in textual authorities such as Avicenna, Hester and Baker relied on "making tryall" and the proof of their experience.

They brought the same confidence in experience to their translation practice. Rather than following the original texts to the letter, Hester "culled out their choise thynges, cut of their superfluities," and editorialized according to his own judgment: "I haue supplied their defectes, I haue laied open their obscuritie, and I haue shadowed their defaultes." As a result, his translation "carrieth in it more plainnesse to the readers, then perhapps thei shall finde in the first writer."[100] In his work of sifting and clarifying, he submitted his source texts to "triall" and "proofs," and only included a passage if "the *experimentum* was confirmed through experience."[101] For a practitioner-translator such as Hester, textual work and practical experience were closely intertwined.

Reflecting on the relationship between the two, Hester writes that practice "sheweth perfectly the use of such thynges, as are but shadowed" in the texts.[102] He even offers his services as a practitioner in recompense for the shortcomings of his translations. In his preface to a work by the French Paracelsian Duchesne, he writes "whatsoeuer I haue offended in the rudenesse of my translation, in the diligent preparing of the medicines, I heere promise to amende." This is no mere metaphor but a genuine offer of in-person demonstration: "if there be any thing else to make, that to this Spagiricall art doth any way belong, the learned, or practitioned shall finde me ready vpon reasonable consideration to performe all."[103] While Hester's experience resonated throughout his translations, this was but a pale echo of what he could do in the workshop. "Such is my grosnesse," he writes, "or such the subtilnesse of the matter it self that I can easilyer by demonstration shew it, then by declaration teach it."[104] Borrowing the language of distillation, Hester juxtaposes the "subtilnesse" of practice with the relative "grosness" of words, problematizing not only the translation between languages, but also the more fundamental translation of gestures into text—a translation that Brunschwig and other how-to writers had grappled with from the start.

In Hester's apt analogy between distillation and translation, the Dutch and English versions of Brunschwig's *Small Book* are simple, unmixed extracts. However limpid the original matter, it takes great skill and detailed attention to make it condense and coalesce in the vessel of a new vernacular. Through a close comparison of the texts, we can see our translators hard at work clarifying the inevitable impurities. Later distiller-translators such as Hester produced more complex composite distillations. They extracted ingredients from a variety of sources and filtered them through their own knowledge and experience before mixing them anew.

Translation, "Profit," and the Commonwealth

Where the English *Cirurgia* is described rather modestly as "profytable for surgyans" in the colophon on the last printed page,[105] in the *Vertuose Boke* Andrew gives the same assertion pride of place on the title page: his translation served the "profyte / helth and cure / of myn evyn crysten in theyr diseases / And in especyall for them of this noble Royalme of Englonde / my natyfe Contrey."[106] In his prologue, he again advertises his translation as "to the reder more profytable" and as a "profytable booke."[107] This language of individual and collective profit became a mainstay of later English translators. As literary historians have noted, humanist scholars saw translation as social action that produced practical effects in readers and in society. In this context, they understood "profit" not only as material benefit, but also in its medieval sense as a remedy or cure.[108]

London's artisanal translators of surgery and chymistry eagerly adopted this language of books "both learned and profitable,"[109] produced for "profite and gaine,"[110] or "common commoditie or a generall gayne."[111] The surgeon John Read speaks of translation in terms of importing "treasure that is in other languages . . . to distribute in english to them that are unlearned," so that they may contribute to the "better maintenance of their common wealth."[112] In a medical context, the reading of "profit" as a cure for an ailing society resonates all the more powerfully. Read writes that surgery itself is "maymed" when it is separated from physic, implying that his translation will hasten the mutilated discipline's recovery.[113] In his translation of Galen, Thomas Gale rhetorically asks, "for the commoditie of the common wealth of my countrye, what greater profit may come unto many in the time of nede, then to cure his grevous woundes," eliding "profit" with the healing of individual injuries as well as the collective body of the commonwealth.[114]

In the chymical literature, too, translators emphasized the collective benefit of their work. Hester "polished and filed as nye as I could . . . to comfort my whole countrey,"[115] and for Baker, even the style of translation became a question of profit, "for if it were not permitted to translate but word for word, then I say, away with all translations, which were great losse to the common weale."[116] In chymistry as in surgery, common "profit" included medical cures, but it also developed in a different direction in the writings of another vernacular expert. Hugh Plat was the Cambridge-educated son of a wealthy London brewer who published a series of books on practical knowledge such as the *Jewell House of Art and Nature* (1594). Its "diuers rare and profitable inuentions" included a lantern that would not blow out in the wind and bread that would keep during long sea voyages, as well as a section on distilling techniques. Plat knew the works of Fioravanti and other "new professors of rare and profitable inventions," but his project was not just one of textual translation.[117] He also collected recipes and techniques in person, notebook in hand, among London's tradespeople, medical practitioners, and other experts. John Hester taught him a trick for catching fish, in exchange for some other "secret."[118] Behind these miscellaneous and entertaining "conceits" lay a serious concern about the material economy of the commonwealth. Nature, Plat perceived, had grown so weak and sickly with old age that "she beginneth to faint under the husbandmans hand," and threatened to wane in her providential supply of material wealth.[119] Plat's project, aptly characterized as "dearth science" by Ayesha Mukherjee, explored chymical theories and practices to manage material scarcity and restore nature's failing strength.[120] Where surgical translations promised collective "profit" by healing the injured body politic, chymical books paved the way to a cure for nature herself.

Conclusion

The translations of Brunschwig's early works chart itineraries of making and communicating knowledge that reveal hitherto overlooked cultures of vernacular text work. They highlight the fine-grained attention of the anonymous Dutch translator of the *Cirurgia* who checked Brunschwig's German borrowing of Guy de Chauliac against a Dutch edition of the medieval surgeon's work and implemented intricate philological clarifications as well as larger emendations of text and image. Reorganizing the book's structure, this text worker replaced entire chapters with Gersdorff's gentler treatments and stronger images. Far from a betrayal of the original, these kinds of changes were in the spirit of Brunschwig's project of compiling learned texts and contemporary artisanal practice.

All vernacular text workers in this story—the two anonymous Dutch translators and Laurence Andrew, assuming that he translated both English books—treat their source texts with careful attention. They streamline Brunschwig's more rambling passages but maintain, or even enhance, his level of technical detail. Barring the odd (and remarkably rare) mistake, they often exceed the original in clarity of instruction, coherent measurements, or carefully glossed terminology. Andrew in particular tweaks the books' audience appeal, emphasizing Brunschwig's experience in surgery and pitching distillation as a "scyence," but these mainly touch the framing of the works. While he signed his name in only some of his previous translations, in the *Vertuose Boke* he makes a confident appearance in the "prologue of Laurence Andrew the Translatour," calling attention to his labor and his credentials as a vernacular text worker.

The practices and strategies we see in the Brunschwig translations were readily taken up and developed by subsequent generations of vernacular text workers. Unlike Andrew and the Dutch translators, the group around Hill, Baker, and Hester were practicing surgeons or distillers, and had more of a medical agenda—their efforts to build a corpus of English texts served to consolidate their professional identities and advertise their expertise. Nevertheless, their translations suggest that they developed a culture of textual practice that goes back to the English editions of Brunschwig's works. Like Andrew, they paid close attention to the "many wordes of arte," as Baker put it, practicing a vernacular philology of precise technical language that was no less purposeful and painstaking than the efforts of learned humanists. Andrew's focus on experience, too, was taken up and developed by the Elizabethan practitioner-translators. For the surgeons, experience was something to be balanced between public-facing texts and closed-off practical training that allowed them to control the boundaries of their elite group. For a distiller such as Hester, the experiential trial and error of chymical practice became the model for the textual work of distilling the useful from the dross and recombining it into new and potent products. Taking another a page out of Andrew's "profytable" book, compilers of medical and chymical how-to literature increasingly insisted upon the public profit of vernacular text work, as a cure for the body politic, or for the commonwealth's ailing soil. Overall, the new kinds of textual practices sparked by Brunschwig's books proved influential. As they developed, they ever more closely resembled the artisanal practices of cutting and distilling that they translated.

✵ 10 ✵

Reading and Writing Medicine

ANNOTATIONS IN BRUNSCHWIG'S BOOKS

Brunschwig's books are rich material texts not just because of their deliberate and controversial move towards new imagined or intended "common" readers. Spreading far and wide in numerous editions and translations, they prompted many real readers to reach for their pen and make the text their own in a variety of ways. These material traces survive in sufficient numbers to tell us something of the ways in which readers engaged with Brunschwig's books. While a comprehensive census was impossible, I examined close to 150 individual copies of Brunschwig's works from collections in Europe, the UK, the US, and Canada, in person where possible, and occasionally through digital images. For the Dutch translations, I was able to draw on Andrea van Leerdam's thorough study of readers' engagement with early medical print in the Low Countries.[1] Around 58 percent of all surviving copies bear handwritten annotations, a slightly higher proportion than has been observed for sixteenth-century books overall, but in line with the generally higher annotation rates of books on medicine and other practical topics.[2] But the picture changes slightly if we look at Brunschwig's works individually. Discounting the plague treatise due to its small sample size, the distillation books are clearly in the lead with 67 percent and 69 percent, respectively. Looking at the English translations, fewer copies survive, but of those that do, up to 90 percent are annotated. Six copies each survive of the Dutch translations; all six copies of the *Cirurgia* and three copies of the *Small Book* are annotated, making an overall annotation rate of 75 percent. Some of Brunschwig's books, then, were annotated at a much higher than average rate, and even come close to the rate of 70 percent among copies of Vesalius's *Fabrica*, arguably one of the most well-known medical books of the sixteenth century.[3]

Historians over the past three decades have turned to annotations or "marginalia" in printed books to uncover past habits of reading and processing information. The earlier studies turned quite naturally to prominent figures such as Gabriel Harvey and John Dee, who left behind identifiable and heavily scribbled-in personal libraries.[4] More recently, scholars have explored more ordinary reading habits and how they fit into everyday life, from casual browsing to opinionated commentary or censorship.[5] As Bill Sherman has noted, early modern readers' annotations have proved overwhelmingly diverse, idiosyncratic, and delightfully resistant to generalizing narratives, although it is possible to discern patterns shaped by training and genre.[6] There is a considerable body of scholarship on reading and writing in the history of science. Ann Blair and Richard Yeo, for instance, have explored humanists' textual practices for making and organizing knowledge from annotating to commonplacing and notebook keeping, and Elaine Leong has explored paper technologies for practical knowledge and recipes.[7] The recent census of Vesalius's *Fabrica* demonstrates what an aggregate analysis of hundreds of annotated copies can tell us about the wider "material reception" of a single medical work.[8] On the whole, the historiography skews towards Latinate readers from the late Renaissance onwards. The many surviving annotations in Brunschwig's works are a valuable opportunity to explore less well-known readers and kinds of "book use" in the early sixteenth century and in a largely vernacular context.

A number of terms have been used for handwritten interventions in books. Contemporary terms such as "gloss," "scholia," or "postil" carry specific and often learned connotations that do not quite capture what Brunschwig's readers are doing. The nineteenth-century term "marginalia" does not reflect the many scribblings elsewhere on the page or in the book.[9] I have opted for "annotations" as a capacious and value-neutral term, bending its literal meaning only slightly to include any handwritten intervention, without judging whether it directly "annotates" or responds to the printed text. This includes hastily drawn crosses to mark a particular passage as well as lovingly rubricated headings or entire handwritten recipes.

Working with annotations raises methodological questions.[10] In some cases, the evidence tells us something about the ways in which a particular copy was used by an individual reader or institution. With many copies, however, there are limits to reconstructing such individual stories. Even where ownership inscriptions are present, books were handed down through generations and sometimes annotated by more than one person. In some copies, it is difficult to tell how many annotators left their mark, since one well-practiced early modern hand might look virtually the same as another, and two distinct-looking entries might in fact be the same person using italic script for Latin and cursive for the vernacular. Some inscriptions include a date, and the majority of the rest can be roughly located in the sixteenth or seventeenth century based on the script, but they do not yield a fine-grained chronology of changes in reading over time. It is clear, then, that care must be taken in interpreting annotations as reflections of reading habits—for all we know, the person who left a single mark in their copy might have dipped into the volume at random for two minutes, or read it cover to cover. With these caveats in mind, however, even terse anonymous annotations, analyzed in the aggregate, can reveal patterns of engagement.

I begin with an overview of the kinds of annotations we find in Brunschwig's books and what they can and cannot tell us. Studying ownership inscriptions and other signs of provenance, I characterize readers and reading contexts of Brunschwig's works and assess how they reflect his stated target audience. I then turn to the ways in which annotators appropriated and (re)organized the books' contents. From these discontinuous reading practices, themes and areas of focus emerge that attracted the most annotations. In particular, I argue that readers' lively engagement with Brunschwig's distilled cures for "the stone" speaks to a widespread interest in material understandings of health and disease. Finally, I explore annotations relating to testing and efficacy in Brunschwig's experience-driven writings.

Crosses, Keywords, and Jokes: Readers' Material Interventions

Surveying the bewildering range of material traces early moderns left in their books, William Sherman has suggested that we should understand them as evidence of "book use" rather than "reading."[11] Indeed, in some cases we can be fairly sure that reading was the one thing the annotator was not doing, at least at that very moment—when they were trying the nib of their pen, practicing writing their name in an elegant script, or jotting down an unrelated thought in the first empty space that came to hand. From mindless doodling to the deep engagement of a Dee or a Harvey, early modern annotators went about using their books in myriad ways, engaging with the text to varying extents and at different levels. Scholars have suggested ramifying classifications of the kinds of interventions they deal with.[12] For the purposes of this chapter, I broadly distinguish four groups: annotations that seem to bear little or no relation to the text; edits and corrections to the source text; visual finding aids, from nonverbal marks such as underlining to marginal keywords; and responses to the contents of the text, be they critical, playful, or supplying relevant additional information.

The annotations in Brunschwig's books reflect a wide spectrum of book use. Just over a handful of copies have pen trials and other scribblings with no obvious relation to the text, sometimes in combination with inscriptions of ownership. On the endpaper of an English edition of Brunschwig's "Thesaurus," one Jordayne Crosland wrote their name and turned it into the anagram "nor can a soldier dy," followed by a short poem. Other than a short list of medical conditions taken from the beginning of the text and an illegible scribble on one page, these are the only annotations in the copy.[13] Paper was expensive, and many people resorted to blank spaces in books to keep track of the household budget, births and deaths, or medical recipes, mingling domestic and textual economies in what sometimes amounted to a veritable family archive.[14] The blank spaces in Brunschwig's books are generally filled with things more closely related to their contents—mostly medical recipes, but some annotations point to a wider context of medical and humanist education. In a copy of the *Apoteck* (a reprint of the collection of simple remedies at the end of the *Large Book*), the endpapers are covered with Latin quotations and the annotator's attempts at a German translation between the lines. As well as an exercise in translating Latin, this was a collection of elegant phrases from Ovid and Ecclesiastes about the dignity and limitations of medicine, of the kind aspiring scholars were taught to collect in commonplace books to impress learned patients or colleagues.[15]

Another common kind of annotation concerned the book's text and its integrity. Renaissance readers of Lucretius amended the source text with scrupulous philological attention to preserve the author's ancient Latin.[16] In around a third of the copies of Vesalius's *Fabrica,* typos in the printed text are corrected by hand, and for around 20 percent of copies, the annotator went through the list of errata appended to the book to amend the typos in the body of the text.[17] It is perhaps not surprising that Brunschwig's rambling vernacular did not command the same kind of philological attention. In only one case, an English edition of the *Cirurgia,* do we find manual corrections that appear to be based on the errata list.[18] However, readers did occasionally correct foliation and indices to facilitate their navigation of the text, as I will show. Terminology is of course essential when it comes to identifying body parts, diseases, and medicinal plants—as we saw in chapter 9, translators of Brunschwig's works engaged in a kind of vernacular philology to ensure an appropriate rendering of technical language. Some readers, too, translated Latin terms in the margin or, more frequently, provided a Latin gloss for Brunschwig's German description. Plant names often varied by region, and readers might note in passing that *schelkraut* (celandine) was called *chaelidonium* in Latin, "and *maienkraut* by others,"[19] or collect alternative local names in the index of plants in the *Small Book*.[20] Its Dutch translation is often similarly annotated; in one case a distinctly French hand noted down the names of plants and remedies in French.[21]

Cross-references are relatively rare in the surviving annotations. On page 1 of a copy of the *Apoteck*, a remedy for an eye condition is annotated "see p. 18 ♂," where the ♂ mark is repeated to indicate another eye remedy.[22] In a copy of the *Small Book*, similar references to other folios connect separate entries on different kinds of poppy.[23] References to other texts occur in only one case. An English reader of the *Homish Apothecarye* noted related treatments in a number of medical books, often with precise page ranges: for constipation, see "Rich store house 1 part : 9 : 2 part : 31 : to 35," and on urinating with ease, "Fletcher of urin : 128 : 129," referring to *A Rich Store-House or Treasury for the Diseased*, reprinted many times from 1596, and John Fletcher's *Differences, Causes, and Iudgements of Vrine* (1598).[24] Other works include Thomas Sydenham's *Practice of Physick*, which dates the annotations to sometime after the publication of that work in 1695. While this is an exceptional case, we can speculate that this English translation of Brunschwig's work was part of a "living library" of medical texts that were read and annotated in conversation with each other.[25]

Early modern readers were not generally shy about expressing their approval, criticism, or other commentary in the margins of their books. At Cambridge University Library, the classmark "Adv." designates an entire collection of heavily annotated volumes that document "adversarial" readers actively engaged in debates with and about the text.[26] From the Vesalius census, we know that early modern annotators were often drawn into the author's battle with Galen, while a number of modern annotations retrace art historians' debates about the production of the book's famous woodcut images.[27] In Brunschwig's surviving books, such rich dialogue with the contents is relatively rare even in heavily annotated copies. Some readers, as I will show, made notes on the efficacy of medical remedies, perhaps from personal testing. One reader of the *Small Book* clearly doubted Brunschwig's claim that nettle water prevents cold hands and feet, and scornfully scribbled "this may believe who will" in the margin.[28] In another copy, a handwritten note points out, perhaps from painful experience, that the distilled water of human feces "stinks cruelly foul."[29] Perhaps it was the same outspoken annotator who provided a witty addition to Brunschwig's entry on pig's blood. As if simply adding one more medical application, it states "pig's blood, in a gut casing and eaten at ten o'clock in the morning (but first boiled) is good for the hunger."

This parody of a medical recipe—for a hearty second breakfast of black pudding—prompted a second annotation, directly below, in a darker ink and possibly a different hand: "my journeyman thinks this is true."[30] Rather than engaging in an intellectual dialogue with the text, these annotations connect Brunschwig's books to the everyday lives of readers who were not disembodied questing minds, but people who recoiled at the stench of human feces, got hungry, and enjoyed a joke.

Besides these discursive responses, the vast majority of annotations in Brunschwig's books are visual finding aids to highlight and retrieve specific passages. This includes marks such as underlining or marginal symbols—a curly brace, a vertical line, a simple cross or tick mark, or a stylized pointing hand known as a manicule. Phrases such as "nota bene" (often in abbreviated form, "NB" or "Nõ") serve much the same function, attracting attention without specifying anything about the annotated content. Alternatively or in addition, many readers wrote keywords in the margin to remind themselves of the specific disease or remedy under discussion. These nondiscursive or even nonverbal annotations mark but a fleeting moment of a reader's attention, and spell out little of their motives or intentions. In the aggregate, however, they can reveal what kinds of information readers looked for in Brunschwig's books, and how they went about obtaining it.

Surgeons, Monks, and Women: Material Practices of Ownership

Inscriptions of ownership can begin to give us an idea of how and by whom Brunschwig's books were read. The *Cirurgia* has by far the highest proportion, with six out of ten annotated copies bearing owners' names. For the *Large Book*, the proportion is smaller nearly by half (three out of nine copies), the *Small Book* and the *Thesaurus* are similar with around 24 percent of copies bearing a name, and the *Apoteck* comes in last with only two out of eleven copies. Of those copies including a geographic indication, most are located in the Southern German lands. No doubt the farthest-flung book, in time as well as place, is a late edition (1646) of the *Apoteck* that was dated August 5, 1762, in Sveaborg.[31] Its few annotations do not tell us much about the book's use in a Swedish sea fortress built as a defense against Russia, but it is worth noting that Brunschwig's book was still printed in the seventeenth century and still read (or at least annotated) in the eighteenth.

The *Cirurgia* stands out among Brunschwig's works because it is explicitly addressed to surgical specialists. In one surviving copy, this target audience is vindicated by no less than three practitioners, all from the Bavarian town of Kitzingen. On the front pastedown, Johannes Heyring inscribes himself as "citizen and barber at Kitzingen," with the date 1597. Below, in a later hand, one Valentin Plindneicher claims to be a "barber and surgeon [*wundtartzt*]" of the same town. On the back pastedown, Hans Hirsing also states that he is a surgeon in Kitzingen, and adds in a barely legible and partly crossed-out note that he hoped to "become a master there," and that he had "gone to school" and learned from "good fellows."[32] There are less than a handful of annotations in the body of the text, mostly in German, and they do not reveal much beyond an apparent interest in healing powders. Still, the inscriptions on the pastedown leaves tell a story of multiple artisanal readers of this copy, which was perhaps handed down to succeeding generations of surgeons. At least in this one copy, we can see the *Cirurgia* being read exactly as Brunschwig intended, by aspiring surgeons as part of their training.

Philip Fulder was a different kind of reader. He wrote his name in an elaborate italic hand on the pastedown leaf in the back of his copy of the *Cirurgia*.[33] We do not know who he was, but if the annotations in the volume are his, he was one of the most conscientious readers of Brunschwig whose traces have survived, showing serious engagement with the text in many chapters throughout the book. He underlined passages and marked them in the margin, sometimes with a generic "nota bene," but more frequently with a key word or phrase, often in Latin. In the chapter on wounds and their causes, he copied out the Latin technical terms (*ulcus*, *punctura*, *incisio*, etc.) from the text into the margin. Elsewhere, he glossed Brunschwig's German with Latin phrases that could not be found in the printed text, such as the anatomical terms *dura mater* and *pia mater*, or *frenesis* for a condition that Brunschwig described in German as "when the sick person goes mad."[34] For practical instructions, on the other hand, Philip used plain German, noting for example that "stapling is done in five ways: first, note the following."[35] On the same page as his name, there are four handwritten recipes, also in German. One of the chapters with the most underlined passages and marginal keywords is Brunschwig's discussion of assessing wounds in a legal context, to determine how they affect a citizen's livelihood to work out the proper punishment or compensation (see chapter 3). Comfortable with Latin and medical terminology, Philip appears more learned than a reader of the *Cirurgia* needed to be, but he shows a keen interest in the practical side of the surgeon's work, including his role as a forensic witness. His annotations suggest that he was a physician or a surgeon with some academic training, perhaps with a municipal appointment in a German city that required him also to act as wound inspector.

Turning to Brunschwig's distillation books, there is at least one owner who can be identified. A copy of the *Small Book* inscribed "Hanns Risheymer zu Seeholtzn" likely belonged to one Hans Rißheimer, who still has a street in Munich named after him.[36] He was a civil administrator, inspector of the mills in the Würm Valley, and sometime owner of Schloss Seeholzen, a castle (no longer extant) on a peninsula in the River Würm, a few kilometers south of Munich. In his book, Hans signs in a bold, decorative script as the proud master of Seeholzen, complete with a drawing of a coat of arms (which I have been unable to trace), perhaps marking the book as part of a nascent library worthy of his castle.

In its subsequent fortuna, Rißheimer's copy ended up in a religious institution. Underneath his inscription, a second hand noted, in German, "the book belongs to the Andechs monastery on the holy mountain."[37] The Benedictine foundation can still be found no more than twenty-five kilometers from the former Schloss Seeholzen. This is one of five distillation books held in monasteries, alongside two at St. Quirinus in Tegernsee, one in Himmerod Abbey, and one at the convent of St. Peter in Salzburg. Their traces of use are suggestive. In three of the copies, for example, it is clear that they were acquired soon after their publication. Himmerod Abbey inscribed their copy with the date 1503, three years after it was published.[38] The monastery of St. Quirinus in Tegernsee lost even less time: the ownership inscriptions are dated 1501 in the *Small Book*, and 1513 in the *Large Book*, indicating that both were bought in the year following their publication.

The Tegernsee copies offer some insight into the material treatment books like Brunschwig's might receive in a monastic context. While the Latin inscription in the *Small Book* states that it was "bought in its binding in 1501 under Abbot Heinrich,"[39] the *Large Book* was "bought and bound in 1513," suggesting that it came in loosely gathered printed sheets and was custom-bound by the monks.[40] Additional labor was then spent on rubricating the books. Skilled hands not only filled in the gaps—left by the printer for just this purpose—with elaborate initials in red ink, but also marked the beginnings of chapters and sections with red underlining and paragraph symbols. In the *Small Book*, the rubricator ran out of steam less than a quarter of the way through the volume's over four hundred pages, but the much longer *Large Book* is lavishly rubricated throughout, with a vivid blue as well as a red ink, and occasional ornamental flourishes. The investment in time and skilled labor suggests that Brunschwig's books were a valued part of the Tegernsee library. The Andechs copy, too, has rubrication throughout, but here it extends to some of the annotations as well as the printed text, specifically the chapter headings copied out into the margin of the disease register by a practiced hand. For one thing, this suggests that some annotations were deemed an integral part of the text, and just as valuable. For another, it illustrates that annotations were part of a larger culture of material interventions in printed books that included binding and rubrication as well as handwritten words.

These monastic copies show some of the most extensive and consistent annotation activity. In the Himmerod volume, for example, the herbal section is annotated throughout, suggesting that it was read in its entirety. Here and in the Tegernsee and Salzburg copies, the annotations reflect a degree of academic learning. Similar to Philip Fulder's annotations in the *Cirurgia,* these monastic readers regularly use Latin keywords to gloss diseases that the printed text only names in German, including technical terms such as *ictericia* for jaundice. This points to readers with some academic training, for whom these specialist terms functioned as a snappier visual aid or mnemonic than the corresponding German phrase—perhaps a brother in charge of the infirmary, or even a university-trained physician. None of these copies, however, are annotated exclusively in Latin. They include the German version of nota (*merck*), but also keywords for techniques ("distilling honey") and symptoms ("evil stomach") that were perhaps closer to daily life than medical theory and thus more recognizable in German.[41] While these annotators used their Latinate framework to orient themselves in the text, they paid no less attention to Brunschwig's pragmatic vernacular.

In the distillation books, and especially the *Large Book,* Brunschwig expressed a hope that these would appeal to learned as well as "common" readers. He did not mention monks or monasteries specifically, but they fit his audience claims on two counts: they were places of medical learning and practice, and they were often isolated, which was Brunschwig's favorite argument for medical self-sufficiency. While the odds of book survival no doubt favor volumes that are kept safe in monastic libraries, the five copies discussed here show that Brunschwig's distillation books engaged the (often detailed and sustained) attention of at least partly learned readers. More importantly, they show the careful attention paid to an individual copy's materiality and its possibilities, from the binding to interlocking rubrication and annotation.

The copy of the *Large Book* owned by the convent of St. Peter in Salzburg raises the question of women as users and owners of Brunschwig's books. As we have seen, Brunschwig mentions women as a source of practical knowledge, and the title image of the *Small Book* suggests that they had a place in his projects. However, the surviving annotations shed little new light. If the annotations of the Salzburg copy stem from the *Petersfrauen* of the convent, there is nothing particularly distinctive about them. They include both Latin and German keywords, and speak to an interest in turpentine-based remedies. A number of medical recipes in German are added on flyleaves at the back, which is not uncommon although we do not see it in the other monastery copies.

Three other copies bear ownership inscriptions by women. An eighteenth-century owner of the *Vertuose Boke,* the English translation of the *Small Book,* asserted her ownership with some force. Her name appears in a large italic hand on the verso of the title page, and later in the book, on a page facing a previous owner's ornate inscription and as if to counteract it, she repeated "Anna Corselles hur booke." On the final page, she noted "This Book is two hundr[ed] years since the Printing of, only Deduct 19 yea[rs], that is one hundred & eighty one yeares ago, this Being 1708."[42] Her distinctive hand left no other annotations, so all we know for certain is that she was keen to mark her ownership and her place in the book's long history.[43]

Back in the German lands, one Elisabetha Fischerin signed her name in a copy of the *Thesaurus* that bears only one further annotation—a cross-reference to another page—so we can only speculate that the book may have aided her in caring for the well-being of the household in her role as *hausmutter.*[44] The final example is a copy of the *Cirurgia,* inscribed on the front pastedown "this book belongs to Anna Maria von Nippenburg," and further down the page, "Maria von Wolmershausen, née von Nippenburg." Little is known about Anna Maria, but her second inscription likely relates to her marriage to Hans Philip von Wolmershausen, a minor Franconian nobleman, who died in 1581.[45] There are no annotations in this copy, except for two baffling interventions in the woodcut illustrations where the surgeon's instrument bag was blacked out. We can therefore only speculate how Anna Maria might have used this book. As Alisha Rankin has shown, for high-born German women the gendered medical duties of a *hausmutter* often extended to their entire estate, but those duties were typically pharmaceutical more than surgical. Distillation in particular chimed with female virtues of purity and charity, so Brunschwig's other books would have been a more obvious choice for Anna Maria.[46] But perhaps she was involved in overseeing the surgical care of her servants and subjects, or perhaps the topic simply piqued her curiosity as she played her part as a medical *landesmutter.*

Customized Indexing and Discontinuous Reading

As we have seen, a majority of annotations in Brunschwig's books serve readers' orientation on the page and through the text. Many annotators went even further in appropriating and reorganizing information, by taking a pen to the tables of contents, alphabetical indices, and other devices that allowed readers to find their way around. The Grüninger press spared no efforts in providing such finding aids—in fact, the first edition of the *Small Book* was rather overburdened with over two hundred pages' worth of alphabetical plant lists and two cross-referencing head-to-toe indices of diseases and remedies. As the preface to the second edition freely admits, "the previously printed book of distillation had a register at the back, which took up nearly as much space again as the book, and was unnecessarily printed twice; then it referred back into the book proper. This has now been left out and a new register made, easier to understand and faster to find."[47] Although these navigation devices were far from perfect, they were recognized as key components, worthy of the painstaking labor of indexing and improving the layout for future editions. This indicates that Brunschwig's books were not necessarily intended to be read cover to cover. In particular, sections such as the herbal in the *Small Book* or the collection of medical recipes in the *Large Book* consisted of short, self-contained textual units. Chapters on specific herbs and individual recipes could be read independently of each other, and while their sequence in the book (alphabetical or head to toe) helped locate individual units, it did not itself generate meaning. As noted in my discussion of recipes in chapter 7, literary scholars have characterized such texts as "discourse colonies" whose independent units form a concerted, highly functional whole.[48] Such texts—miscellanies, encyclopedias, or recipe collections—rose to great popularity over the course of the early modern period, embodying and facilitating discontinuous reading habits.[49] This was the norm rather than the exception. People were used to reading the Bible in weekly selections from the Old and New Testament, and the idea of reading a book cover to cover would have seemed alien to most before the rise of the novel in the eighteenth century.[50] Medical books in particular were perused in discontinuous chunks—readers of Vesalius, for example, freely underlined, excerpted, and skipped over information according to their needs and interests.[51]

The annotations in Brunschwig's books suggest that many readers proceeded in just such a targeted way, and used the provided indices to do so. One simple but effective way in which annotations could be used to facilitate navigation was by correcting the frequently inconsistent folio numbers printed at the top right of each recto page. The folio numbers given in the indices, too, are occasionally unreliable. In the Tegernsee copy of the *Large Book*, one annotator went through five densely printed pages of an alphabetical index with pen in hand, marking each entry with a cross to indicate that the folio number had been checked, and correcting it where necessary.[52] After about a page and a half, there are no more crosses, but the emendations of folio numbers continue. This requires looking up the indicated folio for each entry and, if it turns out to be wrong, hunting around for the correct folio. As one quickly realizes when one retraces the annotator's steps to verify even just a few entries, this is a weary task indeed. To warrant such trouble, a functioning index must have been of great utility to the reader.

A number of readers went beyond correcting the indices and customized them to their interests and needs. As the preface to the second edition of the *Small Book* indicates, the previously clunky set of indices was replaced with a more streamlined version. Under the heading for a disease, it no longer listed the relevant plant names and their use, but merely a folio number and key letter that helped locate the passage on the page. For some readers, this was perhaps a little too streamlined, and they looked up the entry to write the plant names back into the index. This also gives us a clear indication where their focus lay. In two copies of the *Vertuose Boke*, the plant names are written back in for several diseases in the index, sometimes for every single entry.[53] Conversely, for some distilled waters in the main body of the herbal section, all its indications against various diseases are indicated by keywords in the margin—perhaps these waters were readily available to the reader, and they wished for a quick overview of their uses. The result is a highly customized copy geared towards the reader's interest in certain diseases and certain remedies, but also their way of retaining, visualizing, and navigating that information.

Some readers went one step further and compiled their own navigation devices. The 1561 edition of the *Homish Apothecarye*, for example, does not come with a printed index, and one annotator—the same who noted cross-references to the *Rich Store-House* and other medical works—created their own index on a flyleaf at the back of the book. While the work was neatly begun, it was abandoned after the letter "D."[54] In another copy of the same edition, what may have been a nascent index was aborted after a single entry.[55] In another case, we can glimpse an annotator reading against the grain of the book's structure. In their copy of the *Large Book*, they not only corrected the printed table of contents, but constructed their own topical index on an empty flyleaf.[56] For most parts of the book, it provides a selection of topics that broadly follows Brunschwig's structuring principles, such as chapters dedicated to certain distillation techniques or to the treatment of specific diseases. But for the book's second part, which deals with the manufacture of compound remedies, the annotator uses their own categories. Where Brunschwig listed his compound remedies according to disease indications, this reader indexed them according to their form—as oils, electuaries, pills, etc. In this part of the book at least, the reader seems more interested in the physical forms of remedies and how to make them than in Brunschwig's nosological categories, and customized their copy accordingly with a bespoke navigation tool. Readers of early vernacular print, it turns out, were just as confident (if perhaps less rigorous) in marking their books on their own than a Harvey or a Dee.

These are prime examples of annotations that are best understood as different modalities of book use rather than signs of "reading." They show not only what people read, but also how they chose to navigate and organize the text as they used, corrected, and modified the printed tools provided, or constructed their own according to their priorities. In this way, annotations highlight the range of discontinuous pathways readers used to navigate Brunschwig's more miscellaneous writings, and the topics they chose to prioritize.

Reading for the Stone

In just over a third of all annotated copies, the annotations converge on one or more clear themes. These could be particular remedies, such as natural and artificial balsam, which were marked by several readers of the *Cirurgia* and the *Large Book*. More frequently, annotators focused on particular diseases. In some copies, they homed in on a single medical complaint. In a copy of the *Large Book* with annotations on only three pages, one of them shows markings against three separate passages on uneasy sleep, especially in children.[57] In a copy of the *Homish Apothecarye*, all three annotations refer to diseases of the mouth, including two handwritten recipes for waters used for gargling.[58] One particularly single-minded annotator of the *Small Book* marked nothing except nine separate passages throughout the book that deal with hernias, including three that are not listed in the printed index.[59]

More significantly, there are common themes across different copies and annotators. In six exemplars of the *Small Book* and one *Apoteck*, there is a noticeable density of annotation around passages concerning, in general terms, "the diseases and sickness of women," and more specifically complaints relating to menstruation, labor, and childbirth, but also treatments to improve fertility and lactation.[60] None of these annotations are in copies with ownership inscriptions by women, and they include two books owned by monasteries. Another distinct focus in eight annotated copies are diseases of the eyes, from serious wounds in the *Cirurgia* to distilled waters that promised to improve eyesight or clear a cloudy lens. The most frequent topic, both in terms of annotated copies and the number of annotations, is "the stone"—a set of painful but common complaints caused by stone-like concretions in the urinary tract. At best, they manifested as sand-like "grit" that could still be passed in the urine; at worst, they were firmly lodged kidney or bladder stones.

Ten of the examined books show consistent engagement with the topic, from the vernacular reader of the *Apoteck* who repeatedly noted *für den stein* (for the stone) in the margin, to the annotators of the Himmerod and Tegernsee monasteries who used technical Latin keywords such as *calculus* or *lapis* as well as the vernacular *für den gryeß* (for the grit).[61] In the Tegernsee copy of the *Small Book* and in a copy of its English translation, entries relating to the stone are highlighted in the index. In the latter case, the annotator took the time to look up about half of the entries under "For to withdryve and consume the stone" and noted the name of the corresponding plant against the listed folio number, as well as manually adding an entry that had perhaps been missed in the compilation of the index.[62]

It is worth noting that these annotations are concentrated in the distillation books and their offshoots. In the one copy of the *Cirurgia* that bears stone-related annotations, these too take the form of recipes for distilled waters, added by hand on a blank flyleaf. One likely reason for the density of annotation is that Brunschwig's distilled remedies against the stone offered welcome alternatives to the usual surgical interventions. Lithotomy, or "cutting for the stone," was invasive, painful, and risky—so much so that many practitioners wanted nothing to do with it. In fifteenth-century Italy, many general surgeons and learned doctors explicitly refused to perform the operation. Alongside procedures such as couching cataracts or repairing hernias, lithotomy was left in the hands of a distinct group of practitioners who specialized in a single procedure.[63] While "empirics" hawking a single nostrum were generally viewed with suspicion, these surgical specialists attracted clients from all echelons of society and sometimes enjoyed municipal appointments and tax exemptions.[64] We know less about lithotomy in the German lands, but here, too, it was performed by specialists. For instance, the author and learned surgeon Jakob Ruf from Zurich was known as a *lithotomus* or *steinschnyder* (stonecutter), rather than a general *scherer*, and wrote a manual on the "ophthalmic art" in the 1540s. Another surgeon, Kaspar Stromayr, penned a *Practica* that described cutting for stones and hernias as well as removing cataracts.

Neither work was published, serving instead to pass on technical knowledge to sons and apprentices.[65] In Augsburg, the wealthy Fugger family founded the *Schneidhaus* (cutting house), something of a specialist clinic for lithotomy and hernia surgery where the poor were treated for free and often traveled from far afield.[66] In Brunschwig's home town, the archives reveal that one of the local charitable hospitals employed a surgeon named Lamprecht Röderer who specialized in cutting for the stone as well as curing hernias and eye diseases.[67]

Brunschwig himself emphasizes the specialist experiential skill required for the procedure in his short passage on the stone in the *Large Book.* "As for the cutting," he writes, "one can hardly write about it properly, because it should be cut by masters who have seen it before and learned it, so I will not write of it here."[68] In the *Cirurgia*, Brunschwig makes no mention of the stone, indicating perhaps that he was not one of those masters experienced in lithotomy. Nor does he advertise the remedies in his distillation books as straightforward substitutes for the operation. In the index of the *Small Book*, he writes "in truth I have rarely seen a solid, hard stone expelled except by cutting." However, when the concretion is not yet "complete and solidified," then "it may be consumed with these waters."[69] The index contains five entire sections on stone-related treatments, from "grit in the loins, bladder, or kidney" to waters that "expel and consume" the dreaded calculus, with a total of just over a hundred indications of different distilled waters.[70] Where cutting was risky and painful even if one could find a skilled specialist, internal remedies that promised to dissolve the stone without additional pain or bloodshed were likely an appealing option. Since surgical treatments for hernias and eye complaints were similar specialisms, sometimes performed by the same practitioner, this may also explain the clusters of annotations around these topics in Brunschwig's works.

In addition to the distillation books' therapeutic potential, it was perhaps also their underlying account of stone-related complaints that drew readers' attention, at a time when questions about specific pathological matter took center stage in changing perceptions of the body and its workings. As we have seen, Brunschwig explained distillation in terms of matter, its qualities such as "subtlety," and its movement in the body. As I suggested in chapter 4, this influenced later, more explicitly "chemical" accounts of human physiology and pathology of Paracelsus and his followers, notably Thurneisser's model of human health as a distilling alembic. In these bodily imaginaries, mineral concretions such as the stone came to play a central part. Paracelsus posited an entire category of "tartaric" diseases that were caused by the precipitation of mineral deposits at different locations in the body, wherever one's internal alchemy had gone wrong. Medicines taken internally, especially acidic ones, could help dissolve such tartaric blockages.[71] Thurneisser's diagnostic distillation of urine, which mirrored the patient's internal distillations, also focused on the appearance of mineral substances that might indicate deposits in specific parts of the body.

Brunschwig's notion of bodily distillation is not fully spelled out in his writings, and he certainly does not emphasize the stone as a new pathological paradigm. Nonetheless, it is worth exploring how Brunschwig thinks of stones inside the body in material terms. In a short passage in the *Large Book,* he describes their formation in traditional Galenic terms of humors and primary qualities. Stones or grit form as a concretion of accumulated phlegmatic matter, through an excess of either heat or cold.[72] In his proposed cures, however, his thinking goes beyond the Galenic framework. As noted, Brunschwig uses terms such as "expel" and "consume" to describe the action of his distilled waters on the stone. The material evidence of these effects was readily visible in the body's excretions. Water of willow leaves, for example, was "good for the grit, which it drives out from you, as you can readily see in the chamber pot that there is much sand in it."[73] To support his claim that oak leaf water could tackle even solid concretions, Brunschwig suggests a simple test: "receive the urine in a glass, and you will find something resembling sand in it; thus you see that the stone is driven from the person."[74] Working with the trappings of traditional uroscopy (the "glass" likely refers to a *matula* or urine flask), Brunschwig steers the focus away from humoral qualities and towards the material root of the problem: localized concretions, their dispersal and expulsion from the body.

In his discussion of another potent remedy, a solution of lye and ashes of bean straw, Brunschwig obliquely comments on its efficient properties when he expresses his concerns that, with prolonged use, "the bladder might be punctured by its sharpness."[75] That same "sharpness" seems a likely candidate for what makes this medicine effective in dissolving the stone, in the same way that the "subtlety" of distilled waters underpins their penetrating but harmless action within the body. We thus get a glimpse of a material interaction—what we would now call a chemical reaction—between a remedy and a specific pathological substance. Such interactions could be observed outside the body, as Brunschwig suggests in a sort of in vitro test of a distilled water: "To test it, put a stone in this water; it will crumble into sand in three days." Brunschwig admits that he "cannot quite confirm this," and that in practice one may well have to take the water for several weeks, depending on the nature and size of the stone, "since a smaller one will be dispersed much more readily than a larger one."[76] His level-headed qualifications notwithstanding, he draws a clear analogy between a solid stone crumbling to sand in a dish of distilled water before one's eyes, and the same water's action on a stone in the body. By no means limiting his etiology to Galenic humors solidified by excessive heat or cold, Brunschwig thinks of the stone as a specific pathological substance whose material properties and interactions were substantially the same as those of other mineral substances that could be observed outside the body.

These accounts of the stone and its treatment resonate all too readily with later Paracelsian and other "chemical" views of the body. However, my point here is not to make a teleological claim about Brunschwig as a proto-Paracelsian. Rather, his works (and his readers' responses) suggest a more subtle mingling of Galenic and (al)chemical, distillation-based views of the body, in the still under-investigated early decades of the sixteenth century, before Paracelsus's radical polemic pitted chemical medicine against learned orthodoxy. While Paracelsus got the adversarial response he no doubt hoped for, recent scholarship has shown that the schism was often upheld in rhetoric more than in practice. Accounts of the body in terms of matter and movement rather than a balance of the four humors were widespread, and not necessarily seen as incompatible with Galenic theory, even when the therapeutic focus shifted towards specific diseased matter.[77]

However, less attention has been paid to pre-Paracelsian dynamics between learned humoral and alchemical medicine. Brunschwig's treatment of the stone suggests that what later became known as a rival "chemical" (or "chymical") paradigm was not as radical as Paracelsus liked to make out. Brunschwig thinks about the stone both as the product of orthodox humors, and as a specific pathological substance whose movements and transformations in the body could be understood in analogy to external material phenomena, such as a stone dissolving in a dish of distilled water. Viewed in this way, as a rich account that addressed the body's inside and outside, the mineral and the organic, as well as surgical, humoral, and chemical approaches, it is easy to see how Brunschwig's writings on the stone might have influenced later medical reformers in the same way it evidently captured the attention of his audience. Brunschwig's books and their readers' marks thus document a vernacular medical discourse that subscribed to the basic tenets of Galenic theory while also finding great appeal in artisanal and chemical perspectives on specific material changes, within and without the body.

Experience in the Margins

As I have shown in chapter 7, Brunschwig based much of his authority on experiential knowledge, and invited his readers to make their own experiences. This raises the question to what extent such knowledge making is reflected in the surviving annotations. Some annotations cluster around Brunschwig's accounts of successful cures. In a copy of the *Small Book*, for example, one reader underlined an anecdote about a treatment for breast cancer in a female patient that caused the abscessed breast to regrow "as good as the other one."[78] In the same copy, several sections in the chapter on speedwell are similarly marked. They describe the herb's remarkable effects on animals. Bears, we are told, lose weight and become manly and brave upon eating the herb, and it also helps them digest a large meal of ants. Gray water snakes like to impregnate eels, and the mongrel brood is apt to kill the parent—unless the eels abort them first by eating speedwell. These colorful natural histories, Brunschwig writes, were "experienced by a shepherd who over twenty years got to know many herbs and the nature of many an animal."[79] Another reader marked a similar anecdote in which a toad is repeatedly stung by a poisonous spider, but survives each time by eating plantain leaves—until the herb is removed and the toad succumbs to the venom. The reader's marginal note *mirabile dictu* (marvelous to tell) might be read as incredulity, but more likely expresses the early modern view that nature sometimes revealed her powers more clearly in her wondrous quirks than in her ordinary course.[80]

Readers were also attracted to passages where Brunschwig writes more explicitly of trial and experimentation. One reader of the English *Cirurgia* underlined Brunschwig's account of another surgeon's boasting of a powder that could heal a broken bone in eight days—"thys to prove a dogges legge was broken" and successfully treated with the powder in the stated time.[81] A German annotator underlined similar statements about a the efficacy of remedies in the *Small Book*, such as "proven for the pestilence by master Henchen Ungerech in Frankfurt," the endorsement of lavender water by one Hans Pfarrer of Würzburg, or simply Brunschwig's own assertion that St. Benedict's thistle is "proven for the pestilence."[82] As we have seen, anecdotes of cures and witness reports were one of Brunschwig's key strategies for bolstering the credibility of experiential insights, and they clearly made an impression on some of his readers.

A few annotators noted their own assessments of individual remedies, although it is not always clear from a terse endorsement such as "a good piece" whether this was a conclusion drawn from further reading, reports from others, or personal testing.[83] The Latin *probatum* (proven), written against one recipe for the cough, has more explicit connotations of being tried and tested.[84] In any case, such annotations, made in direct response to the printed text, are relatively rare—perhaps surprisingly rare if we consider the richness of "raw material" for testing provided in Brunschwig's books. By contrast, in recipes added to flyleaves and other blank spaces in the books, experiential claims are proportionally much more common—for the one occurrence of a "probatum" note on thousands of printed pages, there are four on the less than a hundred pages of added manuscript material. This suggests an explanation for the relative scarcity of such annotations in the printed texts: perhaps readers were more likely to employ references to experiential testing in their own manuscript collections than in the margins of their books.

As Elaine Leong has shown in her study of early modern recipe collections and testing regimes, these often extended over many levels. Recipes might be acquired in the form of letters from acquaintances or copied from printed books, first into some form of rough notebook. They could then be subjected to "paper trials" involving additional reading or exchanges with local experts before a hands-on attempt was made. Only then was a recipe likely to make it into frequent use and a fair copy of a recipe book to preserve for future generations. The bulk of a recipe's trial was thus recorded in multilayered and bespoke paper technologies rather than in the margins of printed books, and it is in manuscript notebooks such as Johanna St. John's that we see experiential terms such as "tryall," "approved," or "experience" in operation.[85] With this in mind, the relative dearth of such statements in the margins of Brunschwig's books does not necessarily mean that readers did not follow the author's invitation to try for themselves, or that they did not record their experiences—they may simply have done so in separate manuscript collections, of which the recipes added on flyleaves perhaps give us a glimpse. For all we know, the series of crosses, notas, and other marks against individual remedies that we find in many copies are but the grist for the mills of annotators' experimental notebooks and workshops.

Conclusion

The lively and varied annotation activity explored here suggests that Brunschwig's project of crafting a new kind of medicine in print was successful on several counts. His ambition to reach new audiences is borne out by the range of annotation profiles in his books, from learned students of medicine and alchemy to vernacular artisans and the "common man." He could hardly have hoped for a clearer vindication of his target audience than the copy of the *Cirurgia* that was used by three generations of barbers from Kitzingen during their professional training. However, as Philip Fulder's copy shows, the work also captured the sustained attention of more learned practitioners with a good grasp of medical theory. Other annotations point to readers and contexts that Brunschwig had not explicitly anticipated. As it turns out, several monasteries acquired his distillation books, often soon after they were published, and treated them to lavish bindings and rubrication. Here they were read and annotated extensively, sometimes using Latin keywords, and with an emphasis on organizing and navigating medical information. If Brunschwig did not mention monastic readers as a target market, they readily fit his sales pitch, which promises medical self-sufficiency to country dwellers and emphasizes the imperative of charitable care. Finally, there are many anonymous annotators who bring less prior knowledge and training to their reading and prefer annotating in the vernacular. For example, behind the copies of the *Small Book* where the disease indications for individual distilled waters are written back into the table of contents, we can imagine a "common" reader such as a *hausvater* or *hausmutter*, creating a handy reference of applications for those remedies they had ready on the shelf.

In addition to the range of readers, the annotations reflect the kind of engagement Brunschwig anticipated: not passive absorption, but the practical, embodied, epistemically engaged reading of the kind of virtual apprentice discussed in chapter 6. To be sure, at first glance the annotations suggest little in the way of direct exchange with the printed text. Brunschwig's readers certainly show no signs of scholarly "adversarial" reading. Where they do inscribe their own commentary or contrasting voice, these appear firmly rooted in everyday life—commenting on the foul stench of distilled feces rather than their efficacy, or humorously subverting a medical recipe into a second breakfast of black pudding. At a closer look, however, the annotations fulfill important epistemic functions. Marks, crosses, and keywords helped readers to visually negotiate the page, while interventions in printed navigation devices and the creation of bespoke indices document targeted, discontinuous reading practices and a commitment to knowledge organization and retrieval. While explicit references to experiential trying and testing are relatively rare in the examined copies themselves, they suggest a link to paper technologies that typically went beyond the margins of the printed book. Brunschwig's readers read actively and with purpose, treating his books as dynamic repositories of practical knowledge.

Brunschwig's project is perhaps most clearly vindicated in the many annotations that show readers coming to his works looking for solutions to specific medical problems, especially hernias, cataracts, and the stone. If, as I have argued, readers latched onto these passages because they offered alternative treatments to risky and painful surgery, this suggests that Brunschwig was successful in framing a new, distillation-based medicine for the common man. As I have shown in chapter 4, he made a case for the safety and efficacy of distilled remedies even in the hands of practitioners with little or no formal training, based on the waters' subtlety and gentle penetration, within a framework that chimed with widespread imaginaries of the body in terms of matter and its movements. It is highly indicative, therefore, that a majority of annotations are clustered around entries on the stone where such questions of pathological matter come to a head. This suggests that Brunschwig's artisanal and distillation-based imaginary of the body resonated with many of his readers, creating a broad vernacular base for "chemical" understandings of disease associated with more prominent writers later in the century.

Overall, this study of readers' material practices reveals a wide-ranging and dynamic readership of early vernacular print that historians are only beginning to explore.[86] The annotations in Brunschwig's books overwhelmingly vindicate his intended new readership for medical print, but they also help us paint a more nuanced picture of this emerging audience. Brunschwig's turn towards "common" vernacular readers was no doubt pioneering and controversial, but it should not lead us to think of such readers in reified categories. What we know of his audience, intended and real, clearly challenges Miriam Usher Chrisman's thesis that the flourishing Strasbourg print market further entrenched distinct "lay" and "learned" cultures of reading.[87] If anything, the surviving annotations in Brunschwig's books show the opposite of a "two cultures" situation; they map out a range of readers from erudite to common, and within that spectrum they often blur the boundaries between learned and lay, Latinate and vernacular. The majority of the German copies are annotated in a combination of Latin and German. Some annotators show a preference for the vernacular with a working knowledge of Latin technical terms. Others clearly had some academic training, but resort to German for particulars of everyday medical practice—one annotator, for instance, used the Latin *calculus* as a keyword for the solid stone, but the German *grieß* for its less well-defined manifestation as "grit." Few copies are annotated exclusively in Latin or German, and as we have seen, even the monastery copies used both languages or, in the case of the Andechs copy, only German. The Dutch copies show more of a preference for vernacular annotations, but some annotators switch between both languages, sometimes in the middle of a sentence.[88] Of the eighteen annotated English copies, only one contains any Latin. Rather than a well-defined category of vernacular readers, the annotations suggest a spectrum of hybrid readers—a motley group, embodied by the middling figure of the striped layman, but one receptive to Brunschwig's artisanal imaginary of matter and the human body.

✷ CONCLUSION ✷

Crafting Knowledge in Early Modern Europe

This book began with Brunschwig's assertion that medicine was craft. From his surgical textbook to his plague treatise and books on distillation, I have argued that he was motivated by an imperative to codify a new kind of expertise for new vernacular readers, including guild-trained surgeons, professional purveyors of medicines, and above all, the "common man." Brunschwig's books pushed the boundaries of the print medium to convey embodied knowledge. They also pushed the social boundaries of who should know what in the early days of vernacular print, fostering new agencies for non-learned practitioners and patients. At the heart of this project lay an expertise that offered new ways of knowing and manipulating the materialities of bodies and natural substances. The force of that expertise, as I have shown, lay in its eclectic but purposeful merging of different forms and traditions of knowledge—in the way it combined alchemical and medical ideas with practical, sensory skill as well as experiential knowledge in the form of recipes or secrets.

Previous scholarship has done much to characterize this kind of expertise as the product of a "mindful hand," a form of "ingenuity," a "material imaginary," or, in the case of eighteenth-century chemistry, a "hybrid" expertise.[1] However, historicizing the formation of such knowledge at the intersection of different traditions remains a challenge. The idea of hybridity is appealing because it suggests a genuine blend or fusion, but it tends to gloss over the dynamic interactions between different knowledges. These dynamics take center stage in other models, including the idea of "trading zones" that brought together scholars and artisans, and the figure of the go-between, brokering relations between different sites and cultures of knowledge.[2] Such accounts have helped us make sense of many different contexts, but they can be limiting in their tendency to ask questions about movements and exchanges between essentially reified cultures of, for instance, theoretical and practical knowledge.

As this study has shown, Brunschwig was more than a go-between. While he sometimes speaks with different authorial voices, the image he fashions of himself is a genuine fusion of the experienced artisan and the serious student of learned texts. This is borne out by the substance of his expertise. Brunschwig revives the medieval ideal of the rational surgeon precisely because it models a unity of learned principles and manual skill. His material imaginary of nature and the body was not simply "informed by" learned theories on one side and practical experience on the other. It was a nexus of deep connection between alchemical matter theory, medical concepts of material flow and obstruction within the body, and the embodied knowledge of working with difficult materials, complex techniques, and specialist tools. It is through these connections that Brunschwig formulated new treatments for the common man as well as a framework that explained how and why they worked.

The hybrid nature of Brunschwig's expertise is reflected in his intended readership. It included a wide spectrum from the learned to the barely literate, but it also broke a new middling ground by focusing on the "common man." This new reader was just as "hybrid" a figure as Brunschwig himself. The woodcut illustrations in the books underscore this by alluding to the "striped layman," a trope coined by local humanists to describe a troublesome social upstart. To them, his stripes signified a dangerous half-knowledge, a pretense that could not hold together. In Brunschwig's books, the striped layman becomes the ideal reader and practitioner, his contrasting stripes symbolizing different forms of knowledge joined into a new hybrid expertise. This challenges contemporary views on the education of layfolk, but also the boundaries drawn by modern historians around epistemic cultures. At the local level, as we have seen, it casts doubt on Miriam Usher Chrisman's characterization of early Strasbourg print as an early modern two-culture situation in which "lay" and "learned" had nothing to say to each other.[3] Although Brunschwig's texts were (almost) entirely written in German and indeed challenge the exclusiveness of learned authority, we have seen that they explode any such distinction, both in their content and in their intended audience. My study of readers' traces in surviving copies shows that the majority of copies and annotators rarely stick to either Latin or the vernacular, using instead a hybrid language to navigate Brunschwig's hybrid knowledge.

Brunschwig was more than a broker between cultures; he was something of a one-man trading zone unto himself. His eclectic approach challenges us to rethink how we historicize different forms of knowledge and their intersections. In concluding this book, I suggest that the idea of "braiding knowledge" offers a productive model. Projit Bihari Mukharji has articulated "braided sciences" as a framework to make sense of the radical reimagining of the body in late nineteenth-century Ayurvedic medicine, which braided together diverse traditions from East and West.[4] In *Braiding Sweetgrass*, Robin Wall Kimmerer uses a similar image to explore the relationships between contemporary Indigenous knowledge and Western science.[5] This idea of creating something new from existing strands, I suggest, aptly captures how early modern experts crafted new knowledge from different cultures and traditions—intertwining and yet individually recognizable, diverse but serving a common purpose. In what follows, I lay out how this perspective enables us to understand Brunschwig's unique project, how it advances histories of science and medicine, and how it might open up new conversations.

Crafting Medicine, Braiding Knowledge

The metaphor of braiding helps us understand a corpus like Brunschwig's as a unified project, even if not all strands are visible at all times, passing over and under as they weave in and out of his books. It also avoids a positivist fallacy that plagues histories of science, including the early scholarship on Brunschwig. Instead of measuring texts and practitioners of the past in terms of "originality" or "derivativeness," the idea of braiding offers an elegant way out of this false dichotomy. We need not pretend that Brunschwig single-handedly spun new knowledge out of thin air to acknowledge his significant feat of fashioning existing threads into a new way of knowing for new kinds of knowers.

Brunschwig opens up a new understanding of the human body by drawing together the teachings of humoral medicine, the alchemy of distillation, and the sensory expertise of artisanal body workers. The surgeon's practiced senses read the surface of the body for signs of leprosy, but also probed its unseen depths, from anatomical structures to hidden fractures and projectiles lodged in deep wounds. As a forensic expert, the surgeon's expertise went deeper than skin and flesh, judging injuries to honor and livelihood inflicted upon the social bodies of his fellow citizens. Through the distiller's art, Brunschwig reimagined the body in material terms, as a system of moving fluids and spirits. This account offered new rationales for the efficacy and safety of distilled medicines in the hands of the common man, especially since it resonated with a widespread popular imagination of the body's inner workings in terms of fluxes and obstructions. Brunschwig, I suggest, reimagines this "fluid" body as a "distilling body"—a concept that clearly gained currency in the early modern period, although its history—from Avicenna's *homo alembicus* to Thurneisser's body-alembic and beyond—remains to be written.

Brunschwig's braided account of distillation also spells out a new way of knowing the natural world. He draws on Aristotelian matter theory, the alchemical notion of quintessence, and the artisan's embodied knowledge of materials and processes to conceptualize how subtle healing powers can be extracted from coarse substances. In so doing, he offers rare insights into the epistemic relationship between the artisan-expert and nature. Distillation opened up new means of sensing and manipulating materials' most elusive, "subtle" components, and offered control over the unstable nature of earthly matter. It allowed the practitioner to discipline even the toughest, most obstinate materials. Ultimately, however, distillation was an intimate embodied give-and-take, and one could only hope to master nature by becoming her patient apprentice. In Brunschwig's books, distillation emerges as a braided technology for knowing the material world. The impact of his account on wider cultures of natural inquiry remains to be fully elucidated. For instance, as I have begun to show elsewhere, distillation played a key part in Girolamo Cardano's project of formulating a new cosmology centered on the idea of intellectual and material "subtlety."[6]

Ultimately, Brunschwig's approach offered new understandings not only of nature and the body, but also of the interface between the two. The distillation of medicinal substances produced forms of natural matter so refined that they penetrated the material fabric of the body safely and efficiently. Mediating between inside and outside materialities, distillation fostered a joint understanding of the inner microcosm of the body and the outer macrocosm of the material universe. We can gain a more fine-grained understanding of this overarching framework by thinking of it as braided knowledge, as this invites questions about the specific functions of different strands in keeping the whole together. In Brunschwig's project, even the core threads of alchemy, medicine, and craft are not always braided entirely symmetrically. Medical concerns brought the alchemy of distillation into homes and workshops in the first place, and both relied on the artisan's way of knowing bodies and materials. But I argue that alchemical concepts and techniques played the key part in weaving together artisanal and medical approaches into a joint material imaginary of nature and the body. Brunschwig's case suggests that we can (and should) recenter alchemy in histories of science and medicine—not so much as the driving force of an ostensibly radical new medicine formulated by Paracelsus and others, but as a set of material ideas and practices that created new connections between diverse ways of knowing.

Crafting Text: The Labor of Braiding and Unbraiding

Thinking of Brunschwig's expertise as braided knowledge calls our attention to a diversity of expertise, but also to the significant work that goes into braiding them together. Brunschwig was well aware, for example, that rational theory and practical experience did not naturally combine. To tie the loose strands of experiential insight into a larger project, his "vernacular empiricism" combined strategies of legitimation from medical *experimenta*, recipe culture, and occult traditions, ultimately recruiting his readers as "virtual witnesses" avant la lettre. As recent scholarship has begun to show, the diversity of different early modern cultures of experiential knowledge making defies any notion of a unified scientific "empiricism."[7] Brunschwig brings further complexity to the narrative by combining a remarkable diversity of approaches within a single project.

The work of braiding in Brunschwig's case was in large part a work of capturing and communicating different forms of knowledge in print. This was especially challenging in relation to what I have called the "material sensorium"—the expert's fine-tuned sensory knowledge of hidden injuries, complex distillations, and invisible material qualities. Reflecting the local artisanal discourse that emerges from the Strasbourg archives, Brunschwig conceives of such embodied knowledge as a "habit" that was formed by repeated practice. He reflects lucidly on the difficulties of capturing the artisan's gestural and sensory habits in writing, and he does his best to stretch those limits. His cautionary anecdotes serve as vicarious experience to his reader-apprentices. Where the material phenomena of the body or the distillation process are beyond the unaided senses, he provides specific tests to determine hidden skull fractures or a distilled water's remaining medicinal power.

In close collaboration with his publisher Johann Grüninger, Brunschwig's books develop textual and visual strategies for printing braided knowledge. A generation or two before Brunfels and Vesalius, Brunschwig thinks carefully about the limitations of images in conveying sensory knowledge of plants and bodies. In contrast, technical knowledge of instruments, glassware, and furnaces is confidently presented in a well-crafted interplay of detailed woodcuts and printed text. Overall, Brunschwig's books represent an innovative endeavor to guide his readers' senses and experience in a virtual apprenticeship in print.

Crucially, the printing of braided knowledge also involved providing tools to help readers navigate the many aspects of Brunschwig's expertise. Treating a particular disease might require nosological indications as well as information on medicinal herbs and surgical or distilling techniques. This was a lot to juggle in practice, and the books offer detailed tables of contents and indices to help readers make relevant connections between different sections. They anticipate and enable discontinuous reading, opening up a range of different paths through the books. As the annotations in surviving copies show, readers made full use of this potential. They corrected folio numbers in the indices or customized them by collecting the plants or diseases they were most interested in. Some even went further and compiled their own finding aids, teasing apart Brunschwig's categories and imposing their own. The work of braiding and unbraiding knowledge continued beyond Brunschwig's books and into the material practices of his readers.

The work of translators, too, gains significance if we pay attention to the ways in which they spliced together texts, images, and technical know-how from different sources, including their own experience. In tracing the itineraries of Brunschwig's books across languages, I have begun to uncover hitherto neglected cultures of what I call "vernacular text work"—the transmission and adaptation of specific, local, and practical knowledge between vernacular contexts—by otherwise obscure specialists such as Laurence Andrew. While they were not expert practitioners themselves, they grasped the importance of clear and precise technical language, and practiced a vernacular philology of sorts to realize this in their translations. In the English book market, the Brunschwig translations provided models for the text work of an overlapping community of Elizabethan surgeons and distillers. This next generation of practitioner-translators confidently braided their expertise as medical artisans into their textual practices, excising excess matter from their source texts or distilling them through the alembic of their own experience. These early modern cultures of vernacular text work—and others that remain to be explored—provide a new perspective on the role of translation in histories of knowledge, especially regarding the asymmetries of appropriation, silenced experts, and hidden translators.[8]

Finally, thinking about diverse braids of knowledge encourages us to create braided scholarly methodologies, combining a plurality of approaches that match the eclectic, multifaceted nature of our most interesting sources. Brunschwig's books, as we have seen, call for intellectual, social, and material approaches as well as histories of print, translation, and reading. Looking ahead, they challenge us to integrate strands of inquiry that have thus far remained largely separate—on the one hand, scholarship on the embodied knowledges of craftspeople and other practitioners,[9] and on the other, conversations that seek to historicize the body itself, from academic theories to popular perceptions within histories of the emotions or global comparative approaches.[10] As this study of Brunschwig has shown, knowledge *of* the body and knowledge *through* the body were intertwined in early modern material imaginaries. The challenge for historians is to deploy diverse methodologies to explore this relationship between what the body could know and what was known about the body.

Overall, I hope that this detailed study of Brunschwig's braided expertise opens up conversations in histories of science and medicine. It offers a model for studying the interactions between different knowledges and traditions without either essentializing them or underplaying their real and significant differences. Brunschwig was well aware of the fault lines between surgeons and physicians, learned scholars and the common man, or rational medicine and empirical cures. But he shows how new knowledge can form across and between these kinds of categories without tearing them down in the process. This allows us to bring nuance to our main historical narratives without relinquishing our ambitions to ask big questions about surgery, (al)chemical medicine, or empirical inquiry.

Acknowledgments

This book was a long time in the making, and I owe a debt of gratitude to many people. I first came across Brunschwig during research for my master's thesis in history and philosophy of science at Cambridge, supervised by Elaine Leong and Jenny Rampling, who continued to advise on the project as it grew. Lauren Kassell has been the most unwaveringly supportive mentor anyone could wish for. From the draft of my first article on Brunschwig to the final touches to the conclusion of this book many years later, her clear insight and generous guidance have been invaluable. Sachiko Kusukawa went far beyond the call of duty in offering her expertise on images and early print. Alisha Rankin and Peter Jones read an early version of the project with generosity of knowledge and attention, and the finished book owes much to our continued conversations in the intervening years. Many friends and colleagues offered careful readings and incisive criticism on talks, draft chapters, and the book prospectus, including Ulinka Rublack, Pamela Smith, Bill Sherman, Claudia Stein, Annemarie Kinzelbach, Anton Bruder, and the members of the Early Science and Medicine Reading Group at Cambridge. I am deeply grateful for your collective knowledge, advice, and wisdom.

At various stages, the research for this book was supported by a Wellcome Trust Doctoral Award, a postdoctoral position at Columbia University, the Huntington Library Molina Fellowship in History of Medicine and Allied Sciences, a Frances Yates Research Fellowship at the Warburg Institute, and a postdoctoral fellowship at the Herzog August Bibliothek Wolfenbüttel. Scholarship would be nowhere without financial and institutional support, but I would also like to thank the excellent communities at my various temporary homes for their companionship and support, especially during the weirder and darker episodes of the COVID pandemic.

This book would not have been possible without the support and encouragement of friends and family to keep me sane. Thank you, Emma and Toby, for taking in a stray scholar, and to Grace for always being there. To Lia, who has known me the longest, many thanks for your generous mentorship and our conversations about life, work, tea, and podcasts. I was thrilled to work with my talented sister Josi on the cover design for this book. Finally, and most importantly, I am grateful to my partner, Qian, without whose love and understanding I would be nowhere.

Notes

Introduction

1. Pierre Bachoffner, "Jérôme Brunschwig, chirurgien et apothicaire strasbourgeois, portraituré en 1512," *Revue d'histoire de la pharmacie* 81, no. 298 (1993): 269–78; Pierre Bachoffner, "Du nouveau à propos de Jérôme Brunschwig," *Revue d'histoire de la pharmacie* 60, no. 212 (1972): 32–33; Pierre Bachoffner and Théodore Vetter, "Brunschwig, Jérôme," in *Nouveau Dictionnaire de Biographie Alsacienne*, ed. Jean-Pierre Kintz (Strasbourg: Fédération des sociétés d'histoire et d'archéologie d'Alsace, 1985), 393–94; Gerhard Eis, "Brunschwig, Hieronymus," in *Neue Deutsche Biographie* (1955), 688.

2. Archives de la ville et de la communauté urbaine de Strasbourg (hereafter AVCUS) AA 298/46: "Jeronimy Brunzwig 1 streit ax."

3. See chap. 4.

4. AVCUS KS 35 I, fol. 68r: "des erfarnen Hieronymi Brunschwigs vnsers bürgers seligen dochter."

5. Hieronymus Brunschwig and Marsilio Ficino, *Liber de arte distulandi Simplicia et Composita* [Medicinarius]. *Das nüv buoch der rechten kunst zuo distillieren. Ouch von Marsilio Ficino vnn anderen hochberoempten Ertzte natürliche vnn guote künst* (Strasbourg: Johann Grüninger, 1509), fol. 2v: "der bücher groß vnnd clein die ich gesehen vnnd gelesen hab / in etlichen alten liberien vnnd von guotenn meystern mit worheit ich wol sprechen mag ob drei tusent."

6. On surgical manuscripts, see, e.g., Gundolf Keil and Rolf Müller, "Deutsche Lanfrank-Übersetzungen des 14. und 15. Jahrhunderts," in *Medizingeschichte in unserer Zeit: Festgabe für Edith Heischkel-Artelt und Walter Artelt zum 65. Geburtstag*, ed. Hans-Heinz Eulner, Gunter Mann, Gert Preiser, Rolf Winau, and Otto Winckelmann (Stuttgart: Enke, 1971), 90–110. On alchemy, Udo Benzenhöfer, *Johannes de Rupescissa Liber de consideratione quintae essentiae omnium rerum, deutsch: Studien zur Alchemia medica des 15. bis 17. Jahrhunderts, mit kritischer Edition des Textes* (Stuttgart: Steiner Verlag Wiesbaden, 1989).

7. Gundolf Keil, "Hieronymus Brunschwig, Buch der Cirurgia," *Beiträge zur Geschichte der deutschen Sprache und Literatur* 93 (1971): 475–77; Gundolf Keil, "Brunschwig, Hieronymus," in *Die deutsche Literatur des Mittelalters: Verfasserlexikon*, ed. Wolfgang Stammler, Karl Langosch, and Kurt Ruh, vol. 1 (Berlin: Walter de Gruyter, 1978), 1074–75.

8. See, for example, Lawrence I. Conrad et al., *The Western Medical Tradition: 800 BC to AD 1800, Vol. 1* (Cambridge: Cambridge University Press, 1995), 296; Roy

Porter, *The Greatest Benefit to Mankind: A Medical History of Humanity* (London: Harper Collins, 1997), 187.

9. Gustav Klein, *Das Buch der Cirurgia des Hieronymus Brunschwig* (Munich: Kuhn, 1911). See also Henry Sigerist, *Hieronymus Brunschwig and His Work: A Fifteenth Century Surgeon* (New York: Ben Abrahamson, 1946).

10. Keil, "Hieronymus Brunschwig, Buch der Cirurgia"; Keil, "Brunschwig, Hieronymus."

11. Robert Forbes, *A Short History of the Art of Distillation: From the Beginnings up to the Death of Cellier Blumenthal* (Leiden: Brill, 1948).

12. For a recent overview of Paracelsus scholarship, see Bruce T. Moran, *Paracelsus: An Alchemical Life* (2019). Paracelsus is also central to Pamela H. Smith, *The Body of the Artisan: Art and Experience in the Scientific Revolution* (Chicago: University of Chicago Press, 2004). On Fioravanti, see in particular William Eamon, *The Professor of Secrets: Mystery, Medicine, and Alchemy in Renaissance Italy* (Washington, DC: National Geographic Society, 2010).

13. See most recently Pamela Long, *Artisan/Practitioners and the Rise of the New Sciences, 1400–1600* (Corvallis: Oregon State University Press, 2011); Pamela H. Smith, *From Lived Experience to the Written Word: Reconstructing Practical Knowledge in the Early Modern World* (Chicago: University of Chicago Press, 2022).

14. Alisha Rankin, "How to Cure the Golden Vein: Medical Remedies as Wissenschaft in Early Modern Germany," in *Ways of Making and Knowing*, ed. Pamela H. Smith, Harold Cook, and Amy Meyers (Ann Arbor: University of Michigan Press, 2014), 109–33; Tillmann Taape, "Distilling Reliable Remedies: Hieronymus Brunschwig's *Liber de Arte Distillandi* (1500) Between Alchemical Learning and Craft Practice," *Ambix* 61, no. 3 (2014): 236–56; Tillmann Taape, "Common Medicine for the Common Man: Picturing the 'Striped Layman' in Early Vernacular Print," *Renaissance Quarterly* 74, no. 1 (2021): 1–58.

15. Joachim Telle, *Pharmazie und der gemeine Mann: Hausarznei und Apotheke der frühen Neuzeit* (Wolfenbüttel: Herzog August Bibliothek, 1988); Joachim Telle, "Medizin für den gemeinen Mann," in *Bibliotheca Palatina. Katalog zur Ausstellung vom 8. Juli bis 2. November 1986, Heiliggeistkirche Heidelberg*, ed. Elmar Mittler (Heidelberg: Braus, 1986), 335–40.

16. Alix Cooper, "Latin Words, Vernacular Worlds: Language, Nature, and the 'Indigenous' in Early Modern Europe," *East Asian Science, Technology, and Medicine* 26 (2007): 17–39; Mary Fissell, *Vernacular Bodies: The Politics of Reproduction in Early Modern England* (Oxford: Oxford University Press, 2004); Hannah Murphy, *A New Order of Medicine: The Rise of Physicians in Reformation Nuremberg* (Pittsburgh: University of Pittsburgh Press, 2019).

17. Sigerist, *Brunschwig and His Work*, 9–14.

18. On the publication history of the *Cirurgia*, see Karl Sudhoff, *Deutsche medizinische Inkunabeln. Bibliographisch-literarische Untersuchungen* (Leipzig: Barth, 1908), 51–61. On the Schönsperger reprint, see also Klein, *Das Buch der Cirurgia*, iv.

19. On the translations, see chap. 9 in this book.

20. Sudhoff, *Deutsche medizinische Inkunabeln*, 185–88.

21. Hieronymus Brunschwig, *Liber de arte distillandi de simplicibus* [Small book of distillation]. *Das buch der rechten kunst zü distilieren die eintzigen ding* (Strasbourg: Johann Grüninger, 1500), fol. C1v: "Aber dz groß buoch ich ouch sehen will lassen zuo siner zyt den gelerten diß wercks."

22. Annarita Pogliani, "Il successo editoriale *del Büchlein von den ausgebrannten Wässern* di Michael Puff aus Schrick," in *La letteratura tecnico-scientifica nel Medioevo germanico: Fachliteratur e Gebrauchstexte*, ed. Letizia Vezzosi (Alessandria: Edizioni dell'Orso, 2009), 193–212.

23. Key examples of the vast literature include William Eamon, *Science and the Secrets of Nature: Books of Secrets in Medieval and Early Modern Culture* (Princeton, NJ: Princeton University Press, 1994); Elaine Leong and Alisha Rankin, *Secrets and Knowledge in Medicine and Science, 1500–1800* (Farnham: Ashgate Publishing, 2011); Pamela Long, *Openness, Secrecy, Authorship: Technical Arts and the Culture of Knowledge from Antiquity to the Renaissance* (Baltimore: Johns Hopkins University Press, 2001); Smith, *From Lived Experience to the Written Word.*

24. Bruce Moran, *Distilling Knowledge: Alchemy, Chemistry, and the Scientific Revolution* (Cambridge, MA: Harvard University Press, 2005).

25. Fissell, *Vernacular Bodies.*

26. Murphy, *New Order of Medicine*, 21–43, 147–71.

27. Miriam Usher Chrisman, *Lay Culture, Learned Culture: Books and Social Change in Strasbourg, 1480–1599* (New Haven: Yale University Press, 1982).

28. The term "Strasbourg school" was first used by Friedrich Wieger, *Geschichte der Medicin und ihrer Lehranstalten in Strassburg vom Jahre 1497 bis zum Jahre 1872* (Strasbourg: Karl Trübner, 1885), 4; more recently, see Vivian Nutton, "Humanist Surgery," in *The Medical Renaissance of the Sixteenth Century*, ed. Andrew Wear, Roger Kenneth French, and Iain M. Lonie (Cambridge: Cambridge University Press, 1985), 91–92.

29. A noted exception with respect to Gersdorff is Melanie Panse, *Hans von Gersdorffs "Feldbuch der Wundarznei": Produktion, Präsentation und Rezeption von Wissen* (Wiesbaden: Reichert, 2012).

30. Michael McVaugh, *The Rational Surgery of the Middle Ages* (Florence: Sismel, 2006).

31. Harold Cook, *Trials of an Ordinary Doctor: Joannes Groenevelt in Seventeenth-Century London* (Baltimore: Johns Hopkins University Press, 1994); Margaret Pelling, *Medical Conflicts in Early Modern London: Patronage, Physicians, and Irregular Practitioners, 1550–1640* (Oxford: Oxford University Press, 2003); John Henry, "Doctors and Healers: Popular Culture and the Medical Profession," in *Science, Culture, and Popular Belief in Renaissance Europe*, ed. Stephen Pumfrey, Paolo Rossi, and Maurice Slawinski (Manchester: Manchester University Press, 1991), 191–221.

32. Sandra Cavallo, *Artisans of the Body in Early Modern Italy: Identities, Families and Masculinities* (Manchester: Manchester University Press, 2007); Celeste Chamberland, "Honor, Brotherhood, and the Corporate Ethos of London's Barber-Surgeons' Company, 1570–1640," *Journal of the History of Medicine and Allied Sciences* 64, no. 3 (2009): 300–332; Robert Jütte, "Zur Sozialgeschichte der Handwerkschirurgen im 16. Jahrhundert," in *Paracelsus und Salzburg*, ed. Heinz Dopsch and Peter Kramm (Salzburg: Gesellschaft für Salzburger Landeskunde, 1994), 45–60; Annemarie Kinzelbach, "Erudite and Honoured Artisans? Performers of Body Care and Surgery in Early Modern German Towns," *Social History of Medicine* 27 (2014): 668–88; Margaret Pelling, "Appearance and Reality: Barber-Surgeons, the Body and Disease," in *London 1500–1700: The Making of the Metropolis*, ed. A. L. Beier and Roger Finlay (London: Longman, 1986), 82–112; Sabine Sander, *Handwerkschirurgen* (Göttingen: Vandenhoeck & Ruprecht, 1989).

33. J. Andrew Mendelsohn, Annemarie Kinzelbach, and Ruth Schilling, *Civic Med-*

icine: Physician, Polity, and Pen in Early Modern Europe (Abingdon: Routledge, 2020); Murphy, *New Order of Medicine.*

34. Laurence Brockliss and Colin Jones, *The Medical World of Early Modern France* (Oxford: Clarendon Press, 1997), 214–25; Mary C. Erler, "The First English Printing of Galen: The Formation of the Company of Barber-Surgeons," *Huntington Library Quarterly* 48, no. 2 (1985): 159–71, https://doi.org/10.2307/3817541; Toby Gelfand, *Professionalizing Modern Medicine: Paris Surgeons and Medical Science and Institutions in the 18th Century* (London: Bloomsbury Academic, 1980); McVaugh, *Rational Surgery*; Nutton, "Humanist Surgery"; Richard Palmer, "Physicians and Surgeons in Sixteenth-Century Venice," *Medical History* 23, no. 4 (1979): 451–60; Linda Payne, "'A Spedie Reformation:' Barber-Surgeons, Anatomization, and the Reformation of Medicine in Tudor London," in *Paracelsian Moments: Science, Medicine, and Astrology in Early Modern Europe*, ed. Gerhild Scholz Williams and Charles D. Gunnoe (Truman State University Press, 2002), 71–92; Owsei Temkin, "The Role of Surgery in the Rise of Modern Medical Thought," *Bulletin of the History of Medicine* 25, no. 3 (1951): 248–59.

35. For a detailed overview of different ways of understanding the body, see Claudia Stein, *Negotiating the French Pox in Early Modern Germany* (Aldershot: Ashgate, 2009), 27–66. On the fluid body, see, e.g., Ulinka Rublack, "Fluxes: The Early Modern Body and the Emotions," *History Workshop Journal* 53, no. 1 (2002): 1–16. On chemical interpretations, see, e.g., John Christie, "The Paracelsian Body," in *Paracelsus: The Man and His Reputation, His Ideas and Their Transformation*, ed. Ole Peter Grell (Leiden: Brill, 1998), 269–91.

36. Mark Jenner and Patrick Wallis, *Medicine and the Market in England and Its Colonies, c. 1450–c. 1850* (Basingstoke: Palgrave Macmillan, 2007).

37. Elaine Leong, "Collecting Knowledge for the Family: Recipes, Gender and Practical Knowledge in the Early Modern English Household," *Centaurus* 55 (2013): 81–103; Alisha Rankin, *Panaceia's Daughters: Noblewomen as Healers in Early Modern Germany* (Chicago: University of Chicago Press, 2013).

38. Mary Fissell, "The Marketplace of Print," in Wallis and Jenner, *Medicine and the Market in England and Its Colonies*, 108–32. See also William Eamon, "Pharmaceutical Self-Fashioning or How to Get Rich and Famous in the Renaissance Medical Marketplace," *Pharmacy in History* 45, no. 3 (2003): 123–29.

39. Leah DeVun, *Prophecy, Alchemy, and the End of Time: John of Rupescissa in the Late Middle Ages* (New York: Columbia University Press, 2009); Robert Multhauf, *The Origins of Chemistry* (London: Oldbourne, 1966); Michela Pereira, "Heavens on Earth: From the *Tabula Smaragdina* to the Alchemical Fifth Essence," *Early Science and Medicine* 5 (2000): 131–44.

40. Moran, *Distilling Knowledge*; Tara Nummedal, *Alchemy and Authority in the Holy Roman Empire* (Chicago: University of Chicago Press, 2007); Tara Nummedal, *Anna Zieglerin and the Lion's Blood: Alchemy and End Times in Reformation Germany* (Philadelphia: University of Pennsylvania Press, 2019); Jennifer M. Rampling, *The Experimental Fire: Inventing English Alchemy, 1300–1700* (University of Chicago Press, 2020).

41. On the term "chymistry" as a historical category, see William R. Newman and Lawrence M. Principe, "Alchemy vs. Chemistry: The Etymological Origins of a Historiographic Mistake," *Early Science and Medicine* 3, no. 1 (1998): 32–65.

42. Edgar Zilsel, "The Sociological Roots of Science," *American Journal of Sociology* 47 (1942): 544–62.

43. Eamon, *Science and the Secrets of Nature.*

44. Long, *Openness, Secrecy, Authorship*; Long, *Artisan/Practitioners*; Smith, *Body of the Artisan*; Smith, *From Lived Experience to the Written Word.*

45. Pamela H. Smith, *Secrets of Craft and Nature in Renaissance France: A Digital Critical Edition and English Translation of BnF Ms. Fr. 640* (New York: Making and Knowing Project, 2020), https://edition640.makingandknowing.org/.

46. The historical literature on Paracelsus is vast. For a highly readable recent introduction, see Moran, *Paracelsus.* On the medical and alchemical context, see, e.g., Walter Pagel, *Paracelsus: An Introduction to Philosophical Medicine in the Era of the Renaissance* (Basel: Karger, 1958). On Paracelsus as a key advocate of artisanal epistemology, see Smith, *Body of the Artisan*, 59–94.

47. Charles Webster, *Paracelsus: Medicine, Magic and Mission at the End of Time* (New Haven: Yale University Press, 2008), 55.

48. Mark Clarke, Bert De Munck, and Sven Dupré, *Transmission of Artists' Knowledge* (Brussels: KVAB Press, 2012); Smith, *From Lived Experience to the Written Word*; Matteo Valleriani, *The Structures of Practical Knowledge* (Cham: Springer, 2017).

49. Sachiko Kusukawa, *Picturing the Book of Nature: Image, Text, and Argument in Sixteenth-Century Human Anatomy and Medical Botany* (Chicago: University of Chicago Press, 2012). See also Andrea van Leerdam, *Woodcuts as Reading Guides: How Images Shaped Knowledge Transmission in Medical-Astrological Books in Dutch (1500–1550)* (Amsterdam: Amsterdam University Press, 2023).

50. Peter Dear, *Discipline and Experience: The Mathematical Way in the Scientific Revolution* (Chicago: University of Chicago Press, 1995); Peter Dear, "The Meanings of Experience," in *The Cambridge History of Science, vol. 3, Early Modern Science*, ed. Lorraine Daston and Katharine Park (Cambridge: Cambridge University Press, 2006), 106–31.

51. Lorraine Daston and Elizabeth Lunbeck, *Histories of Scientific Observation* (Chicago: University of Chicago Press, 2011); Gianna Pomata and Nancy Siraisi, *Historia: Empiricism and Erudition in Early Modern Europe* (Cambridge, MA: MIT Press, 2005).

52. Elaine Leong and Alisha Rankin, "Testing Drugs and Trying Cures: Experiment and Medicine in Medieval and Early Modern Europe," *Bulletin of the History of Medicine* 91, no. 2 (2017): 157–82; Alisha Rankin, *The Poison Trials: Wonder Drugs, Experiment, and the Battle for Authority in Renaissance Science* (Chicago: University of Chicago Press, 2021).

53. For a full review of the vast literature on translation, see chap. 9. As recent examples of studies in the history of medicine and early modern studies, see Tara Alberts, Sietske Fransen, and Elaine Leong, "Translating Medicine, ca. 800–1900: Articulations and Disarticulations," in *Translating Medicine Across Premodern Worlds*, ed. Tara Alberts, Sietske Fransen, and Elaine Leong, *Osiris* 37 (2022): 1–21 (Chicago: University of Chicago Press); Marie-Alice Belle and Brenda M. Hosington, *Thresholds of Translation: Paratexts, Print, and Cultural Exchange in Early Modern Britain* (Basingstoke: Palgrave Macmillan, 2018).

54. William H. Sherman, *Used Books: Marking Readers in Renaissance England. Philadelphia* (University of Pennsylvania Press, 2008). Further literature on reading and annotating practices is discussed in chap. 10.

Chapter One

1. Lorenz Fries, *Uslegung der merkarthen* (Strasbourg: Grüninger, 1525), fol. 14 r: "dweyl dise stat ein straß ist in Lotringen / Frankrych / Burgund / Lampartenn / Schwitzerland / Brißgaw / Schwabenland / Niderland / und allenthalben hyn / ligt mitten in einer landstrassen zuo wasser und land."

2. A rich portrait of early modern Strasbourg is given in Thomas A. Brady, *The Politics of the Reformation in Germany: Jacob Sturm (1489–1553) of Strasbourg* (Atlantic Highlands: Humanities Press International, 1997), 40–60.

3. Sabine von Heusinger, "Von 'antwerck' bis 'Zunft.' Methodische Überlegungen zu den Zünften im Mittelalter," *Zeitschrift für historische Forschung* 37 (2010): 37–71.

4. Hieronymus Brunschwig, *Liber de arte distillandi de simplicibus* [Small book of distillation]. *Das buch der rechten kunst zü distilieren die eintzigen ding* (Strasbourg: Johann Grüninger, 1500), fol. 15v: "eim solichen vngelerten eynfeltigen artzet / der anfang / mittel noch dz end / nye geleret gehoert / oder im nit wissen ist / du din leben so gar lichtiglichen vnderwürffig machst vnn nit ingedencken bist ein ydem weckmann nit zuo nennen ist eynen werck meister des wercks / das er nye gelert hat / vnn nit getrüwen bist eym dz mynste werck dyns huses das er nye gelert hat dir zuo machen ein gewandt von guotem tuoch / oder schuoch von guotem leder / ich geschwyg eins genßstals von guotem holtz . . . das tuoch / leder oder holtz zuo uerhauwen würd / das doch mit cleynem gelt wider zekauffen ist / vnn wagst lyb vnd leben / das mit keynem gelt oder golt zebezalen ist."

5. Hieronymus Brunschwig and Marsilio Ficino, *Liber de arte distulandi Simplicia et Composita* [Medicinarius]. *Das nüv buoch der rechten kunst zuo distillieren. Ouch von Marsilio Ficino vnn anderen hochberoempten Ertzte natürliche vnn guote künst* (Strasbourg: Johann Grüninger, 1509), fol. 36v: "das artzney edler sei wan alle andere künst das sicht man. wol. Würt ein meister siech / dz sein kunst nicht gehelffen mag / darumb begert er der artznei das er sein kunst vnn sein leben da mit behalt."

6. Heusinger, "Von 'antwerck' bis 'Zunft.'"

7. On the political and social roles of the Strasbourg guilds, see Sabine von Heusinger, "The Topography of Sacred Space and the Representation of Social Groups—Confraternities in Strasbourg," in *Politics and Reformations: Communities, Polities, Nations, and Empires: Essays in Memory of Thomas A. Brady, Jr.*, ed. Thomas A. Brady and Christopher Ocker (Brill, 2007); Sabine von Heusinger, "Die Handwerksbruderschaften in Straßburg," in *Städtische Gesellschaft und Kirche im Spätmittelalter*, vol. 62, ed. Sigrid Schmitt and Sabine Klapp (Stuttgart: Steiner, 2008), 123–40; Sabine von Heusinger, *Die Zunft im Mittelalter: zur Verflechtung von Politik, Wirtschaft und Gesellschaft in Straßburg* (Stuttgart: Steiner, 2009), 13–15.

8. On late medieval shifts in the guilds' political influence, see Martin Alioth, *Gruppen an der Macht: Zünfte und Patriziat in Strassburg im 14. und 15. Jahrhundert: Untersuchungen zu Verfassung, Wirtschaftsgefüge und Sozialstruktur* (Basel: Helbing & Lichtenhahn, 1988). For a more detailed discussion of the literature, see chap. 3.

9. On the processions, see Heusinger, "The Topography of Sacred Space"; Heusinger, "Handwerksbruderschaften," 127–37; Heusinger, *Die Zunft im Mittelalter*, 115–55.

10. Kathy Stuart, *Defiled Trades and Social Outcasts: Honor and Ritual Pollution in Early Modern Germany* (Cambridge: Cambridge University Press, 2006).

11. Hieronymus Brunschwig, *Liber de arte Distillandi de Compositis* [Large book of

distillation]. *Das buch der waren kunst zü distillieren die Composita vnn simplicia / vnd dz Buoch thesaurus pauperum / Ein schatz der armen* (Strasbourg: Johann Grüninger, 1512), fol. 27v: "hinder dem flecken genant Oberkirch / in Straßburger bistum."

12. Brunschwig, *Small Book*, fol. 6v: "hagenower krüg."

13. Brunschwig, *Large Book*, fol. 27r: "als da ist die erd von Syburgk dem stetlin / gelegen im bistum von Koeln."

14. Brunschwig, *Large Book*, fol. 27r: "Aber Sigen ist ein dorff by dem stetlin obgemelt / da macht man die kochhefen vnd nit die krueg / Wan man die krueg wil lassen machen / so muoß man kummen vmb vnser lieben frowen tag der eren / oder xiiii. tag darvor / Oder vmm sant Margarethen tag / wan sie arbeiten sunst im iar nit mer / dann der meister seint nit me dann zwölff / vnd die sechß arbeiten vmb vnser lieben frowen tag / vff sechß wochen / vnn dier andern sechß vmb sant Martins tag sechß wochen."

15. Brunschwig, *Small Book*, fol. 6v: "die besten glesser sint die von den bemischen glaser oder von den Venedischen schyben stick gemacht sint."

16. Brunschwig, *Large Book*, fol. 16v: "Aber schwerlich werden sie gemacht / oder selten funden einen glaser der sie machen kan."

17. Brunschwig, *Large Book*, fol. 25r: "calcenieren als lang der kalck vnd stein brent / das ist hie zu Straßburg vff .xii. oder xiiii. tag."

18. Brunschwig, *Small Book*, fol. 3v: "ob aber du ein ziegler haben macht so bring im die form und heiß dir stein machen."

19. Archives de la ville et de la communauté urbaine de Strasbourg (AVCUS) 1MR 13, fol. 266r: "Item daz man eigen modele zü allen stucken mache vnd habe darnach alle wergk gemacht werden sollent."

20. Brunschwig, *Small Book*, fol. 11v: "dyn geschlagen bly . . . dar in geschriben mit eim grifel / oder gestochen von eym golt schmit oder geschlahen von dem steheln kegel der geschrifft des text by eim drucker."

21. Brunschwig, *Large Book*, fol. 130r: "nim die gülden / vnd mach sie glüendig / vnd schlage sie also dünn als du magst / als die golt schmid thuon wann sie vergülden."

22. Brunschwig, *Large Book*, fol. 124r: "aber besser von eim golt schleher."

23. Hans Sachs, *Eygentliche Beschreibung Aller Staende Auff Erden / Hoher Vnd Nidriger / Geistlicher Vnd Weltlicher / Aller Künsten / Handwercken Vnd Haendeln / Etc. Vom Groesten Biß Zum Kleinesten / Auch von Irem Vrsprung / Erfindung Vnd Gebreuchen* (Frankfurt: Georg Raben for Sigmund Feyerabend, 1568), fol. S1r: "Der Kupfferschmidt"; "Brennhuet zum Wasser brennen nuetz."

24. Patrick Wallis and Catherine Wright, "Evidence, Artisan Experience, and Authority in Early Modern England," in *Ways of Making and Knowing*, ed. Pamela Smith, Harold Cook, and Amy Meyers (Ann Arbor: University of Michigan Press, 2014), 138–63. On inspections in early modern Dijon and their abuse, see James Farr, *Hands of Honor: Artisans and Their World in Dijon, 1550–1650* (Ithaca, NY: Cornell University Press, 1988), 19–29. On inspections in London barber shops, see Celeste Chamberland, "Honor, Brotherhood, and the Corporate Ethos of London's Barber-Surgeons' Company, 1570–1640," *Journal of the History of Medicine and Allied Sciences* 64, no. 3 (2009): 300–332.

25. AVCUS III 11/8, fol. 1r: "geferwete Cristall." This version of the goldsmiths' ordinance is dated 1482.

26. AVCUS III 11/8, fol. 3r: "besehen durch den meister vnd das geriht goltschmyde antwercks . . . vnd durch dieselben erkant das es werschafft sy."

27. Cf. "währschaft" in Jacob and Wilhelm Grimm, *Deutsches Wörterbuch* (Leipzig, 1854–1961), vol. 23, col. 980, http://www.woerterbuchnetz.de/DWB?lemma=waehrschaft.

28. Bert De Munck, "Guilds, Product Quality and Intrinsic Value. Towards a History of Conventions?" *Historical Social Research/Historische Sozialforschung* (2011): 103–24; Bert De Munck, "The Agency of Branding and the Location of Value. Hallmarks and Monograms in Early Modern Tableware Industries," *Business History* 54, no. 7 (2012): 1055–76.

29. AVCUS III 11/8, fol. 3v: "mit siner eigen hant"; goldsmiths' regulations dated 1543, ed. in Meyer, "Die Strassburger Goldschmiedezunft," 89: "welcher ein stück fürbrecht, das er nit mit eygener handt gemacht hette, . . . dem soll syn lade zuogethan un in jar und tag in disser stat Straßburg zuo arbeiten nit zuogelossen werden."

30. AVCUS 1MR 28, fol. 355r: "Divile die buhsse drie die ersten schusse bestanden sÿ das er dann werschafft geton habe."

31. AVCUS 1MR 28, fol. 354r: "das sie die harre bestentlich sin oder werschafft tün möge . . . Man giesse sie dann anderwerbe."

32. The pharmacy regulations are edited in Peter Dilg, "Die Strassburger Apothekerordnung (um 1500)," *Pharmazeutische Zeitung* 124 (1979): 2668–69: "der Stat zwen artztete, dor zu der eltestt und berichtest under den appoteckern"; "sich verstott umb die matterialien."

33. Dilg, "Strassburger Apothekerordnung," 2669: "was sy erfinden nit nutze oder toigentlich sin hin we thun, ußschutten oder verbrennen."

34. Dilg, "Strassburger Apothekerordnung," 2669: "wie all materielien, wurzten, krutter, wasser und ander ding als Lattwergen, Syruppen und ander vermuschungen In den Appotecken sin und gehalten werden sollent und wie lang yedes werschafft thun und nutz sin moge."

35. Brunschwig, *Small Book*, fol. C2r: "als oft geschicht das die heiden oder sarraceni reubarbara oder lignum aloes in wasser siedent oder beyssen/vnnd das wasser nützen vnd den reubarbara wieder dörrent vnn verkouffentz für guot/nit erkannt von eynem vnerfarenden appotecker."

36. Elma Brenner and François-Olivier Touati, *Leprosy and Identity in the Middle Ages* (Manchester: Manchester University Press, 2021); Carole Rawcliffe, *Leprosy in Medieval England* (Woodbridge: Boydell Press, 2006).

37. On the *Blatterhaus*, see Pierre Bachoffner, "Notes sur le Gaïac au XVIe et au début du XVIIe siècle," *Veröffentlichungen der internationalen Gesellschaft für Pharmazie* 36 (1970): 17–25; Jacques Héran, *Histoire de la médecine à Strasbourg* (Strasbourg: La Nuée Bleue, 1997), 86; Ernest Wickersheimer, "Le Gaïac à Strasbourg Au XVIe Siècle," *Analecta Medico-Historica* 1 (1966): 55–66; Otto Winckelmann, *Das Fürsorgewesen der Stadt Strassburg* (Leipzig: Heinsius, 1922), part 1, 94–165.

38. Winckelmann, *Fürsorgewesen*, part 1, 5–27.

39. The information on Strasbourg practitioners is collated from my own research in the Strasbourg archives and the appendix file "Personendatei" in Heusinger, *Die Zunft im Mittelalter*. The key figures are also mentioned in Friedrich Wieger, *Geschichte der Medicin und ihrer Lehranstalten in Strassburg vom Jahre 1497 bis zum Jahre 1872* (Strasbourg: Karl Trübner, 1885); Winckelmann, *Fürsorgewesen*.

40. Dilg, "Strassburger Apothekerordnung."

41. AVCUS VI 368/6, 1502: "den Appotegkern by unns mit den verlegnenn Ma-

terialien merclichenn betrogenn werd*en*"; "uns furgenommen, die Appotegkenn zü besichtigenn"; den hochgelerten he*r*n Fuchsenn uwe*r*n physicum."

42. Dilg, "Strassburger Apothekerordnung."

43. Hannah Murphy, *A New Order of Medicine: The Rise of Physicians in Reformation Nuremberg* (Pittsburgh: University of Pittsburgh Press, 2019).

44. Sabine Schlegelmilch, "Promoting a Good Physician: Letters of Application to German Civic Authorities, 1500–1700," in *Civic Medicine: Physician, Polity, and Pen in Early Modern Europe*, ed. J. Andrew Mendelsohn, Annemarie Kinzelbach, and Ruth Schilling (Abingdon: Routledge, 2020), 88–109.

45. Wieger, *Geschichte der Medicin*, 32–37.

46. Héran, *Histoire de la médecine à Strasbourg*, 73, 93; Wieger, *Geschichte der Medicin*, 33–37.

47. Brady, *Politics of the Reformation.*

48. On Brant, see Hellmut Rosenfeld, "Brant, Sebastian," *Neue Deutsche Biographie* 2 (1955): 534–36, http://www.deutsche-biographie.de/pnd118514474.html.

49. On the early Strasbourg humanists, see Brady, *Politics of the Reformation*, 72; Miriam Usher Chrisman, *Lay Culture, Learned Culture: Books and Social Change in Strasbourg, 1480–1599* (New Haven: Yale University Press, 1982), 81–102.

50. Winckelmann, *Fürsorgewesen*, 21–27.

51. Johann Geiler von Kaysersberg, *Die Emeis* (Strasbourg: Johann Grüninger, 1516), fol. 8r: "wispeln durcheinander vnn weiß doch yetliche was sy thuon sol."

52. Geiler von Kaysersberg, *Die Emeis*, fol. 29r: "Ein bader bedarff des schuomachers / ich bedarff des schuomachers auch."

53. Geiler von Kaysersberg, *Die Emeis*, fol. 9r: "ir speisen vns so solten wir euch das zeitlich ersetzen mitt dem geistlichen."

54. Thomas A. Brady, "The Themes of Social Structure, Social Conflict, and Civic Harmony in Jakob Wimpheling's Germania," *Sixteenth Century Journal* 3, no. 2 (1972): 65–76.

55. On Strasbourg printers, see Chrisman, *Lay Culture, Learned Culture*, 3–36. On Grüninger, see also Jürgen Schulz-Grobert, *Das Straßburger Eulenspiegelbuch* (Max Niemeyer Verlag, 1999), 93–94, 105–10.

56. Chrisman, *Lay Culture, Learned Culture*, 34.

57. Paul Oskar Kristeller, *Die Strassburger Bücherillustration im XV. und im Anfange des XVI. Jahrhunderts* (Leipzig: Seemann, 1888), 7–50.

Chapter Two

1. The first printed book on surgery appears to be an Italian translation of Guglielmo's text, *La ciroxia vulgarmente fata* (Venice: Philippo di Pietro, 1474). Guy de Chauliac's *Cirurgia magna* was first printed in Middle French translation (Lyon: Barthélemy Buyer, 1478), and then in Italian (Venice: Niccolo Girardengo, 1480).

2. Karl Sudhoff, *Deutsche medizinische Inkunabeln. Bibliographisch-literarische Untersuchungen* (Leipzig: Barth, 1908), 51–61.

3. Hieronymus Brunschwig, *Boek der Wundenartzstedye* (Rostock: Ludwig Dietz, 1518); Hieronymus Brunschwig, *The Noble Experyence of the Vertuous Handy Warke of Surgeri* (London: Petrus Treveris, 1525); Hieronymus Brunschwig, *Dits dat hantwerck der cirurgien ende leert alle wonden, gehouden gesteken gheslaghen* (Utrecht: Jan Berntsz,

1535). On the Low German edition, see Chiara Benati, *Dat boek der Wundenartzstedye und der niederdeutsche chirurgische Fachwortschatz* (Göppingen: Kümmerle, 2012).

4. On German translations of the rational surgeons circulating in manuscript, see Gundolf Keil and Rolf Müller, "Deutsche Lanfrank-Übersetzungen des 14. und 15. Jahrhunderts," in *Medizingeschichte in unserer Zeit: Festgabe für Edith Heischkel-Artelt und Walter Artelt zum 65. Geburtstag*, ed. Hans-Heinz Eulner et al. (Stuttgart: Enke, 1971), 90–110. See also entries "Bruno von Longoburgo," "Guy de Chauliac," and "Lanfrank von Mailand" in Wolfgang Stammler, Karl Langosch, and Kurt Ruh, *Die deutsche Literatur des Mittelalters: Verfasserlexikon*, vol. 1 (Berlin: Walter de Gruyter, 1978).

5. Manfred Gröber, *Das wundärztliche Manual des Meisters Hans Seyff von Göppingen (ca. 1440–1518): der Cod. med. et phys. 20 8 der Württembergischen Landesbibliothek Stuttgart* (Göppingen: Kümmerle, 1998).

6. Hieronymus Brunschwig, *Dis ist das buch der cirurgia* [Cirurgia]. *Hantwirckung der wund artzny* (Strasbourg: Johann Grüninger, 1497), fol. 2v: "allen den ginnnen die da begerent zuo leren yn chirurgia"; "iungen angonden meister vnd knecht der scherrer vnn wundartzet."

7. Jan Frederiksen, "Johannes (Hans) von Gersdorff," in *Die deutsche Literatur des Mittelalters: Verfasserlexikon*, ed. Wolfgang Stammler, Karl Langosch, and Kurt Ruh, vol. 4 (Berlin: Walter de Gruyter, 1978), 626–30; Melanie Panse, *Hans von Gersdorffs "Feldbuch der Wundarznei": Produktion, Präsentation und Rezeption von Wissen* (Wiesbaden: Reichert, 2012).

8. Vivian Nutton, "Humanist Surgery," in *The Medical Renaissance of the Sixteenth Century*, ed. Andrew Wear, Roger Kenneth French, and Iain M. Lonie (Cambridge: Cambridge University Press, 1985), 91–92. The term was first used by Friedrich Wieger, *Geschichte der Medicin und ihrer Lehranstalten in Strassburg vom Jahre 1497 bis zum Jahre 1872* (Strasbourg: Karl Trübner, 1885), 4. See also Jacques Héran, *Histoire de la médecine à Strasbourg* (Strasbourg: La Nuée Bleue, 1997), 74.

9. Gundolf Keil, "Hieronymus Brunschwig, Buch der Cirurgia," *Beiträge zur Geschichte der deutschen Sprache und Literatur* 93 (1971): 475–77; Gustav Klein, *Das Buch der Cirurgia des Hieronymus Brunschwig* (Munich: Kuhn, 1911); Henry Sigerist, *Hieronymus Brunschwig and His Work: A Fifteenth Century Surgeon* (New York: Ben Abrahamson, 1946).

10. Heidi Hausse, *The Malleable Body: Surgeons, Artisans, and Amputees in Early Modern Germany* (Manchester: Manchester University Press, 2023); Panse, *Hans von Gersdorffs Feldbuch der Wundarznei*.

11. Margaret Pelling, "Appearance and Reality: Barber-Surgeons, the Body and Disease," in *London 1500–1700: The Making of the Metropolis*, ed. A. L. Beier and Roger Finlay (London: Longman, 1986), 82–112.

12. Annemarie Kinzelbach, "Erudite and Honoured Artisans? Performers of Body Care and Surgery in Early Modern German Towns," *Social History of Medicine* 27 (2014): 672. A similar figure (one in 460) was found in Norwich. See Margaret Pelling and Charles Webster, "Medical Practitioners," in *Health, Medicine, and Mortality in the Sixteenth Century*, ed. Charles Webster (Cambridge: Cambridge University Press, 1979), 225. See also Annemarie Kinzelbach, "Arztsein, Handwerk und Politik. Die Wund- und Stadtärzte, Barbiere, Bader, Chirurgen in Nördlingen," *Jahrbuch des Historischen Vereins für Nördlingen* 35 (2017): 249–90.

13. Robert Jütte, "Zur Sozialgeschichte der Handwerkschirurgen im 16. Jahrhun-

dert," in *Paracelsus und Salzburg*, ed. Heinz Dopsch and Peter Kramm (Salzburg: Gesellschaft für Salzburger Landeskunde, 1994), 49.

14. Jütte ("Sozialgeschichte der Handwerkschirurgen") has argued that the oft-cited dearth of medical practitioners in early modern Germany may apply to physicians, but not surgical practitioners, who were fairly readily available even in rural areas, with an overall figure of five hundred to six thousand inhabitants per practitioner across the empire (49–50). On surgeons as the main port of call for most of the population, see also Gundolf Keil, "Die medizinische Versorgung durch Bader und Wundärzte zur Zeit des Paracelsus," *Salzburger Beiträge zur Paracelsusforschung* 29 (1995): 50–74.

15. For example, Toby Gelfand, *Professionalizing Modern Medicine: Paris Surgeons and Medical Science and Institutions in the 18th Century* (London: Bloomsbury Academic, 1980).

16. Thomas Schlich, "Introduction: What Is Special About the History of Surgery?" in *The Palgrave Handbook of the History of Surgery*, ed. Thomas Schlich, 1–24 (Cham: Springer, 2018); Faith Wallis, "Pre-Modern Surgery: Wounds, Words, and the Paradox of 'Tradition,'" in Schlich, *Palgrave Handbook of the History of Surgery*, 49–70.

17. Michael McVaugh, *The Rational Surgery of the Middle Ages* (Florence: Sismel, 2006).

18. Nutton, "Humanist Surgery"; Linda Payne, "'A Spedie Reformation:' Barber-Surgeons, Anatomization, and the Reformation of Medicine in Tudor London," in *Paracelsian Moments: Science, Medicine, and Astrology in Early Modern Europe*, ed. Gerhild Scholz Williams and Charles D. Gunnoe (Truman State University Press, 2002), 71–92; Andrew Wear, "Early Modern Europe," in *The Western Medical Tradition: 800 BC to AD 1800*, vol. 1, ed. Lawrence I. Conrad, et al. (Cambridge: Cambridge University Press, 1995), 293–98.

19. Nutton, "Humanist Surgery," 92–96.

20. Paolo Savoia, *Gaspare Tagliacozzi and Early Modern Surgery: Faces, Men, and Pain* (Routledge, 2020), 44–82.

21. Laurence Brockliss and Colin Jones, *The Medical World of Early Modern France* (Oxford: Clarendon Press, 1997), 220–24.

22. Robert Ralley, "Medical Economies in Fifteenth Century England," in *Medicine and the Market in England and Its Colonies, c. 1450–c. 1850*, ed. Patrick Wallis and Mark Jenner, 24–46 (Basingstoke: Palgrave Macmillan, 2007).

23. Payne, "A Spedie Reformation."

24. John Henry, "Doctors and Healers: Popular Culture and the Medical Profession," in *Science, Culture, and Popular Belief in Renaissance Europe*, ed. Stephen Pumfrey, Paolo Rossi, and Maurice Slawinski (Manchester: Manchester University Press, 1991), 191–221; Margaret Pelling, *Medical Conflicts in Early Modern London: Patronage, Physicians, and Irregular Practitioners, 1550–1640* (Oxford: Oxford University Press, 2003).

25. Hausse, *Malleable Body*, 19–52.

26. My survey of Strasbourg practitioners is based on the archival materials consulted during my research and on the "Personendatei" appendix to Sabine von Heusinger, *Die Zunft im Mittelalter: zur Verflechtung von Politik, Wirtschaft und Gesellschaft in Straßburg* (Stuttgart: Steiner, 2009).

27. Archives de la ville et de la communauté urbaine de Strasbourg (AVCUS) AA 298, 119.

28. Otto Winckelmann, *Das Fürsorgewesen der Stadt Strassburg* (Leipzig: Heinsius, 1922), part 1, 104–9.

29. AVCUS HA 1316, ed. in Winckelmann, *Fürsorgewesen*, part 2, 71: "keinem andern zugelossen werden solle, neben im die armen im huse (dern kranckheiten so in sinem wissen und erfarenheit stond) zu arzeneien."

30. AVCUS StA varia coll. 1 (no. 849), ed. in Otto Winckelmann, *Fürsorgewesen*, part 2, 14: "kunftiger arzt hie zü Straszburg."

31. AVCUS StA varia coll. 1 (no. 849), 14: "ob es ouch güt und nutz si, das man gestat vnd zülasze, jederman hie sich anzünemen arznien die zü geben und andre ding, so den doctor der arzni zühört, züzulassen, als getouften juden, scherern, alten und sunst torochten wibern, lantfere, zü latin all genant emperici, und zum letsten dem henker, als hie zu Straszburg geschicht, als ich globlich bericht bin."

32. Pelling, *Medical Conflicts*.

33. Johann Widmann, *Tractatus Clarissimi Medicinarum Doctoris Joannis Widman* (Strasbourg: Grüninger, 1497), 15: "anderswo in groszen stetten hat man gesworen hebammen."

34. Widmann, *Tractatus*, 14: "das doch arzni zü triben, nemlich libarzni, nemen gestatt soll werden dann allein den doctor der erzni und sunst dheinem emperico"; "übel und schaden, der us ir unwissenteit entstot."

35. Widmann, *Tractatus*, 15: "der scherer nit witer handelt, dann er die ader schlecht und anders tut, so im doctores heiszent und befelhend."

36. AVCUS 1AH 112, fol. 124v: "Er soll ouch des Spittals scherer Inn allem sinem anligen vnd geschäfften den siechen Rat züthün / beholffen vnnd beroten sin noch synem besten vermögen vnnd verstentniß / ouch den getrüwlich zü vnderwisen / wann vnnd wie er eim yeden lassen / schrepffen vnd baden / oder sanguisugas ansetzen soll / damit das niemans durch vnwissenheit versumpt oder verderbt werde." A document with virtually the same wording, dated 1515, is cited by Brucker, *Zunft- und Polizei-Verordnungen*, 289. AVCUS 1AH 114 no. 1, fol. 10v–11r, also has this text.

37. Michael Stolberg, "Learning from the Common Folks. Academic Physicians and Medical Lay Culture in the Sixteenth Century," *Social History of Medicine* 27 (2014): 649–67.

38. Widmann, *Tractatus*, t.p. On Widmann, see Wilhelm von Heyd, "Widmann, Johann," in *Allgemeine Deutsche Biographie* (Leipzig: Duncker und Humblot, 1897), 355–57, https://www.deutsche-biographie.de/pnd102507368.html#adbcontent.

39. For a short overview of surgeon-physician relations across Europe, see Nancy Siraisi, *Medieval and Early Renaissance Medicine* (Chicago: University of Chicago Press, 1990), 178–80.

40. Brunschwig, *Cirurgia*, fol. A2v: "vil der menschen schaden nement von den iungen anganden mester de scherer vnnd wund ertzet die sich vnder stant des daß sie nit gelert han weder anfang mittel noch end nit wissent."

41. Brunschwig, *Cirurgia*, fol. 32r: "vnerfarnden cirurgicus oder scherer"; Ibid., fol. 19v: "etlich doericht wund artzet . . . uß vnwissenheit die hinderung der heilung."

42. Brunschwig, *Cirurgia*, fol. 9r–v: "darum das sin wirckung mit der hand geschicht dz der phisicus oder lib arzt nit endüt."

43. Brunschwig, *Cirurgia*, fol. 9v: "zuo dem Cirur. als einem hantwircker." See also Sabine von Heusinger, "Von 'antwerck' bis 'Zunft.' Methodische Überlegungen zu den Zünften im Mittelalter," *Zeitschrift für historische Forschung* 37 (2010): 37–71.

44. Brunschwig, *Cirurgia*, fol. A1v: "in der practick vnn mit der handt geuebt."

45. For Brunschwig, see Pierre Bachoffner, "Jérôme Brunschwig, chirurgien et apothicaire strasbourgeois, portraituré en 1512," *Revue d'histoire de la pharmacie* 81, no. 298 (1993): 269–78. On Gersdorff, see Ralf Vollmuth, "War Klaus von Matrei der Lehrer Hans von Gersdorffs?" *Sudhoffs Archiv* (1996): 109–17. For an overview of surgery highlighting the importance of battle experience, see Roy Porter, *The Greatest Benefit to Mankind: A Medical History of Humanity* (London: Harper Collins, 1997), 186–90.

46. Brunschwig, *Cirurgia*, fol. 29v–30r: "als einem geschach in dem Burguner krieg von blomunt der do geschossen ward mit einem pfil / vnd das ysin darinnen bleib vnnd das holtz hervß kam / Zuo dem etwan mancher guoter meister berüfft ward die vnder stunden im das isin vß zuo ziehen / do wolt es in nit folgenn. Do wart hanß meyer vonn Straßburg ein scherer vnnd wund artzet dar zuo berüfft als er soelche irrunge gesehenn het. Do nam der die zang in sin hant vnnd mercket das das ysin wider hacken het / do want er die zang vmm das die geschicklikeit des wider hacken gekert wart als er in was gangen / von stund an gieng das pfil isin hervß Von allen anderen meistern Ritter vnd knechtenn er gelobet wart. Doch sie nit westen wie er es her vß gethon het."

47. A barber named Hans Meier is listed as a member of the confraternity of bathers and barbers in 1478, AVCUS AMS II 119/9, cited in Heusinger, *Die Zunft im Mittelalter*, app. "Personendatei," entry no. 2128. The name "hanß meiger" also appears on a list of members of that fraternity who participated in the Burgundy campaigns (AVCUS AA298, item 71), although he is listed as a bather (*bader*) rather than barber (*scherer*); he may be the same person, or a different professional who coincidentally bore the same (fairly common) name.

48. The two main rival views held that gunshot wounds were compounded either by the poison of the powder, or by the burn caused by the shot. Important German sources are the work of Johannes Beris, who was the first to treat gunshot wounds as poisoned, and his pupil, Heinrich of Pfalzpaint. See entries by Gundolf Keil for "Beris, Johannes" (vol. 1, col. 725–26), and "Heinrich von Pfalzpaint (Pfolspeunt)" (vol. 3, col. 856–62), in Stammler, Langosch, and Ruth, *Verfasserlexikon*. For a brief overview of gunshot treatment in fifteenth-century surgery, see also Ralf Vollmuth, "Wann einer mit einer puchssen geschossen wirt. Eine Mehrschritt-Therapie für Schußwunden aus der zweiten Hälfte des 15. Jahrhunderts," *Sudhoffs Archiv* (1998): 102–4.

49. Brunschwig, *Cirurgia*, fol. 31r: "das pulver die wund vergifft hat"; Hans von Gersdorff, *Feldtbuoch der wundtartzney* (Strasbourg: Johann Grüninger, 1517), fol. 40r: "bitz der brant gelescht ist."

50. Gersdorff, *Feldtbuoch*, fol. 40r: "dann ich es oft gebrucht hab im Burgundischen kryeg bey meinem mister Mulartzt genant / hertzog Sigmunds von Ostereich loblicher gedaechtnuß chirurgicus vnd wundartzt"; fol. 42r: "namlich in dreyen feldtschlachten / Granße / Murten / vnd Nanße."

51. On the identity of Gersdorff's master, see Vollmuth, "Klaus von Matrei"; on Matrei's *Arzneibüchlein*, 111.

52. Guy de Chauliac, *Guigonis de Caulhiaco (Guy de Chauliac), Inventarium sive Chirurgia Magna. Volume One: Text*, ed. Michael McVaugh (Leiden: Brill, 1997), 9: "Si enim neque geometria neque astronomia neque dyalectica neque aliqua alia doctrina bonorum egerent medici, prompte coriarii, carpentatores, fabri, eciam ceteri, dimittentes artes suas currerent ad medicinam et efficientur medici."

53. Brunschwig, *Cirurgia*, fol. 9v: "nit alein in der artzeni. Sunder och in andern natürlichen künsten der Phylosophia."

54. Brunschwig, *Cirurgia*, fol. 9v: "Er sol ouch vast dar in lessen vnd studieren wan zehen mal über lessen vnd dar vff gemercket mit fliß verfaßt danocht kum die vernunft."

55. McVaugh, *Rational Surgery*, 89–134. On the consolidation of surgical *intentiones*, see pp. 65–66.

56. McVaugh, *Rational Surgery*, 91–94.

57. Gersdorff, *Feldtbuoch*, fol. A1r: "Von dem aderlasszen in einer gemein schoene regelen vnd leer"; "Von den xii. zeychen des himels / ire würckungen vnd natur."

58. McVaugh, *Rational Surgery*, 48.

59. Brunschwig, *Cirurgia*, fol. 39r–v: "Dar vmb soltu wissen / die krafft / tugent vnd eygenschafft eines ieden heilenden dings Wann wor vmb was in eim land bald heilt / heilt im andern langsam / In einem gelidt / im ander nit. In einer complex. in der ander nit."

60. Domenico Bertoloni Meli and Cynthia Klestinec, "Renaissance Surgery Between Learning and Craft," *Journal of the History of Medicine and Allied Sciences* 72, no. 1 (2017): 1–5; Cynthia Klestinec, "Translating Learned Surgery," *Journal of the History of Medicine and Allied Sciences* 72, no. 1 (2017): 34–50.

61. Cynthia Klestinec, "Renaissance Surgeons: Anatomy, Manual Skill and the Visual Arts," in *Early Modern Medicine and Natural Philosophy*, ed. Peter Distelzweig, Benjamin Goldberg, and Evan Ragland (Springer, 2016), 43–58.

62. Brunschwig, *Cirurgia*, fol. 10r: "uff dass er wisse wo er den menschen schniden oder etzen solle das im kein doetlicher schad von dem schniden oder atzunge uff erstande."

63. Hieronymus Brunschwig, *Das buch der wund Artzney. Handwirckung der Cirurgia von Jyeronimo brunschwick* (Strasbourg: Johann Grüninger, 1513), fol. 10v: "besehen vnd erfaren"; "die hut / fleisch vnn geeder nit geoffenbart werden mag durch die figur würt besehen durch die gesichte der doten körper durch schneidung oder sieden oder beissen in ungeleschtem kalck."

64. Gersdorff, *Feldtbuoch*, 13v: "in beysein ettlicher der gelerten vnd bewerten physicis / doctoribus / chirugicis vnd schereren . . . künstlich declariert durch den erfarnen vnn hochgelerten medicine doctorem Wendelinum hock von Brackenauw / und als bald abcontrafact verzeychnet mit aller gestalt / farben / und worer anzoeig wie du es in nochgonder figuren findest."

65. On the anatomical images in Gersdorff, see also Sachiko Kusukawa, *Picturing the Book of Nature: Image, Text, and Argument in Sixteenth-Century Human Anatomy and Medical Botany* (Chicago: University of Chicago Press, 2012), 8–15.

66. Katharine Park, *Secrets of Women: Gender, Generation, and the Origins of Human Dissection* (New York: Zone Books, 2006).

67. For slightly dated but serviceable overviews of anatomy, see Porter, *Greatest Benefit*, 176–86, and Wear, "Early Modern Europe," 264–92. For an excellent summary of more recent work, especially on dissection, see Hannah Murphy, *A New Order of Medicine: The Rise of Physicians in Reformation Nuremberg* (Pittsburgh: University of Pittsburgh Press, 2019), 76–80.

68. Kusukawa, *Picturing the Book of Nature*.

69. Kusukawa, *Picturing the Book of Nature*, 198–227.

70. Paolo Savoia, "Skills, Knowledge, and Status: The Career of an Early Modern Italian Surgeon," *Bulletin of the History of Medicine* 93, no. 1 (2019): 27–54.

71. Brunschwig, *Cirurgia*, fol. 128r: "dar vmm bitt ich üch gelerten vnn erfanden mit hohem fliß soelches vff zuo nemen im bestenn von mir / wann ich begerenn bin die lere von üch witter zuo erfarn."

72. Brunschwig, *Cirurgia*, fol. 43v: "wan es eim phisico me zu gehoert dan dem cirurgico."

73. Brunschwig, *Cirurgia*, fol. 36r: "darvmm ist etwas dar von zuo schribenn vnd alle vsseren ding die do sint zuo mercken mit der handt."

74. Brunschwig, *Cirurgia*, fol. 36r: "der cirurgicus zuo im nem den phisicus / soelche cur vnn gifftikeit ab zuo stellenn das hertz vnn krefft des menschen zuo behütten."

75. Brunschwig, *Cirurgia*, fol. 36r: "nümer kein guotter cirurgicus sin mag. er sy dan etwas in phisica gelert Des gelichen der phisicus er sy dan ouch in der cirurgi geleret / wan zum dickern mal eim fuor kumpt das dem andern gezimpt zuo tuon."

76. Brunschwig, *Cirurgia*, fol. 43v: "e du zu dem phisicus schickest / wer der mensch wol halben dott."

77. For a more detailed discussion of Brunschwig's audience claims, see Tillmann Taape, "Common Medicine for the Common Man: Picturing the 'Striped Layman' in Early Vernacular Print," *Renaissance Quarterly* 74, no. 1 (2021): 1–58.

78. This has been shown for the Württemberg area in the eighteenth century; see Sabine Sander, *Handwerkschirurgen* (Göttingen: Vandenhoeck & Ruprecht, 1989), 51.

79. Gersdorff, *Feldtbuoch*, fol. 18r: "sovil vnd im zuo gehoert. Dann weyter aspecten der planeten zuo ersuochen / oder anderer medicinen handreichungen gehoeren den astrologis vnn medicis zuo."

80. Gersdorff, *Feldtbuoch*, fol. 37v: "dz er vnn ich mit einander gebrucht haben."

81. Gersdorff, *Feldtbuoch*, fol. m3'v: "ein hundert glyd oder zwey abgeschnitten in sanct Anthonien hoff zuo Strasßburg." On amputation see also Hausse, *Malleable Body*, 51–81.

82. Gersdorff, *Feldtbuoch*, fol. 70v: "ein yeder barbierer wol selbs erkennen mag / vnnd rat darzuo erfunden vß geschicklicheit seiner verstaentnüß."

Chapter Three

1. Hannah Murphy, *A New Order of Medicine: The Rise of Physicians in Reformation Nuremberg* (Pittsburgh: University of Pittsburgh Press, 2019).

2. Sandra Cavallo, *Artisans of the Body in Early Modern Italy: Identities, Families and Masculinities* (Manchester: Manchester University Press, 2007). On surgery in Italian cities, see also Paolo Savoia, *Gaspare Tagliacozzi and Early Modern Surgery: Faces, Men, and Pain* (Routledge, 2020), 118–39. On Germany, Robert Jütte, "Zur Sozialgeschichte der Handwerkschirurgen im 16. Jahrhundert," in *Paracelsus und Salzburg*, ed. Heinz Dopsch and Peter Kramm (Salzburg: Gesellschaft für Salzburger Landeskunde, 1994), 45–60; Annemarie Kinzelbach, "Zur Sozial-und Alltagsgeschichte eines Handwerks in der frühen Neuzeit: Wundärzte und ihre Patienten in Ulm," *Ulm und Oberschwaben* 49 (1994): 111–44; Annemarie Kinzelbach, "Erudite and Honoured Artisans? Performers of Body Care and Surgery in Early Modern German Towns," *Social History of Medicine* 27 (2014): 668–88; Annemarie Kinzelbach, "Arztsein, Handwerk und Politik. Die Wund- und Stadtärzte, Barbiere, Bader, Chirurgen in Nördlingen," *Jahrbuch des Historischen Vereins für Nördlingen* 35 (2017): 249–90.

3. Martin Kintzinger, "Status medicorum: Mediziner in der städtischen Gesellschaft des 14. bis 16. Jahrhunderts," in *Städtisches Gesundheits-und Fürsorgewesen vor 1800*, ed. Peter Johanek (Cologne: Böhlau, 2000), 63–91.

4. For a late fifteenth-century example, see Manfred Gröber, *Das wundärztliche Manual des Meisters Hans Seyff von Göppingen (ca. 1440–1518): der Cod. med. et phys. 20 8 der Württembergischen Landesbibliothek Stuttgart* (Göppingen: Kümmerle, 1998), 23–24.

5. Rudolf Maurer (*Baden, schröpfen, amputieren: Die Geschichte der Bader in Baden bei Wien* [Vienna: Verlagshaus der Ärzte, 2004], 8–9) observes that the practice of bath masters and surgeons was indistinguishable in the small spa town of Baden near Vienna.

6. Kinzelbach, "Erudite and Honoured Artisans?" On tensions between bath masters and barbers, also see Gundolf Keil, "Die medizinische Versorgung durch Bader und Wundärzte zur Zeit des Paracelsus," *Salzburger Beiträge zur Paracelsusforschung* 29 (1995): 64–66. In eighteenth-century Württemberg, these categories become more codified in terms of examinations, and more clearly distinguished in medical ordinances, although their practice still overlapped with that of other trades. See Sabine Sander, *Handwerkschirurgen* (Göttingen: Vandenhoeck & Ruprecht, 1989), 54–58, 65–70.

7. Kinzelbach, "Erudite and Honoured Artisans?"

8. Robert Jütte, "Bader, Barbiere und Hebammen: Heilkundige als Randgruppen?," in *Randgruppen der spätmittelalterlichen Gesellschaft*, ed. Bernd-Ulrich Hergemöller (Warendorf: Fahlbusch Verlag, 1994), 90–121.

9. Archives de la ville et de la communauté urbaine de Strasbourg (hereafter AVCUS) III 11/19b, no.1, fol. 1r: "die Badere . . . hett*en*t Ir ettliche scherer knecht gedinget vnd hett*en*t die by Inen Inn Iren husern vnd Badtstuben vnd schürent also dar men*n*ige / do durch sie den scherern Inn Ir antwerck griff*en*."

10. AVCUS III 11/19b, no.1, fol. 1r: "den scherern Inn Ir antwerck"; "das doch dz halb antwerck wer."

11. AVCUS III 11/19b, no.1, fol. 1r: "in den stett*en* die do nit gezunfftet werent."

12. AVCUS III 11/19b, no.1, fol. 1r: "siech oder wunde lüte den möge*n*t sie die zit wol Rath thün."

13. Kristin Zech, "Le 'Schwörbrief' de 1482: L'origine et les conséquences de l'exclusion du Grand conseil pour les baigneurs de Strasbourg," *Revue d'Alsace*, no. 140 (2014): 59–78.

14. Martin Alioth, *Gruppen an der Macht: Zünfte und Patriziat in Strassburg im 14. und 15. Jahrhundert: Untersuchungen zu Verfassung, Wirtschaftsgefüge und Sozialstruktur* (Basel: Helbing & Lichtenhahn, 1988); Sabine von Heusinger, *Die Zunft im Mittelalter: zur Verflechtung von Politik, Wirtschaft und Gesellschaft in Straßburg* (Stuttgart: Steiner, 2009), 13–15.

15. Sabine von Heusinger, "The Topography of Sacred Space and the Representation of Social Groups—Confraternities in Strasbourg," in *Politics and Reformations: Communities, Polities, Nations, and Empires: Essays in Memory of Thomas A. Brady, Jr.*, ed. Thomas A. Brady and Christopher Ocker (Brill, 2007); Heusinger, *Die Zunft im Mittelalter*, 124–36.

16. Zech, "Schwörbrief."

17. Heusinger, *Die Zunft im Mittelalter*, 83–84, 154–55. On the new constitution

of 1482 and the dissolution of the guild of barbers and bath masters, see 208–11, and Alioth, *Gruppen an der Macht*, 404–21.

18. Jean-Yves Mariotte, *Les sources manuscrites de l'histoire de Strasbourg* (Strasbourg: Archives municipales, 2000); Philippe Dollinger, "L'évolution politique des corporations strasbourgeoises à la fin du Moyen Âge," in *Artisans et ouvriers d'Alsace* (Strasbourg: ISTRA, 1965), 127–35. For a recent overview of this argument, see also Zech, "Schwörbrief." On the dissolution of guilds, see Kristin Zech, "Zunftauflösungen als Spiegel politischer Partizipationschancen und -grenzen sozialer Gruppen in der Stadt: Straßburg, Colmar, Schlettstadt," in *Politische Partizipation in spätmittelalterlichen Städten am Oberrhein*, ed. Olivier Richard and Gabriel Zeilinger (Berlin: Erich Schmidt, 2017), 205–40.

19. Kathy Stuart, *Defiled Trades and Social Outcasts: Honor and Ritual Pollution in Early Modern Germany* (Cambridge: Cambridge University Press, 2006), 189–221.

20. Jütte, "Bader, Barbiere und Hebammen"; Kintzinger, "Status medicorum," 78.

21. Jütte, "Bader, Barbiere und Hebammen," 93–96.

22. Zech, "Schwörbrief."

23. Stuart, *Defiled Trades*, 33–66. See also Knut Schulz, *Handwerksgesellen und Lohnarbeiter: Untersuchungen zur oberrheinischen und oberdeutschen Stadtgeschichte des 14. bis 17. Jahrhunderts* (Sigmaringen: Thorbecke, 1985), 120–22.

24. The ordinance is edited in Jean-Charles Brucker, *Strassburger Zunft-und Polizei-Verordnungen des 14. und 15. Jahrhunderts* (Strasbourg: Trübner, 1889), 398–99. See also Stuart, *Defiled Trades*, 28.

25. Jütte, "Sozialgeschichte der Handwerkschirurgen"; Kinzelbach, "Erudite and Honoured Artisans?"

26. Gundolf Keil, "Seyff, Hans," *Neue Deutsche Biographie* 24 (2010): 295–96, http://www.deutsche-biographie.de/pnd100964559.html.

27. Gröber, *Das wundärztliche Manual des Meisters Hans Seyff*, 189–90; Daniel Carlo Pangerl, "Item als man dem kayser Fridrichen sin fuss abschnitt. Die Beinamputation an Kaiser Friedrich III. am 8. Juni 1493 in Linz," *Sudhoffs Archiv* 94, no. 2 (2010): 195–200.

28. Paolo Savoia, "Skills, Knowledge, and Status: The Career of an Early Modern Italian Surgeon," *Bulletin of the History of Medicine* 93, no. 1 (2019): 27–54.

29. This has been shown in detail for the inspection of lepers in early modern Nuremberg; see Fritz Dross, "Vom zuverlässigen Urteilen. Ärztliche Autorität, reichsstädtische Ordnung und der Verlust armer Glieder Christi in der Nürnberger Sondersiechenschau," *Medizin, Gesellschaft und Geschichte* 29 (2010). On these inspections and the medicalization of leprosy diagnosis in early modern Germany, see also Luke Demaitre, *Leprosy in Premodern Medicine: A Malady of the Whole Body* (Baltimore: Johns Hopkins University Press, 2007), 37–93.

30. AVCUS StA varia coll. 1 (no. 849), ed. in Otto Winckelmann, *Das Fürsorgewesen der Stadt Strassburg* (Leipzig: Heinsius, 1922), part 2, 14: "ob scherer die maledy besehen und darin glich mit den doctorn urteilen und sprechen sollen . . . wiewol sie etwass wissen davon haben."

31. Melanie Panse, *Hans von Gersdorffs "Feldbuch der Wundarznei": Produktion, Präsentation und Rezeption von Wissen* (Wiesbaden: Reichert, 2012), 31; Brucker, *Zunft-und Polizei-Verordnungen*, 78–79; Winckelmann, *Fürsorgewesen*, 27.

32. AVCUS 1MR 30, fol. 90r: "in allen ander*n* stetten vn*n* landen gewonheit ist ein

doctor vn*n* zwuen od*er* drig scherer"; "die wil dehein doctor dz begert zuo duon / so ist doch nit notturfft ein doctor zu haben / vil mer ein wund artzet dem solch kranckheit offenbarren ist."

33. AVCUS 1MR 30, fol. 90r: "dz wir wol versorgt wirden mit hanß von gerstdorff scherrer hind*er* dem münster der sin wol bericht vnd dar in erfaren ist"; "der erkentniß der vssern me dan der innern schad." On leprosy as a disease of the whole body, see Demaitre, *Leprosy in Premodern Medicine.*

34. The distinction between *wunden* and *blutrunsen* goes back at least as far as the city's law book of 1322, which, however, only states that the master of the council should see the wound; inspectors are not mentioned. See Wilhelm Wiegand, *Urkundenbuch der Stadt Strassburg*, vol. 4 (Strasbourg: Karl Trübner, 1898), 53–93.

35. AVCUS 1MR3, fol. 154v: "die cleynen verstandt solcher dinge gehept."

36. AVCUS 1MR3, fol. 154v: "die sie zum geschicksten erachten."

37. AVCUS 1MR3, fol. 154r: "nit wol oder füglich gesetzt vnnd gemacht werden mag"; "die an eym ort vil schädlicher sindt dann an dem andern"; "erkandtniß der wunden beseher."

38. Hieronymus Brunschwig, *Dis ist das buch der cirurgia* [Cirurgia]. *Hantwirckung der wund artzny* (Strasbourg: Johann Grüninger, 1497), fol. 14r: "verlierung der glider."

39. Brunschwig, *Cirurgia*, fol. 14r: "der er doch gantz nottorftig wer zuo seiner hantyerung."

40. Brunschwig, *Cirurgia*, fol. 16r: "wz hantwercks oder gewerbes er ist."

41. Brunschwig, *Cirurgia*, fol. 16r–v: "ein schmidt wagner zimerman vnnd was mit den sweren hamer / oder werckgezeug arbeyt. vnn der geleichen die do mit dem gantzen arm arbeiten sind noturftig der achselen bewegung vnn des oebersten geleichs wo das versert würt. das ist den selbigen ein grosse lemung vnn beroubung eins glides."

42. Brunschwig, *Cirurgia*, fol. 15r: "Ist dan der Cirurgicus guot so ist es eyn cleine wund ist er aber vnerfarend so ist es ein grosse wund."

43. Brunschwig, *Cirurgia*, fol. 63v: "wann es ist ein glidt das do vast gebrucht vnd dem menschen noturfftig ist. besunder etlichen hantwercken. als schmid wagner. zimmerman. vnd der gelichen die do arbeitten mit dem gantzen arm."

44. Hieronymus Brunschwig and Marsilio Ficino, *Liber de arte distulandi Simplicia et Composita* [Medicinarius]. *Das nüv buoch der rechten kunst zuo distillieren. Ouch von Marsilio Ficino vnn anderen hochberoempten Ertzte natürliche vnn guote künst* (Strasbourg: Johann Grüninger, 1509), fol. 36v: "das artzney edler sei wan alle andere künst das sicht man wol. Würt ein meister siech / dz sein kunst nicht gehelffen mag / darumb begert er der artznei das er sein kunst vnn sein leben da mit behalt."

45. Brunschwig, *Cirurgia*, fol. 15r: "ein wund dar zuo ein ewig laster wan sye verdacht werden Ebruchs halben vnderstanden abgeschniten zü werden."

46. Brunschwig, *Cirurgia*, fol. 15r: "er mag verdacht werden durch die backen gebrent sy."

47. For an exploration of similar questions in relation to women's appearance see Jill Burke, *How to Be a Renaissance Woman: An Untold History of Beauty and Female Creativity* (London: Profile Books, 2023).

48. Luke Demaitre, "Skin and the City: Cosmetic Medicine as an Urban Concern," in *Between Text and Patient: The Medical Enterprise in Medieval and Early Modern Europe*, ed. Florence Glaze and Brian Nance (Florence: SISMEL, 2011), 97–120; Mar-

garet Pelling, "Appearance and Reality: Barber-Surgeons, the Body and Disease," in *London 1500–1700: The Making of the Metropolis*, ed. A. L. Beier and Roger Finlay (London: Longman, 1986), 89–90; Savoia, *Tagliacozzi*, 118–39.

49. Savoia, *Tagliacozzi*, 92–98.

50. Savoia, *Tagliacozzi*.

51. Brunschwig, *Cirurgia*, fol. 17v: "an sim lib / an siner er / gesuntheit / an siner hantierung."

52. Brunschwig, *Cirurgia*, fol. 14r: "O wie dick hon ich gesehen eynen vnherfarnen Cyrurgycus eynen binden. den ich dann für einen wunden müst erkenen. het yn ein anderer erfarender Cyrrurgicus oder wund Artzet zuo dem ersten gebunden Es wer ein bluotrunß gewessen danocht nit fast groß."

53. Brunschwig, *Cirurgia*, fol. 14r: "alle vnd och yetliche wund Ertzet"; "üwere Eer vnd aller wund ertzot Eer"; "gebent im yeman zuo der da wüße vnd verste die kunst."

54. AVCUS 1MR 3, fol. 63r: "das hynfürtter alle scherer vnnd Ire knecht by Iren eyden, so Inen ein solicher gewundeter oder geblütrunseter heimbrocht oder kom*m*ett oder sie zü eim solichen geholet werden / den selben verwundten / vnnd den oder die Ime ein solichs gethon / mit Iren namen vnnd Zünamen (ob sie die erfaren mögen) Eim Am*m*eister ye zü zitten angeben sollen."

55. AVCUS III 11/19b no. 2, fol. 1r: "ein ied*en* scherer gegen sin pacienten oder verwunthen*n* nit anders zuachten sin dan*n* ein wirth gege*n* sin gast"; "ein trostlich zuflucht vnd vertruwen*n* zu im hat"; "so geburte eim vil bas dem dether . . . hinweg zuv*er*helffen." The document is not dated but refers verbatim to several aspects of the Rat's 1517 decree in AVCUS 1MR 3 fol. 63r.

56. AVCUS III 11/19b no. 2, fol. 1v: "dwÿl wir vns mit des gemeinenn mans gunst begon miessen."

57. AVCUS III 11/19b no. 2, fol. 1v: "blotter oder sunst wund erzt vnd alten wÿberen"; "vngeschulte heilung."

58. AVCUS III 11/19b no. 2, fol. 1v: "die scherer knecht in frembd*en* land*en* do sollich riegen villicht nit sithlich auch in laster kum*m*en mochten*n*."

59. On journeymen confraternities' control of the local work market, see Sabine von Heusinger, "Die Handwerksbruderschaften in Straßburg," in *Städtische Gesellschaft und Kirche im Spätmittelalter*, vol. 62, ed. Sigrid Schmitt and Sabine Klapp (Stuttgart: Steiner, 2008), 123–40; Heusinger, *Die Zunft im Mittelalter*, 85–90. On honor policing, see Celeste Chamberland, "Honor, Brotherhood, and the Corporate Ethos of London's Barber-Surgeons' Company, 1570–1640," *Journal of the History of Medicine and Allied Sciences* 64, no. 3 (2009): 300–332; Stuart, *Defiled Trades*, 207–12.

60. AVCUS III 11/19b no. 2, fol. 1v: "vns das broth vor de*m* mund heimlich abgeschnitte*n* wurd."

61. Michael McVaugh, "Bedside Manners in the Middle Ages," *Bulletin of the History of Medicine* 71, no. 2 (1997): 201–23.

62. Hans von Gersdorff, *Feldtbuoch der wundtartzney* (Strasbourg: Johann Grüninger, 1517), fol. 19v: "dardurch im ein geschrey / oder ein boeßer ruoff enstand"; Brunschwig, *Cirurgia*, fol. 9v: "boeser loemet"; "schand vnd laster."

63. Lanfranco di Milano, *Kleyne Wundartznei des hochberuempten Lanfranci / auß fürbit des wol erfarnen M. Gregorii Flüguß / Chyrurgen vnn Wundartzt zuo Straßburg / durch Othonem Brunfels verteutscht*, trans. Otto Brunfels (Strasbourg: Christian Egenolff, 1528), fol. A1v: "Es haben vnsere vorfaren die Alten / so da beschriben die Policeien . . . gesagt / Wann man in einer Statt habe guotte Oberkeiten / guotte

seelsorger / aertzet / Schulmeyster / vnd Hebammen / so sei solich policei eygentlich wol versorget."

64. Lanfranco, *Kleyne Wundartznei*, fol. A1v: "Aber meins bedunckens / wol noch moechten darzuo gethan werdenn die Wundaertzet"; "gemeynenn nutz . . . nit allein zuo der zeit des Kriegs / sonder auch in vilen andern Siechtagen der menschen."

65. Walther Hermann Ryff, *Die kleyner Chirurgi D. Gualtheri H. Ryff: Das ist, der grund vnnd kern gemeyner eynleytung der gantzen wundartzney* (Strasbourg: Balthasar Beck, 1542), fol. A2r: "wie nützlich vnd hoch von noeten die wundartzney sey . . . zuoerhalten des gemeynen nutzes rechte ordentlich Pollicey."

66. J. Andrew Mendelsohn, Annemarie Kinzelbach, and Ruth Schilling, *Civic Medicine: Physician, Polity, and Pen in Early Modern Europe* (Abingdon: Routledge, 2020).

67. For provenance of a range of Gersdorff copies, see Panse, *Hans von Gersdorffs Feldbuch der Wundarznei*, 178–212.

Chapter Four

1. D. Goltz, "Aber erschrocklich ist es dem Menschenn. Beziehungen zwischen Chirurgie und Pharmazie im Mittelalter," *Pharmazeutische Zeitung* 123, no. 7 (1978): 278–84. The importance of pharmaceutical expertise in surgery is discussed in the chapters by Sandra Cavallo and Michael Stolberg in "Learning to Cut and Mend: Surgical Training and Practice, 1450–1800," ed. Maria Pia Donato, Elaine Leong, and Tillmann Taape (London: UCL Press, under review).

2. Michael McVaugh, *The Rational Surgery of the Middle Ages* (Florence: Sismel, 2006), 184–215.

3. Lorenz Fries, *Spiegel der Artzny des geleichen vormals nie von keinem doctor in tütsch vßgangen* [. . .] (Strasbourg: Johann Grüninger, 1518), fol. 14r: "apotecker . . . dem die armen lüt für sein hauß lauffen wenn sie wöllen fisch kauffen."

4. Only four pharmacy shops were allowed to operate at any one time in Strasbourg. See Peter Dilg, "Die Strassburger Apothekerordnung (um 1500)," *Pharmazeutische Zeitung* 124 (1979): 2664–70.

5. Pierre Bachoffner, "Jérôme Brunschwig, chirurgien et apothicaire strasbourgeois, portraituré en 1512," *Revue d'histoire de la pharmacie* 81, no. 298 (1993): 269–78.

6. Dilg, "Strassburger Apothekerordnung," 2668.

7. Archives de la ville et de la communauté urbaine de Strasbourg (hereafter AVCUS) 1AH 112, fol. 123v: "die dan*n* die wasser Inn dem Spittal distilieren werden."

8. Hieronymus Brunschwig, *Liber de arte distillandi de simplicibus* [Small book of distillation]. *Das buch der rechten kunst zü distilieren die eintzigen ding* (Strasbourg: Johann Grüninger, 1500), fol. 9v: "als etliche closter frowen brennent / als ich zuo Straßburg selber gesehen hab."

9. Hieronymus Brunschwig, *Liber de arte Distillandi de Compositis* [Large book of distillation]. *Das buch der waren kunst zü distillieren die Composita vnn simplicia / vnd dz Buoch thesaurus pauperum / Ein schatz der armen* (Strasbourg: Johann Grüninger, 1512), fol. 39r: "In dem man gemeinklich den wein brenen sol."

10. Nicolas Thomas, "L'alambic dans la cuisine?" in *La cuisine et la table dans la France de la fin du Moyen-Âge. Contenus et contenants du XIVe au XVIe siècle*, ed. Fabienne Ravoir and Anne Dietrich, 35–50 (Caen: CRAHM, 2009).

11. The *Thesaurus pauperum* by Petrus Hispanus, who may be the same person who

later became pope John XXI. Lynn Thorndike, *A History of Magic and Experimental Science*, vol. 2 (New York: Columbia University Press, 1923), 488–98.

12. Robert Multhauf, *The Origins of Chemistry* (London: Oldbourne, 1966), 82–175; Lawrence M. Principe, *The Secrets of Alchemy* (Chicago: University of Chicago Press, 2013), 37–46, 69–70.

13. These are often conflated with distillers of aquavit or other strong spirits rather than simple distilled waters. See, e.g., Robert Forbes, *A Short History of the Art of Distillation: From the Beginnings up to the Death of Cellier Blumenthal* (Leiden: Brill, 1948)., 55–98.

14. Alisha Rankin, *Panaceia's Daughters: Noblewomen as Healers in Early Modern Germany* (Chicago: University of Chicago Press, 2013); Wendy Wall, "Distillation: Transformations in and out of the Kitchen," in *Renaissance Food from Rabelais to Shakespeare: Culinary Readings and Culinary Histories*, ed. Joan Fitzpatrick, 89–104 (Farnham: Ashgate, 2010).

15. Robert Multhauf, "The Significance of Distillation in Renaissance Medical Chemistry," *Bulletin of the History of Medicine* 30, no. 4 (1956): 329–46. For a highly readable introduction to Paracelsus, see Bruce T. Moran, *Paracelsus: An Alchemical Life* (2019). On Fioravanti, William Eamon, *The Professor of Secrets: Mystery, Medicine, and Alchemy in Renaissance Italy* (Washington, DC: National Geographic Society, 2010).

16. Allen G. Debus, *The Chemical Philosophy: Paracelsian Science and Medicine in the Sixteenth and Seventeenth Centuries* (New York: Neale Watson, 1977), 63–126; Bruce Moran, *Distilling Knowledge: Alchemy, Chemistry, and the Scientific Revolution* (Cambridge, MA: Harvard University Press, 2005), 67–98; W. Newman, "From Alchemy to 'Chymistry,'" in *The Cambridge History of Science*, vol. 3, *Early Modern Science*, ed. L. Daston and K. Park (Cambridge: Cambridge University Press, 2006), 497–515; Principe, *The Secrets of Alchemy*, 107–36.

17. Tillmann Taape, "Trade and Industry: Chemical Economies and the Business of Distillation," in *A Cultural History of Chemistry in the Early Modern Age*, ed. Bruce Moran (London: Bloomsbury, 2021).

18. Leah DeVun, *Prophecy, Alchemy, and the End of Time: John of Rupescissa in the Late Middle Ages* (New York: Columbia University Press, 2009); Robert Multhauf, "John of Rupescissa and the Origin of Medical Chemistry," *Isis* 45, no. 4 (1954): 359–67; Michela Pereira, "Heavens on Earth. From the *Tabula Smaragdina* to the Alchemical Fifth Essence," *Early Science and Medicine* 5 (2000): 131–44; Michela Pereira, "Mater Medicinarum: English Physicians and the Alchemical Elixir in the Fifteenth Century," in *Medicine from the Black Death to the French Disease*, ed. Roger Kenneth French et al. (Aldershot: Ashgate, 1998), 26–52; Jennifer M. Rampling, "Depicting the Medieval Alchemical Cosmos: George Ripley's Wheel of Inferior Astronomy," *Early Science and Medicine* 18, no. 1–2 (2013): 45–86.

19. Forbes, *Short History of Distillation*; Astrid Müller-Grzenda, *Pflanzenwässer und gebrannter Wein als Arzneimittel zu Beginn der Neuzeit* (Stuttgart: Deutscher Apotheker Verlag, 1996); Gerhard Pfeiffer, "Technologische Entwicklung von Destilliergeräten vom Spätmittelalter bis zur Neuzeit," 1986.

20. Brunschwig, *Small Book*, fol. C1r: "das subtyl von dem groben / vnd das grob von dem subtilen zuo scheiden."

21. Brunschwig, *Small Book*, fol. C2v: "on kosten," "mit kosten vnd durch das füer."

22. Brunschwig, *Small Book*, fol. 6v: Brunschwig, fol. 6v: "wasser wyn oder safft dz man begeren ist zescheiden von syner drüben vnn vnlutern substantz."

23. Brunschwig, *Small Book*, fol. 7r: "besunder die gebranten wasser so man sie zum andern mal rectificieren . . . will."

24. Müller-Grzenda, *Pflanzenwässer*, 64–67.

25. Brunschwig, *Small Book*, fol. 7v: "vff vnd ab zuo stigen / do durch sie rectificiert werden / vnn gebessert in alle weg."

26. Brunschwig, *Small Book*, fol. 7v: "al zit stigen in der mitten des glases übersich vff / louffende in den armen wider her ab / oder von einem glase in das ander / also wurt es vas schoen / vnn groesser würckung."

27. Brunschwig, *Small Book*, fol. C4r: "putrifficieren vnn digerieren."

28. Brunschwig, *Large Book*, fol. 31r: "underscheidt zwischen digerieren vn circulieren."

29. Brunschwig, *Large Book*, fol. 31v: "vff das das dritteil des glases hussen stand / vnd von dem lufft oder kelte wider abfal / was durch die hitz vnden vbersich vff gestigen ist."

30. DeVun, *Prophecy, Alchemy, and the End of Time*; Multhauf, "John of Rupescissa and the Origin of Medical Chemistry."

31. Brunschwig, *Large Book*, fol. 13v: "wan ie me dz gedistilliert würt / ie mer suptiliert vnd spiritualisiert vnmatterlicher / das ist darumb es füer drefflicher würt."

32. Brunschwig, *Large Book*, fol. 18r: "So man sehen ist durch ein vff und ab stygen der distillation das grob vnd vnlauter so vast gescheiden vnd gesubtiliert . . . Was würt dann geschehen so es tausent mal vnn noch vil mer vff vnd ab styget durch das distillatorium / on zwyfel / das es kumpt zuo so grosser hoher loeblicheit . . . als es ab gescheiden von dem versamleten."

33. Brunschwig, *Small Book*, fol. C1v: "die wasser die do gescheyden sint von dem groben / als von dem corpus die krafft in sel eim yeden vßgezogen vnd das fürbas zuo schicken an das ende do es mer nutz vnnd heilsamkeit bringen ist / als ich her nach zeigen will."

34. Brunschwig, *Small Book*, fol. C2r: "materialischen teil anhangend etlich teil des menschlichen lyb als an dem magen / lebern oder gedermen im die inschnident vnd schindent vnd weich oder bloede machent oder in etlich ander wyse schmertzen vnn pyn infüerent."

35. Brunschwig, *Small Book*, fol. C2r: "ye doch so ist dennocht sorg zehaben irer substantz darin vermyschet blibt."

36. Brunschwig, *Small Book*, fol. C2v: "ein sollich füchte der inbeissung an vilen ertznyen nympt offenbarlich die farb der medicin. dar vmb sie etwas an sich nemen ist die substantz von dem corpus . . . wann warumb die farb ist ein lyplich qualitet."

37. Brunschwig, *Small Book*, fol. C2v: "mer vermyscht wurt von dem corpus der ertzeny in die kochung."

38. Brunschwig, *Small Book*, fol. C2v: "wie vast sie gelütert wuort über alle die vorgenanten sinnlicheit schaden zuo vermyden der subtantz des corpus dem menschlichen lyp / würt dennocht bewert die wort Auicenne vnd Ypocratis / also das die medicin schwecht vnd macht alt wie wol sie gerchtfertiget vnn gedeissempt / vnd durch die inbeissung gegeben sy."

39. Brunschwig, *Small Book*, fol. C2v: "den wegk funden das grob vnd die irdesche substantz zescheiden von dem subtylen durch dise würckung die genant ist die hantwürckung der distillierung."

40. On movement and blockage in ancient medicine, see Vivian Nutton, *Ancient Medicine* (London: Routledge, 2013), 123–24, 171.

41. Mikhail Bakhtin, *Rabelais and His World* (Bloomington: Indiana University Press, 1984), 355–61.

42. Michael Stolberg, "Der gesunde Leib: Zur Geschichtlichkeit frühneuzeitlicher Körpererfahrung," in *"Erfahrung" als Kategorie der Frühneuzeitgeschichte*, ed. Paul Münch (München: Oldenbourg, 2001), 37–57; Michael Stolberg, *Experiencing Illness and the Sick Body in Early Modern Europe* (Springer, 2011), 95–100, 125–34; Alisha Rankin, "Duchess, Heal Thyself: Elisabeth of Rochlitz and the Patient's Perspective in Early Modern Germany," *Bulletin of the History of Medicine* 82, no. 1 (2008): 109–44.

43. Ulinka Rublack, "Fluxes: The Early Modern Body and the Emotions," *History Workshop Journal* 53, no. 1 (2002): 1–16; Olivia Weisser, "Grieved and Disordered: Gender and Emotion in Early Modern Patient Narratives," *Journal of Medieval and Early Modern Studies* 43, no. 2 (2013): 247–73.

44. Michael Stolberg, "You Have No Good Blood in Your Body: Oral Communication in Sixteenth-Century Physicians' Medical Practice," *Medical History* 59 (2015): 63–82.

45. Brunschwig, *Small Book*, fol. C1v: "durch sin subtilithet dester lichter darzuo behender dringen vnd penetrieren mag mit siner tugende vnn krafft . . . in dem menschlichen lyb."

46. Gianna Pomata, *Contracting a Cure: Patients, Healers, and the Law in Early Modern Bologna* (Baltimore: Johns Hopkins University Press, 1998), 129–39.

47. Nancy Siraisi, *Medieval and Early Renaissance Medicine* (Chicago: University of Chicago Press, 1990), 80–81.

48. Sergius Kodera, "The Art of the Distillation of 'Spirits' as a Technological Model for Human Physiology: The Cases of Marsilio Ficino, Joseph Duchesne and Francis Bacon," in *Blood, Sweat and Tears: The Changing Concepts of Physiology from Antiquity into Early Modern Europe*, ed. Manfred Horstmanshoff, Helen King, and Claus Zittel (Leiden: Brill, 2012).

49. Walther Hermann Ryff, *Das New groß Distillier Buoch Wolgegründter Künstlicher Distillation* (Frankfurt: Christian Egenolff, 1545), fol. 2v: "auffgetriebenc daempff . . . rotz oder Phlegma," "durch die naßlöcher."

50. Adam Lonicer, *Kreuterbuoch Künstliche Conterfeytunge der Baeume / Stauden / Hecken / Kreuter / Getreyde / Gewürtze* (Frankfurt: Egenolff heirs, 1564), fol. 1r: "von der feurstatt der lebern / die pfann oder capell des magens erhitzigt / vnd die innerlichen daempff durch natürliche oder eusserliche zuofaellige hitz / ins haupt vnd hirnschal / als ein Distillierhelm / auffsteigen / vnd durch kaelte desselbigen zusammen getriben / zu rotz vnd schleim werden in die nase / als den schnabel des Distillierhelms vnd hals / wiederumb herab fallen durch schnupffen / huosten / vnnd reuspern / außgefegt muoß werden."

51. Walter Pagel, *Das medizinische Weltbild des Paracelsus: Seine Zusammenhänge mit Neuplatonismus und Gnosis* (Wiesbaden: Steiner, 1962), 190–200.

52. Walter Pagel, *Paracelsus: An Introduction to Philosophical Medicine in the Era of the Renaissance* (Basel: Karger, 1958), 192–94.

53. Rublack, "Fluxes," 4–6.

54. Pagel, *Paracelsus*, 195–96; Pagel, *Das medizinische Weltbild des Paracelsus*, 18–20; Michael Stolberg, *Uroscopy in Early Modern Europe* (Farnham: Routledge, 2016), 66–70.

55. Kodera, "Art of the Distillation of Spirits."

56. Charles Webster, *Paracelsus: Medicine, Magic and Mission at the End of Time* (New Haven: Yale University Press, 2008), 55.

57. Brunschwig, *Small Book*, fol. C1r–v: "das vnlieplich lieplicher zuo behalten."

58. Brunschwig, *Small Book*, fol. C1v: "in einer lieplichen trincklichen gestalt."

59. Brunschwig, *Small Book*, fol. C1v: "den gynnen die me lust habent in dem tranck oder wasser dann in dem essen / oder keyn anderley ertzny."

60. AVCUS 1AH 112, fol. 124r: "ein schühen oder grugel hetten vnnd nit nemmen oder das zubruchen guttlichen sich vberreden lassen wolten / So soll er die by sinem eyde noch niemans anders darzu tringen."

61. The term was first coined by Harold Cook, *The Decline of the Old Medical Regime in Stuart London* (Ithaca, NY: Cornell University Press, 1986). See also Harold Cook, "Good Advice and Little Medicine: The Professional Authority of Early Modern English Physicians," *Journal of British Studies* 33, no. 1 (1994): 1–31; Harold Cook, *Matters of Exchange: Commerce, Medicine, and Science in the Dutch Golden Age* (New Haven: Yale University Press, 2007), 133–74.

62. Margaret Pelling, *Medical Conflicts in Early Modern London: Patronage, Physicians, and Irregular Practitioners, 1550–1640* (Oxford: Oxford University Press, 2003), 342–43.

63. On reappraisals of the "marketplace" as an analytic concept, see Mark Jenner and Patrick Wallis, *Medicine and the Market in England and Its Colonies, c. 1450–c. 1850* (Basingstoke: Palgrave Macmillan, 2007).

64. Moran, *Distilling Knowledge*, 37–66.

65. Jon Arrizabalaga, John Henderson, and Roger French, *The Great Pox: The French Disease in Renaissance Europe* (New Haven: Yale University Press, 1997), 20–32; Roger Arrizabalaga French, "Coping with the French Disease: University Practitioners' Strategies and Tactics in the Transition from the Fifteenth to the Sixteenth Century," in French et al., *Medicine from the Black Death to the French Disease*, 248–87.

66. Ryff, *New groß Distillier Buoch*, fol. 1v: "Zum andern sindt auch solcke künstliche abzüge der Distillation zuom theyl erfunden worden / darmit man die gifftigkeyt vnd schaedliche boßheyt / new auffkommener kranckheyt / mit new erfundener artzenei Curiert / wie solches vilfeltige exempel allein bezeugen in dem gifftigen gebrechen der Frantzosen."

67. Harold Cook, "The New Philosophy and Medicine in Seventeenth-Century England," in *Reappraisals of the Scientific Revolution*, ed. David Lindberg and Robert Westman (Cambridge: Cambridge University Press, 1990), 397–436.

68. Paracelsus, *An Excellent Treatise Teaching Howe to Cure the French-Pockes*, trans. John Hester (London: Charlwood, 1590), fol. π3r.

69. William Eamon, "Alchemy in Popular Culture: Leonardo Fioravanti and the Search for the Philosopher's Stone," *Early Science and Medicine* 5 (2000): 196–213; William Eamon, "Pharmaceutical Self-Fashioning or How to Get Rich and Famous in the Renaissance Medical Marketplace," *Pharmacy in History* 45, no. 3 (2003): 123–29.

70. Christoph Friedrich and Wolf-Dieter Müller-Jahncke, *Geschichte der Pharmazie von der frühen Neuzeit bis zur Gegenwart* (Eschborn: Govi-Verlag, 2005), 80–85. On householders as practitioners, see Gundolf Keil, "Der Hausvater als Arzt," in *Haushalt und Familie in Mittelalter und früher Neuzeit*, ed. Trude Ehlert (Sigmaringen: Thor-

becke, 1991), 219–43; Rudolf Schenda, "Der 'gemeine Mann' und sein medikales Verhalten im 16. und 17. Jahrhundert," in *Pharmazie und der gemeine Mann: Hausarznei und Apotheke der frühen Neuzeit*, ed. Joachim Telle (Braunschweig: Waisenhaus Verlag, 1982), 9–20.

71. Wall, "Distillation: Transformations in and out of the Kitchen."

72. Elaine Leong, "Collecting Knowledge for the Family: Recipes, Gender and Practical Knowledge in the Early Modern English Household," *Centaurus* 55 (2013): 81–103; Elaine Leong, *Recipes and Everyday Knowledge: Medicine, Science, and the Household in Early Modern England* (Chicago: University of Chicago Press, 2018).

73. Elaine Leong, "Making Medicines in the Early Modern Household," *Bulletin of the History of Medicine* 82 (2008): 145–68.

74. Rankin, *Panaceia's Daughters*, 2–17, 93–127.

75. Alisha Rankin, "How to Cure the Golden Vein: Medical Remedies as Wissenschaft in Early Modern Germany," in *Ways of Making and Knowing*, ed. Pamela H. Smith, Harold Cook, and Amy Meyers (Ann Arbor: University of Michigan Press, 2014), 109–33.

76. Eamon, "Alchemy in Popular Culture"; Friedrich and Müller-Jahncke, *Geschichte der Pharmazie*, 93; Moran, *Distilling Knowledge*, 54–55; Maria Roxo Beltran, "Sixteenth Century Books of Distillation," in *Between the Natural and the Artificial: Dyestuffs and Medicines*, ed. Gérard Emptoz and Patricia Aceves Pastrana, 145–55 (Turnhout: Brepols, 2000).

77. Eucharius Röslin, *Kreutterbuoch von Allem Erdtgewaechs* (Frankfurt: Egenolff, 1533), fol. 1v: "edel buechlin . . . für den gemeynen man"; "das Distillier buoch Hieronymi Braunschwig / welches auch nit kleyns lobs wirdig."

78. Röslin, *Kreutterbuoch von Allem Erdtgewaechs*, fol. 1v: "disen Winter / vnden andern gescheфften überlesen"; Lonicer, *Kreuterbuoch*, fol. aa4r: "solches Buoch ist bei dem gemeinen einfaltigen Man gar angenem / vnn kaeuflich gewesen / daß es offtmals im Truck ernewert worden."

79. Ryff, *New groß Distillier Buoch*, fol. 3v: "vilfaltigen schrifftarbeyten . . . zu gmeyner wolfart am nützlichsten."

80. On readers of how-to literature, see William Eamon, *Science and the Secrets of Nature: Books of Secrets in Medieval and Early Modern Culture* (Princeton, NJ: Princeton University Press, 1994), 93–133; Pamela H. Smith, *From Lived Experience to the Written Word: Reconstructing Practical Knowledge in the Early Modern World* (Chicago: University of Chicago Press, 2022), 151–76. On readers as connoisseurs, see also Sven Dupré, "The Value of Glass and the Translation of Artisanal Knowledge in Early Modern Antwerp," in *Trading Values in Early Modern Antwerp*, ed. Christine Göttler, Bart Ramakers, and Joanna Woodall (Leiden: Brill, 2014), 138–61.

81. Conrad Gessner, *Thesaurus Euonymi philiatri De remediis secretis, liber physicus, medicus et partim etiam chymicus, & oeconomicus* (Zurich: Gessner, 1552), 6: "lingua nostra Germanica conscripta & euulgata primum."

82. Gessner, *Thesaurus Euonymi*, 24: "Brunsuicensis & Ryffii scripta breuius comprehenderet."

83. Tillmann Taape, "Distilling Subtlety: Small-Scale Materialities in Medicine, Craft Knowledge, and the Body," in *Microscopic Records: The New Interdisciplinarity of Early Modern Studies, c. 1400–1800*, ed. Stefan Hanss (UCL Press, in preparation).

84. Gessner, *Thesaurus Euonymi*, 51: "Fimi calor (propter calcem admixtam) medio gradu caliditatem Balnei M. excedere iudicatur à Brunsvicensi."

Chapter Five

1. Pamela H. Smith, *From Lived Experience to the Written Word: Reconstructing Practical Knowledge in the Early Modern World* (Chicago: University of Chicago Press, 2022). See also Pamela H. Smith, *Secrets of Craft and Nature in Renaissance France: A Digital Critical Edition and English Translation of BnF Ms. Fr. 640* (New York: Making and Knowing Project, 2020), https://edition640.makingandknowing.org/.

2. Michela Pereira, "Heavens on Earth: From the *Tabula Smaragdina* to the Alchemical Fifth Essence," *Early Science and Medicine* 5 (2000): 131–44; Lawrence M. Principe, *The Secrets of Alchemy* (Chicago: University of Chicago Press, 2013), 9–26; Hans-Werner Schütt, "Die frühe Alchemie und die Technik der Distillation," in *Brückenschläge: 25 Jahre Lehrstuhl für Geschichte der exakten Wissenschaften und der Technik an der Technischen Universität Berlin: 1969–1994*, ed. Hans-Werner Schütt and Burghard Weiss (Berlin: Engel, 1995), 293–309.

3. Leah DeVun, *Prophecy, Alchemy, and the End of Time: John of Rupescissa in the Late Middle Ages* (New York: Columbia University Press, 2009); Bruce Moran, *Distilling Knowledge: Alchemy, Chemistry, and the Scientific Revolution* (Cambridge, MA: Harvard University Press, 2005), 8–36; Robert Multhauf, *The Origins of Chemistry* (London: Oldbourne, 1966), 201–36; Pereira, "Heavens on Earth"; Principe, *Secrets of Alchemy*, 51–82.

4. See, e.g., Sven Dupré, ed., *Laboratories of Art: Alchemy and Art Technology from Antiquity to the 18th Century* (Cham: Springer, 2014); Elaine Leong, "Making Medicines in the Early Modern Household," *Bulletin of the History of Medicine* 82 (2008): 145–68; Pamela Long, *Artisan/Practitioners and the Rise of the New Sciences, 1400–1600* (Corvallis: Oregon State University Press, 2011); Alisha Rankin, *Panaceia's Daughters: Noblewomen as Healers in Early Modern Germany* (Chicago: University of Chicago Press, 2013); Pamela H. Smith, *The Body of the Artisan: Art and Experience in the Scientific Revolution* (Chicago: University of Chicago Press, 2004); Pamela H. Smith, Harold Cook, and Amy Meyers, *Ways of Making and Knowing* (Ann Arbor: University of Michigan Press, 2014).

5. Hieronymus Brunschwig, *Liber de arte distillandi de simplicibus* [Small book of distillation]. *Das buch der rechten kunst zü distilieren die eintzigen ding* (Strasbourg: Johann Grüninger, 1500), fol. C1v: "ob du krüter wurtzeln oder ander substantz stossest / vnd das safft dar vß druckest / so würt es vnlieplich dar zuo ouch nit bliblichen vnd zerstoerlichen wann durch syne grobe vnnd vngelüterte substantz bald dar nach volgen ist eyne fule zerstoerlicheit."

6. Brunschwig, *Small Book*, fol. C1v: "das vnbehalten zuo behalten werden mag. vnd das grob gelütert würt von siner fülung vnd zerstoerlicheit."

7. Hieronymus Brunschwig and Marsilio Ficino, *Das buch der Gesuntheit* [Medicinarius]. *Liber de arte distillandi Simplicia et composita. Das nüv buoch der rechten kunst zuo distillieren. Ouch von Marsilio Ficino vnn andere hochberoepmter Artzte natürliche vnn gute künst* (Strasbourg: Johann Grüninger, 1505), fol. Ddd4v: "Wan du sichst in dem mayen so es zuo Druckenn noch zuo fücht noch zuo kalt noch zuo warm ist."

8. Brunschwig and Ficino, *Medicinarius*, 1505, fol. Ddd4v: "also sag ich dir von den Menschen / wo eins dy fier Element glych hat so ist er gesund."

9. Nancy Siraisi, *Medieval and Early Renaissance Medicine* (Chicago: University of Chicago Press, 1990), 101–4.

10. This analogy is epitomized in a series of late seventeenth-century engravings;

see *The Four Seasons of Human Life: Four Anonymous Engravings from the Trent Collection* (Rotterdam: Erasmus, 2002), 58–59.

11. Tillmann Taape, "Distilling Reliable Remedies: Hieronymus Brunschwig's *Liber de Arte Distillandi* (1500) Between Alchemical Learning and Craft Practice," *Ambix* 61, no. 3 (2014): 236–56.

12. Pereira, "Heavens on Earth," 136–40.

13. John of Rupescissa, *Ioannis de Rupescissa . . . de consideratione Quintae essentiae rerum omnium* (Basel: anon., 1561), 18: "omnem excessum cuiuscunque qualitatis amputet, & quamcunque qualitatem deperditam restauret."

14. On Rupescissa's concept of quintessence, see DeVun, *Prophecy, Alchemy, and the End of Time*, 60–71; Robert Multhauf, "John of Rupescissa and the Origin of Medical Chemistry," *Isis* 45, no. 4 (1954): 359–67; Multhauf, *Origins*, 210–12; Pereira, "Heavens on Earth." On the relationship between heaven and earth in alchemy, also see Jennifer M. Rampling, "Depicting the Medieval Alchemical Cosmos: George Ripley's Wheel of Inferior Astronomy," *Early Science and Medicine* 18, no. 1–2 (2013): 45–86.

15. On quintessence as "perfected nature," see DeVun, *Prophecy, Alchemy, and the End of Time*, 73.

16. Brunschwig, *Small Book*, fol. C1v: "ein elementisch ding zuo machen glich dem hymelschen."

17. It has proved impossible to trace with certainty any specific versions or copies of Rupescissan texts from which Brunschwig might have worked and borrowed. His mediocre Latin would point to a number of German manuscripts circulating in the fifteenth century, but my comparative reading of their critical edition against Brunschwig's text does not strongly implicate them as a direct source. The only direct reference to a text is the mention of an elusive gloss, allegedly made by a mysterious "commentator on John of Rupescissa" at a public disputation at the university of Padua in 1463, which has not been traced and may be lost. Since we must remain agnostic about the texts to which Brunschwig had access, and in the absence of a critical edition of the original Latin manuscripts, I shall refer to the 1561 printed edition of Rupescissa's text for reference. On the German Rupescissa tradition, see Udo Benzenhöfer, *Johannes de Rupescissa Liber de consideratione quintae essentiae omnium rerum, deutsch: Studien zur Alchemia medica des 15. bis 17. Jahrhunderts, mit kritischer Edition des Textes* (Stuttgart: Steiner Verlag Wiesbaden, 1989); on the elusive Rupescissa commentary, see p. 62.

18. Hieronymus Brunschwig, *Liber de arte Distillandi de Compositis* [Large book of distillation]. *Das buch der waren kunst zü distillieren die Composita vnn simplicia / vnd dz Buoch thesaurus pauperum / Ein schatz der armen* (Strasbourg: Johann Grüninger, 1512), fol. 18r: "ein vff und ab stygen der distillation"; "tausent mal vnn noch vil mer vff vnd ab styget durch das distillatorium"; "vnuermüst von dem zerstörlichen ist. Glychet würt dem himel." The passage is closely borrowed from Rupescissa, *De consideratione*, 31.

19. Astrid Müller-Grzenda, *Pflanzenwässer und gebrannter Wein als Arzneimittel zu Beginn der Neuzeit* (Stuttgart: Deutscher Apotheker Verlag, 1996), 152–53.

20. Rupescissa, *De consideratione*, 40: "In omni re quinta essentia est."

21. Rupescissa, *De consideratione*, 44: "Scientiae quintae Essentiae extrahendae à fructibus, radicibus, herbis, & foliis." Cf. Brunschwig, *Large Book*, fol. 30r: "von allen krüteren stengel / bluomen / somen / vnd wurtzeln."

22. Brunschwig, *Large Book*, fol. 30r: "quintam essentiam / in welicher grosse krafft vnn tugent des selben krutz / samen / bluomen / oder wurtzelen ist."

23. Brunschwig, *Large Book*, fol. 30v: "nun möcht gefragt werden was quinta essentia were so es doch nit mer ist dann ein ander gedistilliert wasser."

24. Brunschwig, *Large Book*, fol. 30v: "es ist war / Es ist nit anders in der gestalt vnn gesicht dan ein ander wasser / aber subtyler vnn edler in der krafft / die in im verborgen vnn vnsichtig by einander in einer cleinen substantz vereinigt ist. Darumb dz von im durch die vile der digerierung / distillierung / vnn circulierung / die fleugmatische / als die wasserige füchtikeit verzert ist."

25. Brunschwig, *Small Book*, fol. 13r: "von den krütern die do sint heisser und druckner substantz"; "wurtzlen di eyner grossen oder füchten oder schlymigen substantz oder complexion"

26. Brunschwig, *Small Book*, fol. 12v: "wan es halber gifdt is durch ir grosse kelte."

27. Brunschwig, *Small Book*, fol. 12v: "in dem dritten iar im sin külung getemperiert ist."

28. Brunschwig, *Small Book*, fol. 12r: "so ist es im dritten iar me confortieren dann kelten."

29. Brunschwig, *Small Book*, fol. 14r: "so weret es wider sin zit."

30. Rupescissa, *De consideratione*, 58: "quintae Essentiae, vt extrahat . . . ex quacunque medicinali re, omnes virtutes & proprietates & naturas, & effectus."

31. Rupescissa, *De consideratione*, 24: "indago stellarum nostrarum ad ornatum caeli nostri."

32. Brunschwig, *Large Book*, fol. 17r: "Wann als der himel voll der sterne ist / also mag man zuo setzen alle andere medicin . . . alle ire crafft vnd tugent ansich ziehen ist / vnd wiederumb von im geben in der wirckung. Als das gestirn des himels thuot / durch die bewegniß der himelen."

33. Brunschwig, *Small Book*, fol. C1v: "wann durch bewegung der natürlichen hymeln ein yedes natürlich wesen geregiert werden muoß on den corpus der hymel. Zuo glicher wise des menschen lyp durch ein erfarenden artzet on den corpus der ertzeny sunder durch die wasser die do gescheyden sint von dem groben."

34. Brunschwig, *Large Book*, fol. 13v: "durch die vorgende distillierung gesubtiliert / vnd düner gemacht ist, darumb würt leichter de matery / vnd das suptieller von den suptilen gescheiden / durch ein geringer vnd cleiner füer." Cf. Benzenhöfer, *Johannes de Rupescissa Liber de consideratione, deutsch*, 104.

35. Brunschwig, *Small Book*, fol. 12r: "darumb not ist alle zit dar an geschriben die datum oder iarzal siner brennung oder distillierung."

36. Brunschwig, *Small Book*, fol. 11v: "ab nagung der müse."

37. Brunschwig, *Small Book*, fol. 11v: "O wie offt vnd dick sin mir abgefület vnd benaget die brieff / ouch ab gerissen ich solichs grossen schaden hett."

38. Brunschwig, *Small Book*, fol. 11v: "nach den buochstaben A b c. in kurtzem zefinden welchens du begeren bist."

39. Brunschwig, *Small Book*, fol. C2r: "ob die guot frisch vnn dar zuo vß gelesen werent vnd nit veralt vnnd ob die sel vnn der geist irer würckung ücht entzogen ist."

40. Archives de la ville et de la communauté urbaine de Strasbourg (hereafter AVCUS) 1AH 112, fol. 198v: "Die gebranten wasser soll er auch versorgen, und in reynen geschirren behalten, weliche auch mit iren brieflin verzeychnet sollen werden, damit man wiss wz eyn iedes sy, und wie alt es sy, . . . sich eyn wasser wie auch andre artznyen so in der apotecken sind, lenger hielten mag dan das ander."

41. Brunschwig, *Large Book*, fol. 23v: "so distillieren die spiritus allein durch den

bad schwamen / vnd die fuechtigkeit / als das wasser oder flegma mag nit dar durch vor dem boum oly gon / vnd distillierest also in einer distillation mer wann sunst in dryen."

42. Brunschwig, *Large Book*, fol. 17r–v: "mit cleiner müw vnd arbeit / vnd on grossen kosten"; "vmm verdruß oder kostens willen"; "versaumniß andere ire gescheflt."

43. Brunschwig, *Large Book*, fol. 17v: "waren vnn gerechten wege der experiment"; "fürstenn oder herren durch ire wolhabung das wol zu wegen bringen mögen . . . dem wircker diser kunst / der dz mit grossem flyß / müw / vnd arbeit den fürsten vnd herren zu wolgefallen vnd im selber zu hilff seiner narung thuon ist / da er vleicht in ein andern weg vil swerer grosser müw vnn arbeit haben müst."

44. Helmut Arntz, *Weinbrenner: Die Geschichte vom Geist des Weines* (Stuttgart: Seewald, 1975), 16–68, 124–34, 168–80.

45. Tara Nummedal, *Alchemy and Authority in the Holy Roman Empire* (Chicago: University of Chicago Press, 2007), 1–15; Jennifer M. Rampling, "John Dee and the Sciences: Early Modern Networks of Knowledge," *Studies in History and Philosophy of Science* 43, no. 3 (2012).

46. Nummedal, "Practical Alchemy and Commercial Exchange in the Holy Roman Empire," in *Merchants and Marvels: Commerce, Science, and Art in Early Modern Europe*, ed. Pamela Smith and Paula Findlen (New York: Routledge, 2002), 201–22.

47. Agricola, *Vom Bergkwerck XII Bücher darinn alle Empter, Instument, Gezeuge unnd alles zu disem handel gehörig, mitt schönen figuren, vorbilder und klärlich beschriben seindt*, 370: "so das starck scheidtwasser gemachet wirdt / muoß man zuo nacht wachen / vnd an der sach sonderlichen fleiß muehe vnd arbeit legen"; "nicht vast kostlich / auch nicht arbeitsam."

48. Brunschwig, *Small Book*, fol. 14r: "schwerlich ist zuoerkennen ein wasser wenn es zerstört ist dz man es hyn schütten soll / so man nit wissen ist die zit siner distillierung."

49. Brunschwig, *Small Book*, fol. 14r: "der gesmack und geruch mercklich gemyndert vnn geschwechert oder gantz verloren ist."

50. Brunschwig, *Small Book*, fol. 14r: "soll das wasser von der hoehen gegossen werden / wan es dan louffen ist wie ein faden . . . so ist gegenwürtig sin zerstörung / ist vß zeschütten."

51. Brunschwig, *Small Book*, fol. 14r: "ein tropffen vff ein dumen nagel gelassen fallen / ob er nit bald über ab fliessen ist / so ist gegenwürtig sin zerstoerung ist vß zeschütten."

52. On the epistemic use of the senses in Brunschwig's distillation books, see chap. 6, and Taape, "Distilling Reliable Remedies," 249–52.

53. Brunschwig, *Large Book*, fol. 13v: "die matery im anfang grob / vmb der grobheit nitt leicht gehorsam der distillierung ist / vm yrer rohen vnd vnreinigkeit der substantz."

54. Brunschwig, *Large Book*, fol. 13v: "ein hert als ein grobe wider wertige substantz."

55. Brunschwig, *Large Book*, fol. 14r: "es sol werden gedigirriert vnd geputrificiert / vff das es werd gehorsamer der distillierung."

56. Brunschwig, *Large Book*, fol. 13v: "Not vnn mitt fleiß zuo mercken ist eyn yedes ding oder substantz was wesen / das ist ein herten / oder groben / oder weichen substantz."

57. Jacob and Wilhelm Grimm, *Deutsches Wörterbuch* (Leipzig: Hirzel, 1854–1961), s.v. "widerwärtig," vol. 29, col. 1365–1373, accessed August 23, 2016, http://www.woerterbuchnetz.de/DWB?lemma=widerwaertig.

58. On matter as an active, even anthropomorphic entity to be wrestled with, see Smith, *Body of the Artisan*, 95–127, 114–15. On materials as agents in historical accounts of practice, see the introduction to Ursula Klein and Emma Spary, *Materials and Expertise in Early Modern Europe: Between Market and Laboratory* (Chicago: University of Chicago Press, 2010), 1–19.

59. Smith, *From Lived Experience to the Written Word*, 231–50. See also Smith, *Secrets of Craft and Nature in Renaissance France.*

60. Brunschwig, *Large Book*, fol. 14v: "das das siessiglich vnd milttiglich geschehe . . . nitt mitt schneller maß / als die natur das nit lyden mag schnelle mutacion oder veranderung / also sol ouch die digestion nitt geyebt sein schnel ab zuo schneiden / sunder also das der kunstener oder wircker der natur / als ein discipel geleichformlich dem selbigen nach gedenck."

61. Brunschwig, *Large Book*, fol. 12v: "die natur mag nicht lyden gehe oder schnelle veranderung. Darumb sol ein ieder meister / als ein werckmeister / wircken sein werck / nachgefolget der naturen / als seinem werckmeister / die da verwürfft gezwencknisse / oder mitt gewalt genettiget."

62. Smith, *Body of the Artisan*, 59–94. On Brunschwig as an exponent of artisanal epistemology, see also Alisha Rankin, "How to Cure the Golden Vein: Medical Remedies as Wissenschaft in Early Modern Germany," in Smith, Cook, and Meyers *Ways of Making and Knowing*, 109–33.

63. John Christie, "Paracelsian Body," in *Paracelsus: The Man and His Reputation, His Ideas and Their Transformation*, ed. Ole Peter Grell (Leiden: Brill, 1998), 269–91; Smith, *Body of the Artisan*, 95–128.

64. Brunschwig, *Large Book*, fol. 122v: "darumm ist es ein dorheit dz wir suochen in disem leben kunst die vnsern leichnam ewigklich behalten sol."

65. Brunschwig, *Large Book*, fol. 122r: "suochen kunst dz wir vnser leben erfristen biß an die zeit die vns got gesetzt hat."

Chapter Six

1. Hieronymus Brunschwig, *Liber de arte Distillandi de Compositis* [Large book of distillation]. *Das buch der waren kunst zü distillieren die Composita vnn simplicia / vnd dz Buoch thesaurus pauperum / Ein schatz der armen* (Strasbourg: Johann Grüninger, 1512), fol. 32v: "das du nit wol weist wann es vß gebrandt ist oder ni / sunder allein nach geratwol / vnd gewonheit deren / die des vil gemacht habent."

2. Paracelsus, "On the Miners' Sickness and Other Miners' Diseases," in *Four Treatises of Theophrastus von Hohenheim Called Paracelsus*, ed. Henry Sigerist, trans. George Rosen (Baltimore: Johns Hopkins University Press, n.d.), 91.

3. Walter Pagel, *Paracelsus: An Introduction to Philosophical Medicine in the Era of the Renaissance* (Basel: Karger, 1958), 50–51; Pamela H. Smith, *The Body of the Artisan: Art and Experience in the Scientific Revolution* (Chicago: University of Chicago Press, 2004), 114–20; Pamela H. Smith, *From Lived Experience to the Written Word: Reconstructing Practical Knowledge in the Early Modern World* (Chicago: University of Chicago Press, 2022), 126–30.

4. Tim Ingold, *The Perception of the Environment: Essays on Livelihood, Dwelling and Skill* (New York: Routledge, 2000), 24, 106.

5. Smith, *From Lived Experience to the Written Word*, 117–37.

6. Sven Dupré, ed., *Laboratories of Art: Alchemy and Art Technology from Antiquity to the 18th Century* (Cham: Springer, 2014); Pamela Long, *Openness, Secrecy, Authorship: Technical Arts and the Culture of Knowledge from Antiquity to the Renaissance* (Baltimore: Johns Hopkins University Press, 2001), 102–42; Smith, *Body of the Artisan*, 31–57; Pamela H. Smith, "Craft Techniques and How-to Books," in *Transmission of Artists' Knowledge*, ed. Mark Clarke, Bert De Munck, and Sven Dupré (Brussels: KVAB Press, 2012), 75–84; Simon Werrett, *Fireworks: Pyrotechnic Arts and Sciences in European History* (Chicago: University of Chicago Press, 2010).

7. Pamela Long, *Artisan/Practitioners and the Rise of the New Sciences, 1400–1600* (Corvallis: Oregon State University Press, 2011); Pamela H. Smith, "In a Sixteenth-Century Goldsmith's Workshop," in *The Mindful Hand: Inquiry and Invention from the Late Renaissance to Early Industrialisation*, ed. Lissa Roberts and Simon Schaffer (Amsterdam: Royal Netherlands Academy of Arts and Sciences, 2007), 33–57; Smith, *From Lived Experience to the Written Word*, 91–116.

8. Hieronymus Brunschwig, *Dis ist das buch der cirurgia* [Cirurgia]. *Hantwirckung der wund artzny* (Strasbourg: Johann Grüninger, 1497), fol. π2v: "allen den ginnen die da begerent zuo leren yn chirurgia"; Hieronymus Brunschwig, *Liber de arte distillandi de simplicibus* [Small book of distillation]. *Das buch der rechten kunst zü distilieren die eintzigen ding* (Strasbourg: Johann Grüninger, 1500), fol. A2v: "die do begeren zuo leren die maß vnn kunst der distillierung."

9. On the scholastic format of the *quaestio* in medieval surgery, see Michael McVaugh, "Surgical Education in the Middle Ages," *Dynamis* 20 (2000): 283–304. Teacher-student dialogue was also used in Gregor Reisch's *Margarita philosophica*, which was printed in Strasbourg in 1493 and thus likely known to Brunschwig; see Miriam Usher Chrisman, *Lay Culture, Learned Culture: Books and Social Change in Strasbourg, 1480–1599* (New Haven: Yale University Press, 1982), 125–26. The surgeon Hans Seyff included in his manual a question-and-answer dialogue representing what an aspiring surgeon might expect to be asked at examination; see Manfred Gröber, *Das wundärztliche Manual des Meisters Hans Seyff von Göppingen (ca. 1440–1518): der Cod. med. et phys. 20 8 der Württembergischen Landesbibliothek Stuttgart* (Göppingen: Kümmerle, 1998), 88.

10. Brunschwig, *Cirurgia*, fol. π2v: "sich behelffent mit den selbigen die disse kunst der hantwirckung nie gesehen noch gedriben habent darzuo nie gehort noch yn geoffenbart ist worden."

11. Brunschwig, *Large Book*, fol. 2v: "für den leyen"; Brunschwig, *Small Book*, fol. C1v: "die gemeynen menschen die die ertzeny nit zuo bezalen habent / oder haben moegent."

12. For an overview of European apprenticeship arrangements, see Bert De Munck, "Apprenticeship, Guilds, and Craft Knowledge," in *Encyclopedia of Early Modern Philosophy and the Sciences*, ed. Dana Jalobeanu and Charles T. Wolfe (Cham: Springer, 2022), 98–104. Surgeons' apprenticeships might last three to six years in the southern imperial cities studied by Annemarie Kinzelbach, "Erudite and Honoured Artisans? Performers of Body Care and Surgery in Early Modern German Towns," *Social History of Medicine* 27 (2014): 668–88. Goldsmiths in fifteenth-century Ulm were appren-

ticed for six years, in Cologne up to eight years. Cf. Hans Meyer, "Die Strassburger Goldschmiedezunft von ihrem Entstehen bis 1681: Urkunden und Darstellung. Ein Beitrag zur Gewerbegeschichte des Mittelalters," *Staats- und socialwissenschaftliche Forschungen* 3, no. 2 (1881): 190.

13. For an overview of a typical artisan career in Strasbourg, see Sabine von Heusinger, *Die Zunft im Mittelalter: zur Verflechtung von Politik, Wirtschaft und Gesellschaft in Straßburg* (Stuttgart: Steiner, 2009), 62. An average duration of two to three years was typical for the entire Upper Rhine region according to Knut Schulz, *Handwerksgesellen und Lohnarbeiter: Untersuchungen zur oberrheinischen und oberdeutschen Stadtgeschichte des 14. bis 17. Jahrhunderts* (Sigmaringen: Thorbecke, 1985), 248–54.

14. Reinhold Reith, "Know-How, Technologietransfer und die Arcana Artis im Mitteleuropa der Frühen Neuzeit," *Early Science and Medicine* (2005): 349–77; Schulz, *Handwerksgesellen*, 249.

15. For an overview of apprenticeship learning, see De Munck, "Apprenticeship, Guilds, and Craft Knowledge." See also Bert De Munck, *Technologies of Learning: Apprenticeship in Antwerp Guilds from the 15th Century to the End of the Ancien Regime* (Turnhout: Brepols, 2007); Bert De Munck, Steven L. Kaplan, and Hugo Soly, *Learning on the Shop Floor: Historical Perspectives on Apprenticeship* (New York: Berghahn Books, 2007).

16. Heusinger, *Die Zunft im Mittelalter*, 65; Schulz, *Handwerksgesellen*, 265–75.

17. Archives de la ville et de la communauté urbaine de Strasbourg (hereafter AVCUS) III 11/8, "Antwürt vf den drytten artickel . . ." The document is undated, but the writing is similar to other later fifteenth- and early sixteenth-century hands in the collection.

18. AVCUS III 11/8, fol. 2r: "vnserem Alten harkomen."

19. AVCUS III 11/8, fol. 1v: "nie gemacht noch machen künden / des gelichen nie erdient eriebt noch gelernet haben."

20. AVCUS III 11/8, fol. 2r: "das wir mit grossen kossten erdient vnn eriebet haben . . . do mit das jeder treibet das er gelernet hatt."

21. Brunschwig, *Large Book*, fol. 2r: "mein reglich vbung / bruch / vnn practica der hantwirckung."

22. Smith, *From Lived Experience to the Written Word*, 123–24.

23. Laetitia Boehm, "Artes mechanicae und artes liberales im Mittelalter. Die praktischen Künste zwischen illiterater Bildungstradition und schriftlicher Wissenschaftskultur," in *Festschrift für Eduard Hlawitschka zum 65. Geburtstag*, ed. Karl Rudolf Schnith and Roland Pauler (Kallmünz: Lassleben, 1993), 419–44; Martin Kintzinger, "Experientia lucrativa? Erfahrungswissen und Wissenserfahrung im europäischen Mittelalter," *Das Mittelalter: Perspektiven mediävistischer Forschung* 17, no. 2 (2012): 95–117; Nancy Siraisi, *Medieval and Early Renaissance Medicine* (Chicago: University of Chicago Press, 1990), 136–52.

24. Brunschwig, *Large Book*, fol. 57r: "durch stet vbung die er getrieben vnn gelert hat."

25. Brunschwig, *Large Book*, fol. 57r: "durch sein practica geübet vnn gelert hat."

26. Brunschwig, *Large Book*, fol. 2v: "die ich mit meinen henden / durch die kunst der Practica gethan vnn geübt hab . . . nit wenig / sunder zum offteren mal."

27. Brunschwig, *Cirurgia*, fol. 87r: "vil vnd offt han ich grosse zufelle gesehen von solichen cleinen wunden die man gantz vnd gar vernicht vnd verachtet vnd mir selbiß

vnd vil andren wider faren ist dar vmb gebirt dir mit grossem flys vnd merer fursichtikeit zu betrachten disse wunden."

28. Brunschwig, *Cirurgia*, fol. 34v–35r: "als ich gesehen hab in einer loeblichen stat von den scherern die der kunst vngeübet waren den zuo kam ein wund in eim arm die ward geapostemiert dy sie vil eyter trug / do vnderstundent sie den eyter allen tag vß zu drucken . . . so lang biß nah das gelidt gantz verdarb."

29. Brunschwig, *Cirurgia*, fol. 35r: "Darvmm han ich dir diß geschriben das du dich wissest in soelchen holen wunden zuo richten vff dz der mensch nit verderb."

30. Brunschwig, *Small Book*, fol. 3v: "rysen oder . . . zerstoert werden durch das füer."

31. Brunschwig, *Large Book*, fol. 34r: "da ward ich geholt zuo einem pacienten / vnnd also das oley vnnder das wasser gon was / vnnd als ich wider kum das oley vff dem wasser stünd . . . nit so vil vernunnft oder fürwitz hett / das ich das oley obe ab geseigt hett . . . also deht ich oley vnnd wasser vnder einander in ein frischen cucurbit / vnd meint das wasser hüpschlich dar von zuo ziehen . . . vnn wart aber dar von geholt / in dem ging das wasser gar nahe alles hervß / vnnd also da wenig wasser me darinnen was / vnd also das vber sich vff distilliert an den wenden des glases ein deil wider ab rinen in das oley was / mit grosser vngestiem das oley sich erhuob im glaß der dunst dz darvon der alembic vff erhuob."

32. Brunschwig, *Large Book*, fol. 34r: "als das liecht kam / der dunst aber vß schluog in das liecht ettwas dar von kam das gantz end zuo ring vmb / in eym ougen blick yttel für vnnd verloschen was nit destmynder meinem dienner vnnd mir alle hor ab den cleidern vnd die augbrauwen vns beyden verbrant das wir in itel füer stunden vnnd nider fiellent / vnnd nitt wußen wo wir worent ouch der dünst der das liecht lascht / vnd nitt lang stuonden wir wider vff / vnd holten ein ander liecht in einer lützernnen dz vns das nymm geschehe / vnn thet den offen vff / vnnd warff den offen voll esch das kein füer me im offen was."

33. Brunschwig, *Large Book*, fol. 34r: "das dir nit geschehe wie mir geschehen ist."

34. Sven Dupré, "Doing It Wrong: The Translation of Artisanal Knowledge and the Codification of Error," in *The Structures of Practical Knowledge, ed. Matteo Valleriani* (Cham: Springer International Publishing, 2017), 167–88; Thijs Hagendijk, Márcia Vilarigues, and Sven Dupré, "Materials, Furnaces, and Texts: How to Write About Making Glass Colours in the Seventeenth Century," *Ambix* 67, no. 4 (2020): 323–45.

35. Heidi Hausse, "Learning from Disaster: Surgical Mishaps and Pedagogy in the Age of Print," in Donato, Leong, and Taape, "Learning to Cut and Mend: Surgical Training and Practice, 1450–1800."

36. On Grüninger's reprint of the *Gart* and his use of the woodblocks in the *Small Book*, see Brigitte Baumann and Helmut Baumann, *Die Mainzer Kräuterbuch-Inkunabeln "Herbarius Moguntinus" (1484), "Gart der Gesundheit" (1485), "Hortus sanitatis" (1491): wissenschaftshistorische Untersuchung der drei Prototypen botanisch-medizinischer Literatur des Spätmittelalters (Stuttgart: Hiersemann, 2010)*, 223–38; Agnes Arber, *Herbals: Their Origin and Evolution* (Cambridge: Cambridge University Press, 1938), 201.

37. Brunschwig, *Small Book*, fol. 210v: "von eim yeden krut syn gestalt vnd geschlecht," "dz etwan mit figuren nit zuo zeigen ist / dem der sie vor nit kennen ist."

38. Brunschwig, *Small Book*, fol. 210v: "die figuren nit anders synd dann ein ougen weid vnd ein anzeigung geben ist die weder schriben noch lesen kündent."

39. William Ashworth, "Emblematic Natural History of the Renaissance" in *Cul-*

tures of Natural History, ed. Nicholas Jardine, James Secord, and Emma Spary, 17–37 (Cambridge: Cambridge University Press, 1996); Sachiko Kusukawa, *Picturing the Book of Nature: Image, Text, and Argument in Sixteenth-Century Human Anatomy and Medical Botany* (Chicago: University of Chicago Press, 2012), 17–18. See also Luuk Houwen, "The Noble Lyfe: An Early English Version of the Hortus Sanitatis," in *Schooling and Society*, ed. Alasdair MacDonald and Michael Twomey (2004).

40. Brunschwig, *Small Book*, fol. 210v: "darumm ist nit zuo achten allein vff die figuren / sunder vff die geschrifft," "dz erkennen durch die gesicht." The verb *erkennen* means to recognize, but also to gain insight or to realize, and is related to *erkenntnis*, meaning knowledge or insight.

41. The skeleton looks similar to a woodcut published by one doctor Helain from Nürnberg in 1493; see Gustav Klein, *Das Buch der Cirurgia des Hieronymus Brunschwig* (Munich: Kuhn, 1911), xi.

42. Hieronymus Brunschwig, *Das buch der wund Artzney. Handwirckung der Cirurgia von Jyeronimo brunschwick* (Strasbourg: Johann Grüninger, 1513), fol. 10v: "durch die gesichte der doten körper."

43. Sachiko Kusukawa, "The Uses of Pictures in the Formation of Learned Knowledge: The Cases of Leonhard Fuchs and Andreas Vesalius," in *Transmitting Knowledge: Words, Images, and Instruments in Early Modern Europe*, ed. Sachiko Kusukawa and Ian Maclean (Oxford: Oxford University Press, 2006), 73–96; Kusukawa, *Picturing the Book of Nature*. See also Swan, "The Uses of Realism in Early Modern Illustrated Botany"; Kemp, "'The Mark of Truth': Looking and Learning in Some Anatomical Illustrations from the Renaissance and Eighteenth Century." Brian Ogilvie has argued that naturalists developed a "technology of observation" as a prerequisite for the "science of describing" nature. Ogilvie, *The Science of Describing: Natural History in Renaissance Europe*, 87–138.

44. Brunschwig, *Cirurgia*, fol. 51v: "diß ist sin form."

45. On medieval surgical illustrations, see Peter Jones, *Medieval Medicine in Illuminated Manuscripts* (London: The British Library, 1998), 76–94, and especially 89–91 on John of Arderne's fistula-in-ano. On the latter, see also Peter Jones, "Image, Word and Medicine in the Middle Ages," in Givens, Reeds, and Touwaide, *Visualizing Medieval Medicine and Natural History*, 1–24.

46. See, e.g., British Library, Sloane MS 6, fol. 177v., depicted in Jones, *Medieval Medicine*, 88; Guy de Chauliac, *Guigonis de Caulhiaco (Guy de Chauliac), Inventarium sive Chirurgia magna. Volume One: Text*, ed. Michael McVaugh (Leiden: Brill, 1997), 189.

47. See, e.g., Geneviève Dumas, *Ymage de Vie* (Montpellier: Presses universitaires de la Méditerranée, 2019), British Library, Sloane MS 3648, fol. 25, depicted in Jones, *Medieval Medicine*, 74; Reginald Bess, "Hieronymus Brunschwig's 'De Arte Distillandi' and the Technical Language of Distillation," *Fifteenth-Century Studies* 11 (1985): 9.

48. Some examples can be found in fifteenth-century xylographs depicting the wounds of Christ, but here the (allegedly) true scale of the image was intended to aid the imagination in religious contemplation. See Peter Schmidt, "Beschrieben, bemalt, zerschnitten: Tegernseer Mönche interpretieren einen Holzschnitt," in *Einblattdrucke des 15. und frühen 16. Jahrhunderts: Probleme, Perspektiven, Fallstudien*, ed. Volker Honemann, Sabine Griese, Falk Eisermann, and Marcus Ostermann (Tübingen: Max Niemeyer, 2000), 245–76. On the use of drawings functioning as geometrical scales as

well as illustrative figures in medieval manuscripts on astronomical instruments, see Catherine Eagleton, "Medieval Sundials and Manuscript Sources: The Transmission of Information about the Navicula and the *Organum Ptolomei* in Fifteenth-Century Europe," in Kusukawa and Maclean, *Transmitting Knowledge*, 41–71.

49. Although there is a well-documented tradition of illustrated technical manuscripts in the fifteenth century, their lavish images were primarily aimed at securing patronage rather than divulging technical knowledge. See Long, *Openness, Secrecy, Authorship*. An exception is the (fairly self-contained) visual tradition of fifteenth-century master gunners who developed a "pictorial language" in which images could function as vehicles of craft knowledge. See Rainer Leng, "Social Character, Pictorial Style, and the Grammar of Technical Illustration in Craftsmen's Manuscripts in the Late Middle Ages," in *Picturing Machines 1400–1700*, ed. Wolfgang Lefèvre (Cambridge, MA: MIT Press, 2004), 85–111.

50. Ian Maclean, *Logic, Signs and Nature: Learned Medicine in the Renaissance* (Cambridge: Cambridge University Press, 2002), 276–85. On the practical aspects of semiology in an institutional context, see also Claudia Stein, "The Meaning of Signs: Diagnosing the French Pox in Early Modern Augsburg," *Bulletin of the History of Medicine* 80, no. 4 (2006): 617–48; Claudia Stein, *Negotiating the French Pox in Early Modern Germany* (Aldershot: Ashgate, 2009), 35–40.

51. Tillmann Taape, "The Body and the Senses in Ms. Fr. 640: Towards a 'Material Sensorium,'" in *Secrets of Craft and Nature in Renaissance France: A Digital Critical Edition and English Translation of BnF Ms. Fr. 640*, ed. Pamela H. Smith (New York: Making and Knowing Project, 2020), https://edition640.makingandknowing.org/#/essays/ann_302_ie_19.

52. Brunschwig, *Cirurgia*, fol. 11r: "Item solt du auch mercken dz du mit keinen dingen macht also wol gebrieffen vnnd tastenn ob das dir die hirnschal zerbrochen zerzwungen vnd entzwei sy / als der mit dem finger sanff dar vff tastet / vnnd ob das bein inwendig ledig sy."

53. Brunschwig, *Cirurgia*, fol. 11r: "doch soltu diß mercken by disen zeichenn. Laß den menschenn sin mund vnd nasen hert zuo halten vnd huost dan sere. Sichstu dan vß dem riß etwz quellen / so ist der riß wol durch gangen durch das zerbrochen bein vnn die hirnschal is zerbrochen."

54. Brunschwig, *Small Book*, fol. 13v: "dan dz du ein finger darin geleiden magst."

55. Brunschwig, *Large Book*, 22v: "das magstu mercken dar by / wann die strich in dem helm / oder alembich gabelen gewinnen / vnd grob breit tropffen in dem helm werden als schwitzet der helm inwenig / so synt die spiritus als die geist heruß."

56. Brunschwig, *Large Book*, fol. 23r: "versuoch dz vff der zungen / so entpfindestu ob es sy wasserecht / oder me krafft vnn geist darinnen seint oder nit."

57. Brunschwig, *Large Book*, fol. 34v: "Das im mundt vff der zungen bewert würt durch das versuochen zuo manig malen / wan eß anfahet zuo sueren oder zuo bitzelen die zung / oder starck zuo werden / dan werd denen gethon das fürsatz glaß / vnd ein anders an sein statt."

58. Evan Ragland, "Chymistry and Taste in the Seventeenth Century: Franciscus Dele Boe Sylvius as a Chymical Physician Between Galenism and Cartesianism," *Ambix* 59, no. 1 (2012): 1–21; Pamela H. Smith, "Vermilion, Mercury, Blood, and Lizards: Matter and Meaning in Metalworking," in *Materials and Expertise in Early Modern Europe: Between Market and Laboratory*, ed. Ursula Klein and Emma Spary (Chicago: University of Chicago Press, 2010), 29–48; Taape, "Body and the Senses."

59. Brunschwig, *Large Book*, fol. 23r: "wann du ein tropffen boum oly lassest daryn fallen / so felt er zu hant an den bodem / vnd stygt nicht mer wider vber sich vff / wie fast du das gefeß mit dem gebranten wein hin vnd her schwenckest vnn rüttelst."

60. Brunschwig, *Large Book*, fol. 276v: "dan wirfestu ein tropffen in ein stil ston wasser so get der tropff bis and den bodem . . . vnn würfesstu ein nadel darein so schwimt sie enbor Item es machet milch zuosamen lauffen."

61. Smith, *From Lived Experience to the Written Word*, 91–116. See also Sven Dupré, ed., *Laboratories of Art: Alchemy and Art Technology from Antiquity to the 18th Century* (Cham: Springer, 2014); Pamela Long, *Openness, Secrecy, Authorship: Technical Arts and the Culture of Knowledge from Antiquity to the Renaissance* (Baltimore: Johns Hopkins University Press, 2001), 102–42; Smith, *Body of the Artisan*, 31–57; Pamela H. Smith, "Craft Techniques and How-to Books," in *Transmission of Artists' Knowledge*, ed. Mark Clarke, Bert De Munck, and Sven Dupré (Brussels: KVAB Press, 2012), 75–84; Simon Werrett, *Fireworks: Pyrotechnic Arts and Sciences in European History* (Chicago: University of Chicago Press, 2010).

Chapter Seven

1. Archives de la ville et de la communauté urbaine de Strasbourg (hereafter AVCUS) KS 35 I, fol. 68r: "des erfarnen Hieronymi Brunschwigs vnsers bürgers seligen dochter."

2. Hieronymus Brunschwig and Marsilio Ficino, *Das buch der Gesuntheit* [Medicinarius]. *Liber de arte distillandi Simplicia et composita. Das nüv buoch der rechten kunst zuo distillieren. Ouch von Marsilio Ficino vnn andere hochberoepmter Artzte natürliche vnn gute künst* (Strasbourg: Johann Grüninger, 1505), fol. 2r: "mit grosser arbeit yetzt wol vff xxx iar zuo samen bracht gelesen / gesehen vnn erfaren."

3. Hieronymus Brunschwig, *Liber de arte distillandi de simplicibus* [Small book of distillation]. *Das buch der rechten kunst zü distilieren die eintzigen ding* (Strasbourg: Johann Grüninger, 1500), fol. A2r: "durch sin teglich hantwürckung"; "erfarenden meystern der ertzney."

4. Hieronymus Brunschwig, *Liber de arte Distillandi de Compositis* [Large book of distillation]. *Das buch der waren kunst zü distillieren die Composita vnn simplicia / vnd dz Buoch thesaurus pauperum / Ein schatz der armen* (Strasbourg: Johann Grüninger, 1512), fol. 3r: "mit studieren lesen / vnd gesehen / durch mein vnd ander erfarend."

5. Jan-Dirk Müller, "Erfarung zwischen Heilssorge, Selbsterkenntnis und Entdeckung des Kosmos," *Daphnis* 15 (1986): 307–42; Alisha Rankin, *Panaceia's Daughters: Noblewomen as Healers in Early Modern Germany* (Chicago: University of Chicago Press, 2013), 57–59.

6. Peter Dear, *Discipline and Experience: The Mathematical Way in the Scientific Revolution* (Chicago: University of Chicago Press, 1995); Peter Dear, "The Meanings of Experience," in *The Cambridge History of Science, vol. 3, Early Modern Science*, ed. Lorraine Daston and Katharine Park (Cambridge: Cambridge University Press, 2006), 106–31.

7. Geoffrey Ernest Richard Lloyd, *Magic, Reason, and Experience: Studies in the Origins and Development of Greek Science* (Indianapolis: Hackett Publishing Company, 1999), 37–49, 126–38.

8. R. Jim Hankinson, "Art and Experience: Greek Philosophy and the Status of Medicine," *Quaestio* 4 (2004): 3–24.

9. Jeremiah Hackett, "Roger Bacon's Concept of Experience: A New Beginning in Medieval Philosophy?" *Modern Schoolman* 86, no. 1/2 (2008): 123–46.

10. Isabelle Draelants, "Expérience et autorités dans la philosophie naturelle d'Albert le Grand," in *Expertus sum: L'expérience par les sens dans la philosophie naturelle médiévale*, ed. Isabelle Draelants and Thomas Bénatouïl (Florence: SISMEL, 2011), 89–121; Leen Spruit, "Albert the Great on the Epistemology of Natural Science," in *Erfahrung und Beweis. Die Wissenschaften von der Natur im 13. und 14. Jahrhundert*, vol. 14, ed. Alexander Fidora and Matthias Lutz-Bachmann (Oldenbourg Verlag, 2007), 61–75.

11. William Eamon, *Science and the Secrets of Nature: Books of Secrets in Medieval and Early Modern Culture* (Princeton, NJ: Princeton University Press, 1994), 45–52; Jeremiah Hackett, "Ego Expertus Sum: Roger Bacon's Science and the Origins of Empiricism," in *Expertus sum. L'expérience par les sens dans la philosophie naturelle médiévale*, ed. Isabelle Draelants and Thomas Bénatouïl (Florence: SISMEL, 2011), 145–73; Hackett, "Roger Bacons Concept of Experience."

12. Hans Blumenberg, *Die Lesbarkeit der Welt* (Frankfurt: Suhrkamp, 1983), 58–67; Clyde Lee Miller, *Reading Cusanus: Metaphor and Dialectic in a Conjectural Universe* (Washington, DC: Catholic University of America Press, 2003), 110–46; Klaus Schreiner, "Laienbildung als Herausforderung für Kirche und Gesellschaft. Religiöse Vorbehalte und soziale Widerstände gegen die Verbreitung von Wissen im späten Mittelalter und in der Reformation," *Zeitschrift für historische Forschung* 11, no. 3 (1984): 257–354.

13. Michael Frede, "The Method of the So-Called Methodical School of Medicine," in *Science and Speculation*, ed. Jaques Brunschwig and Jonathan Barnes (Cambridge: Cambridge University Press, 2005), 1–23.

14. Jonathan Barnes, "Medicine, Experience and Logic," in Brunschwig and Barnes, *Science and Speculation*, 24–68; Hankinson, "Art and Experience."

15. Michael Stolberg, "Learning from the Common Folks. Academic Physicians and Medical Lay Culture in the Sixteenth Century," *Social History of Medicine* 27 (2014): 649–67.

16. Danielle Jacquart, "Theory, Everyday Practice, and Three Fifteenth-Century Physicians," *Osiris* 6 (1990): 140–60; Peter Jones, "Mediating Collective Experience: The Tabula Medicine (1416–1425) as a Handbook for Medical Practice," in *Between Text and Patient: The Medical Enterprise in Medieval and Early Modern Europe*, ed. Florence Glaze, Brian Nance, and Michael McVaugh (Florence: Sismel, 2011), 279–307.

17. Katharine Park, "Observation in the Margins," in *Histories of Scientific Observation*, ed. Lorraine Daston and Elizabeth Lunbeck (Chicago: University of Chicago Press, 2011), 15–44.

18. Gianna Pomata, "Praxis Historialis: The Uses of Historia in Early Modern Medicine," in *Historia: Empiricism and Erudition in Early Modern Europe*, ed. Gianna Pomata and Nancy G. Siraisi, 105–46 (Cambridge, MA: MIT Press, 2005); Gianna Pomata, "Observation Rising: Birth of an Epistemic Genre, 1500–1650," in Daston and Lunbeck, *Histories of Scientific Observation*, 45–80; Gianna Pomata, "A Word of the Empirics: The Ancient Concept of Observation and Its Recovery in Early Modern Medicine," *Annals of Science* 68, no. 1 (2011): 1–25.

19. Elaine Leong, *Recipes and Everyday Knowledge: Medicine, Science, and the Household in Early Modern England* (Chicago: University of Chicago Press, 2018); Elaine Leong and Alisha Rankin, "Testing Drugs and Trying Cures: Experiment and

Medicine in Medieval and Early Modern Europe," *Bulletin of the History of Medicine* 91, no. 2 (2017): 157–82; Rankin, *Panaceia's Daughters*; Alisha Rankin, *The Poison Trials: Wonder Drugs, Experiment, and the Battle for Authority in Renaissance Science* (Chicago: University of Chicago Press, 2021); Stolberg, "Learning from the Common Folks."

20. Rankin, *Panaceia's Daughters*, 59.

21. See for example Pamela Long, *Artisan/Practitioners and the Rise of the New Sciences, 1400–1600* (Corvallis: Oregon State University Press, 2011); Pamela H. Smith, *From Lived Experience to the Written Word: Reconstructing Practical Knowledge in the Early Modern World* (Chicago: University of Chicago Press, 2022).

22. For example Harold Cook, "The New Philosophy and Medicine in Seventeenth-Century England," in *Reappraisals of the Scientific Revolution*, ed. David Lindberg and Robert Westman (Cambridge: Cambridge University Press, 1990), 397–436; Dear, "Meanings of Experience"; Allen G. Debus, *Chemistry, Alchemy and the New Philosophy, 1550–1700: Studies in the History of Science and Medicine* (London: Variorum, 1987); Pamela H. Smith, *The Body of the Artisan: Art and Experience in the Scientific Revolution* (Chicago: University of Chicago Press, 2004), 155–82.

23. Hieronymus Brunschwig, *Dis ist das buch der cirurgia* [Cirurgia]. *Hantwirckung der wund artzny* (Strasbourg: Johann Grüninger, 1497), fol. 48v: "die werck der alten"; "nüwe erfarung."

24. Brunschwig, *Cirurgia*, fol. 122r: "hat disse sal genumen vß dem buoch rasis / doch hat er sie gebessert."

25. Brunschwig, *Cirurgia*, fol. 70v: "tugent vnn kraft die ich gesehen hab durch binden vnn wircken des steins."

26. Guy de Chauliac, *Guigonis de Caulhiaco (Guy de Chauliac), Inventarium sive Chirurgia magna. Volume One: Text*, ed. Michael McVaugh (Leiden: Brill, 1997), 190: "partes enim spermatice in pueris cottidie regenerantur, ut videtis"; Brunschwig, *Cirurgia*, fol. 55v: "Aber in der warheit hon ichs selber gesehenn wie oben statt. Dar vmm ich billich gezeügniß gib Galieno."

27. Nancy Siraisi, *Medieval and Early Renaissance Medicine* (Chicago: University of Chicago Press, 1990), 170–73.

28. Brunschwig, *Cirurgia*, fol. 99r: "ein gutter beriempter scherer vnnd wundartzet gebunden der ein ander minern meister by im het."

29. Brunschwig, *Cirurgia*, fol. 99r: "aber ich hab diß gebend vff ein andern weg gemacht das vil nitzer ist / vnnd dem menschen miner we dünt."

30. Brunschwig, *Cirurgia*, fol. 99v: "ein hültzen instrument mit zweien hacken alß die armbroster bruchent die bogen zü sennen."

31. Brunschwig, *Cirurgia*, fol. 99v: "Ist aber der bruch alt vnd verharttet so werde es vnder wegen gelassen alß die alten philosophi schribent wan es wer besser gesin dem wisen . . . das er gelebt het also hincken dan das er gerad wolt werden vnd starb."

32. Brunschwig, *Cirurgia*, fol. 39r: "dz sie die kunst von nüwem erfunden hant . . . vnn nit gedencken das die alten vil baß künt vnn verstanden hant dan wir. So die doch die cur durch insprechung des almechtigen gotz durch ir philosophi funden habent vnd gewist hant die krafft tugent vnd complex des balsams. War vmm er heilt oder nit. oder wo er zuo bruchen ist oder nit / oder in welchen wunden oder landen er gebruchet mag werden / oder wie man in bruchen sol . . . das doch dem cirur. vast not ist zuo Wissen."

33. Brunschwig, *Cirurgia*, fol. 9v: "natürlichen künsten der Phylosophia."

34. Brunschwig, *Cirurgia*, fol. 39r: "die gantz complex vnn natur zerstoert."

35. Brunschwig, *Cirurgia*, fol. 39r: "Wan du einen sochen cirurgicus fragest wor vmb heilt der balsam oder das plaster die wund oder nit Kan kein ander antwurt geben dan dz er spricht dar vmm dz er guot ist vnn ich das von anderen meystern auch gesehenn hab. oder ich weiß nit wor vmb es in nit geheilt hat. Er ist villeicht nit guot zuo der wunden / er hat mir nie gefelt. O wie doerecht antwurt entpfahestu diner frag."

36. Brunschwig, *Cirurgia*, fol. 48r–v: "als Guido hat gesehenn die zweytrechtikeit vnnder denn meystern. Do folte er nach in syner iugent den iungen. Aber do er betracht in dem alter die werck der alten / vermeid er die nüwe erfarung / wan er gewan sie argwoenig."

37. Brunschwig, *Cirurgia*, fol. 48v: "das durch einenn oder durch dry oder me lere die kunst der guotten vnderwegen gelasen werden das wer boeß."

38. Brunschwig, *Cirurgia*, fol. 48v: "Wan in der erfarung bringt die artzeny schaden / dar vmm ist zuo erwelen der gemeinest vnn best weg. Dar vmm mit hilffe des werckmeysters aller ding so gang nit ab dem weg Galieni."

39. Hieronymus Brunschwig and Marsilio Ficino, *Liber de arte distulandi Simplicia et Composita* [Medicinarius]. *Das nüv buoch der rechten kunst zuo distillieren. Ouch von Marsilio Ficino vnn anderen hochberoempten Ertzte natürliche vnn guote künst* (Strasbourg: Johann Grüninger, 1509), fol. 2v: "vnd von manchen gelerten vnnd leyen man vnnd frowen durch experimennt erfaren hab."

40. Brunschwig, *Small Book*, fol. 15r: "eynen erfarenden vnd gelerten artzet"; fol. 15v: "Ob sie aber keinen artzet noch artzney haben moechten etliche hülff zuo bewisen durch die experiment und erfarung die bewert sint durch würckung der wasser."

41. Brunschwig, *Large Book*, fol. 3v: "broesemlin / welch da sein gefallen durch experiment." On Brunschwig's recipes as *experimenta*, see also Alisha Rankin, "How to Cure the Golden Vein: Medical Remedies as Wissenschaft in Early Modern Germany," in *Ways of Making and Knowing*, ed. Pamela H. Smith, Harold Cook, and Amy Meyers (Ann Arbor: University of Michigan Press, 2014), 109–33.

42. Tillmann Taape, "A Commerce of Secrets: Digital and Performative Approaches to an Early Modern How-to Manuscript at the Making and Knowing Project," in *Tracts of Action: Material, Visual, and Practical Dimensions of Early Modern How-to Books*, ed. Stefan Laube (Leiden: Brill, 2024), 126–51.

43. Joachim Telle, "Das Rezept als literarische Form. Zum multifunktionalen Gebrauch des Rezepts in der Deutschen Literatur," *Berichte zur Wissenschaftsgeschichte* 26, no. 4 (2003): 251–74.

44. The literature on recipes is vast and growing; recent examples include Geneviève Deblock, *Le bâtiment des recettes: Présentation et annotation de l'édition Jean Ruelle, 1560* (Rennes: Presses Universitaires de Rennes, 2015); Sara Pennell and Michelle DiMeo, *Reading and Writing Recipe Books* (Manchester: Manchester University Press, 2013); Thijs Hagendijk, "Unpacking Recipes and Communicating Experience: The Ervarenissen of Simon Eikelenberg (1663–1738) and the Art of Painting," *Early Science and Medicine* 24, no. 3 (2019): 248–82; Leong, *Recipes and Everyday Knowledge*; Smith, *From Lived Experience to the Written Word*.

45. Eamon, *Science and the Secrets of Nature*, 113, 135–38, 147–51. On testing of recipes in academies of secrets, see William Eamon, "From Secrets of Nature to Public Knowledge," in Westman and Lindberg, *Reappraisals of the Scientific Revolution*, 333–66; William Eamon, "Alchemy in Popular Culture: Leonardo Fioravanti and the Search for the Philosopher's Stone," *Early Science and Medicine* 5 (2000): 196–213.

46. Pamela Long, *Openness, Secrecy, Authorship: Technical Arts and the Culture of Knowledge from Antiquity to the Renaissance* (Baltimore: Johns Hopkins University Press, 2001), 79–85; Doris Oltrogge, "Transmission of Artists' Knowledge in Germany," in *Transmission of Artists' Knowledge*, ed. Mark Clarke, Bert De Munck, and Sven Dupré (Brussels: KVAB Press, 2012), 25–30; Pamela H. Smith, "What Is a Secret? Secrets and Craft Knowledge in Early Modern Europe," in *Secrets and Knowledge in Medicine and Science, 1500–1800*, ed. Elaine Leong and Alisha Rankin (Farnham: Ashgate Publishing, 2011), 47–66; Smith, *From Lived Experience to the Written Word.*

47. Peter Assion, *Altdeutsche Fachliteratur* (Berlin: Erich Schmidt, 1973), 5–7; Joachim Telle, "Verzeichnis der Exponate: Bücher," in *Pharmazie und der gemeine Mann: Hausarznei und Apotheke der frühen Neuzeit*, ed. Joachim Telle (Wolfenbüttel: Herzog August Bibliothek, 1988), 60–66.

48. Francisco Alonso-Almeida, "Genre Conventions in English Recipes, 1600–1800," in *Reading and Writing Recipe Books, 1550–1800*, ed. Michelle DiMeo and Sara Pennell (Manchester: Manchester University Press, 2013), 68–90; Telle, "Das Rezept als literarische Form."

49. Leong, *Recipes and Everyday Knowledge.*

50. Mesue's *De medicinis universalibus* was seen as the major authority on compounding drugs. The *Antidotarius* compiled by Nicolaus Salernitanus around 1200 was commonly used as a pharmaceutical reference in medical practice. See, e.g., Thomas E. Keys, "The Earliest Medical Books Printed with Movable Type: A Review," *Library Quarterly* 10, no. 2 (1940): 223–25. The official recipe collection which served in the regulation of pharmacy at Strasbourg from around 1500 was in large part based on the *Antidotarius Nicolai*. Cf. Pierre Bachoffner, "La pharmacie est réglementée," in *Histoire de la médecine à Strasbourg*, ed. Jacques Héran (Strasbourg: La Nuée Bleue, 1997), 92–93.

51. Brunschwig, *Cirurgia*, fol. 118v: "güt grien wunt plaster das die nuwen wund artzet bruchent"; "heilt ein iede frische wund"; "vngelerter scherrer"; "kan er nit anders sprechen dan das es die wunden heilt vnd das von andren gesehen hab bruchen."

52. Brunschwig, *Cirurgia*, fol. 118v: "ist notturfft das ein ieder wund artzet wiß die dugent vnnd krafft der simplicio was do in got in plaster oder salben."

53. Brunschwig, *Cirurgia*, fol. 68r: "die salb werd also gemacht in der apatecken"; "fiat vnguentum secundum artem."

54. Hieronymus Brunschwig, *Liber pestilentialis de venenis epidimie. Das buoch der vergift der pestilentz das da genant ist der gemein sterbent der Trüsen Blatren* (Strasbourg: Johann Grüninger, 1500), fol. 14v: "'So aber der merer teil sollicher recept vnn stück die darin gond nit in tütscher sprach in eim kurtzen begriff gebrucht werden moegen vnd gewonlich in den apotecken gefunden vnn auch gemacht werden / so hab ich sie latinisch . . . gelassen als sie die doctores gewonlich schriben sind dz ein yeder der sollichs recept bedarff / müt durch abschrifft in die apoteck senden vnn die lassen machen."

55. Brunschwig and Ficino, *Medicinarius*, 1509, fol. GG4r: "Ein kurtz apoteckisch underweisung."

56. Brunschwig, *Large Book*, fol. 3v: "Ob dir eincher Doctor schrib oder ordinieret in die Apoteck / du wol verston moechtest / ob er dir wol oder vbel geschriben het."

57. Brunschwig, *Large Book*, fol. 319r: "im sumer so hab ich die bleter von weisem andorn clein gehackt"; "ich habs ouch etwan dür genützet."

58. Brunschwig, *Large Book*, fol. 308r: "Ob du aber also arm bist und den ingber nit hast / so mach mießlin uß iungen nesseln."

59. Brunschwig, *Large Book*, fol. 310v: "dise alle oder ir ein teil."

60. Brunschwig, *Large Book*, fol. 283v: "durch mein hand gethon"; "berümpten Meistern."

61. Leong, *Recipes and Everyday Knowledge*; Smith, *From Lived Experience to the Written Word*; Taape, "Commerce of Secrets."

62. Brunschwig, *Small Book*, fol. 125r: "durch expriment erfaren."

63. Herzog August Bibliothek (HAB), Wolfenbüttel, A: 51.2 Med, fol. 123v: "Ein gudtes Stücklein"; Bayerische Staatsbibliothek (BSB), Munich, Rar. 2166, fol. 351r: "est certe probatum."

64. Brunschwig, *Large Book*, fol. 297r: "bin ich beriefft worden von einem beriempten scherrer."

65. Brunschwig, *Large Book*, fol. 297v: "Da band ich in an den vierorten mit den würg bendeln hert / vnd zoch das bluot."

66. Brunschwig, *Large Book*. fol. 297v: "vß solcher erfarniß hab ich seidher vil ander hendel geton / welich da nottufft [*sic*] den nit oder wenig habenden menschen oder die weit von den artzeten oder von apodecken seind."

67. Brunschwig, *Large Book*, fol. 57r (misfoliated lxx): "dz er erfaren vnn gelert hatt die kranckheit zu erkennen / vnn die maß der artzeny zu geben"; "dz er besunder artzeny / die er durch sein practica geübet vnn gelert hat / wie die kranckheit geendet worden ist."

68. Brunschwig, *Large Book*, fol. 57r (misfoliated lxx): "ein wolgelerter vnerfarender artzet . . . alle zeit zu vermeiden ist"; "vmb seiner vnerkantnüß willen / ee dz er die lert erkennen / der kranck todt ist."

69. Brunschwig, *Large Book*, fol. 57r (misfoliated lxx): "Zu allen zeiten ein alter artzet für ein iungen zu erwelen ist / ob er nit wol gelert / so ist er aber wol geübt / vnn me erfaren dann der iung."

70. Sabine Schlegelmilch, "Promoting a Good Physician: Letters of Application to German Civic Authorities, 1500–1700," in *Civic Medicine: Physician, Polity, and Pen in Early Modern Europe*, ed. J. Andrew Mendelsohn, Annemarie Kinzelbach, and Ruth Schilling (Abingdon: Routledge, 2020), 88–109.

71. Brunschwig, *Large Book*, fol. 297r: "je me er eilet ye me ich mich sumpt / sprechend mir ist lieber ersterb ee dan wir dar kumen / dan dz er sterb so wir gegenwirtig seint."

72. Brunschwig, *Large Book*, fol. 297v: "was woellent ir thuon / sehent ir nit das er ietz anfahen wil zuo sterben Vnn so erstirbt meniglich spricht ir haben in gedötet."

73. Brunschwig, *Large Book*, fol. 297v: "vnn wa im ein messer wer worden / er het im selber die kell vff geschnitten."

74. Brunschwig, *Large Book*, fol. 297v: "es was kein beitten oder wartten biß mir die bendel wurden"; "vier schnieren damit man die kornseck zuo bindt."

75. Brunschwig, *Large Book*, fol. 57r: "der empiricus alle zyt zu vermeiden."

76. Brunschwig, *Large Book*, fol. 57r: "von vssen als dem cirurgico durch stet vbung die er getrieben vnn gelert hat."

77. Brunschwig, *Large Book*, fol. 57r: "zum offern mal ein solicher artzet ein experiment künnen / dz vor hundert doctoribus verborgen / oder inen nit gegenwertig in der vernunfft zu wissen ist / sunder allein im in der vbung / vnn stets in seinem bruch ist."

78. Brunschwig, *Large Book*, fol. 56v: "ietzundigen Emperici / durch ir practica erfaren vnd funden."

79. Brunschwig, *Large Book*, fol. 3r: "diße heimlichkeit / und verborgene kunst and dz liecht zuobringen und zuo offenbaren."

80. Eamon, "From Secrets of Nature to Public Knowledge"; Brian Copenhaver, "Natural Magic, Hermetism, and Occultism in Early Modern Science," in Westman and Lindberg, *Reappraisals of the Scientific Revolution*, 261–301.

81. Draelants, "Expérience et autorités."

82. Eamon, *Science and the Secrets of Nature*. Since Eamon's seminal study, the study of secrets has expanded, see, e.g., Allison Kavey, *Books of Secrets: Natural Philosophy in England, 1550–1600* (Urbana: University of Illinois Press, 2007); Elaine Leong and Alisha Rankin, *Secrets and Knowledge in Medicine and Science, 1500–1800* (Farnham: Ashgate Publishing, 2011); Long, *Artisan/Practitioners*, 33–38.

83. Dear, "The Meanings of Experience," 114–15.

84. Brunschwig, *Large Book*, fol. 10v: "ein secret vnd ein heimlichkeit der naturen."

85. Brunschwig, *Large Book*, fol. 258r: "Ich muß dir hie offenbaren ein secret von dem frosch geoegel / weliches ich selber offt gethon."

86. Brunschwig, *Large Book*, fol. 296r: "vnd redt im dreü wort in dz linck or . . . Max. Nax. Pax." This formula, in various forms, was widespread in medieval and early modern collections of charms. See Claude Lecouteux, *Le Livre des Grimoires* (Paris: Imago, 2002), 79–80.

87. Brunschwig, *Large Book*, fol. 296r: "das hab ich vil von im gesehen / Ich hab es auch selber gethon / vnd halff mir etwan / vnd etwan nit / wan ich het kein sundern glauben daruff."

88. Müller, "Erfarung."

89. Jan-Dirk Müller, "'Curiositas' und 'erfarung' der Welt im frühen deutschen Prosaroman," in *Literatur und Laienbildung im Spätmittelalter und in der Reformationszeit*, ed. Ludger Grenzmann and Karl Stackmann (Stuttgart: Germanische Symposien, 1984), 252–71.

90. Christian Kiening, "Erfahrung und Vermessung der Welt in der frühen Neuzeit," in *Text—Bild—Karte: Kartographien der Vormoderne*, ed. Christian Kiening and Jürg Glauser (Freiburg im Breisgau: Rombach Verlag, 2007), 221–51; Müller, "Erfarung," 312–21; Müller, "'Curiositas' und 'erfarung' der Welt im frühen deutschen Prosaroman"; Verena Schmid Blumer, *Ikonographie und Sprachbild: Zur reformatorischen Flugschrift "Der gestryfft Schwitzer Baur"* (Tübingen: Niemeyer, 2004), 245–50.

91. Sebastian Brant, *Das Narren schyff* (Basel: Johann Bergmann, 1494), fol. l3v: "Von erfarung aller land."

92. Johann Geiler von Kaysersberg, *Des hochwirdigen doctor keiserspergs narrenschiff so er gepredigt hat zuo straßburg* (Strasbourg: Johann Grüninger, 1520), fol. 80r.

93. Brant, *Das Narren schyff*, fol. 3v: "Der allzyt bgert erfaren vil / Vnd sich dar von nit besseren wil"; fol. 4r: "Eyn narr ist wer vil land durchfert // Vnd wenig kunst / noch tugent lert."

94. Geiler von Kaysersberg, *Doctor keiserspergs narrenschiff*, fol. 80v: "xii handtwerck, xiii vnglück."

95. Geiler von Kaysersberg, *Doctor keiserspergs narrenschiff*, fol. 12v: "müssen ire hantweck vor leren vnd erlaubung oder ein zunfft haben ee sie es getoeren treiben . . . aber sie seint frei / die bücher dichten sie haben kein gesatzt das sie bindet / es macht ietzt bücher wer da wil on erfarnis."

96. Brant, *Das Narren schyff*, fol. 7v: "abergloub."

97. Geiler von Kaysersberg, *Doctor keiserspergs narrenschiff*, fol. 86r: "Ich habe es erfaren das / das und das hilfft / ich bin gesund davon worden."

98. Brunschwig, *Cirurgia*, fol. π2r: "von vil herfarnden artzten gelernet."

99. Brunschwig, *Liber pestilentialis*, fol. A2v: "in einer stat dar in ein erschröcklich pestilentz regieret."

100. Brunschwig, *Liber pestilentialis*, fol. A2v: "das dir nit geschehe wie mir geschehen ist."

101. Kiening, "Erfahrung und Vermessung der Welt in der frühen Neuzeit."

102. Müller, "Erfarung," 329–31.

103. On Paracelsus and *erfarung*, see Bruce T. Moran, *Paracelsus: An Alchemical Life* (2019), 94–118; Walter Pagel, *Paracelsus: An Introduction to Philosophical Medicine in the Era of the Renaissance* (Basel: Karger, 1958), 56–58; Smith, *Body of the Artisan*, 82–92, 95–128.

104. Daniel Carey, "The Problem of Credibility in Early Modern Travel," *Renaissance Studies* 33, no. 4 (2019): 524–47.

105. Andrea Frisch, *The Invention of the Eyewitness: Witnessing and Testimony in Early Modern France* (Chapel Hill: University of North Carolina Press, 2004).

106. Müller, "Erfarung," 323f; Katharine Park, "Natural Particulars: Medical Epistemology, Practice, and the Literature of Healing Springs," in *Natural Particulars: Nature and the Disciplines in Renaissance Europe*, ed. Nancy Siraisi and Anthony Grafton (Cambridge, MA: MIT Press, 1999), 357–59.

107. Steven Shapin and Simon Schaffer, *Leviathan and the Air-Pump: Hobbes, Boyle, and the Experimental Life* (Princeton, NJ: Princeton University Press, 1985), 22–79; Steven Shapin, *A Social History of Truth: Civility and Science in Seventeenth-Century England* (Chicago: University of Chicago Press, 1994), 243–309.

108. Steven Shapin, "The Invisible Technician," *American Scientist* 77, no. 6 (1989): 554–63; Shapin, *Social History of Truth*, 355–407.

109. The issue of artisans' contributions to the Scientific Revolution was already raised by Edgar Zilsel, "The Sociological Roots of Science," *American Journal of Sociology* 47 (1942): 544–62, but it has received substantive attention in recent scholarship, e.g., Deborah Harkness, *The Jewel House: Elizabethan London and the Scientific Revolution* (New Haven: Yale University Press, 2007); Long, *Artisan/Practitioners*; Smith, *Body of the Artisan*.

110. Frisch, *The Invention of the Eyewitness*, 76–116. See also Smith, *Body of the Artisan*, 31–57.

111. Brunschwig, *Large Book*, fol. 281r: "als ich selber gesehen hab an einem Ersamen Ammeister der loblichen stat Straßburg / vnd an dem edlen iuncker Wilhelm von Bach. Her hanß Ingollt ein treffelicher kouffman. vnd andere mer."

112. Brunschwig, *Large Book*, fol. 123r: "als ich Iheronymus Brunschweick selber gesehen hab einen Vicarien der hohen stifft Straßburg by hundert iaren im alter / diß wasser nutzen so er schwach vnd kranck was."

113. Brunschwig, *Large Book*, fol. 273v: "Hans pfarrer von Würtzburg."

114. Brunschwig, *Large Book*, fol. 297v: "als mir herr Burckhart zuo den kartüselern selber gesagt hatt."

115. Brunschwig, *Large Book*, fol. 31r: "einen alten priester."

116. Brunschwig, *Large Book*, fol. 208v: "von eim geistlichen menschen mir geoffenbart ist."

117. AVCUS VI 386/6, 1523: "domit die geschicht vnnd grundt der worheit gentzlich erfaren werde / Bitt ich den Strengen*n* hern*n* Hanns Bork Ritter / Doctor Michell meyger Statt artzt / Meister hansenn von Constantz vnnd Jacob scherer gegen der pfennig bruck deshalb zubefrogen vnnd zuuerhören*n*."

118. Brunschwig, *Cirurgia*, fol. 14r: "die da minner künen gelernt oder herfarn hon"; "ufnemen zuo einer lere und ungestrafft lassen"; "grüntlicher erfarung."

119. Brunschwig, *Cirurgia*, fol. 16v: "Dan hab ich gesehen etwan manchen."

120. Brunschwig, *Cirurgia*, fol. 29v: "er mir selbs mit sinem mund hat geseit wie er es her vß gethon het."

121. Brunschwig, *Large Book*, fol. 2v: "mit meinen henden / durch die kunst der Practica gethan vnn geübt hab . . . mit meinen augen gesehen und mit meinen oren gehört hab / nit wenig / sunder zum offteren mal."

122. Brunschwig, *Large Book*, fol. 210v–211r: "Desgleichen hab ich ouch gesehen an eim ritter / der do was by dem künig von Sicilien . . . / der do was haben finstere oder dunckele ougen / vnn gar nahe blint wz / der allen tag trincken was vß einem hültzen becher / der gemacht was von dem holtz ebhoew / genant in latin Edera / gewunnen vnd getreyet von eim Trechseler / zwischen den zweyen vnser lieben frowen tagen von eim eichboum kam wiederumb sein gesicht," "Deß gleichen so det ouch her Melchior bluom / burgermeister der statt Franckfurt / der mir fünfzehen gülden darumb geschenckt hat. Deß gleichen ein begin oder ley schwester in der stat Straßburg was / die ietzo nit wol den schein sach / vnd vß eim sollichen becher trincken was / in kurttzen vergangenen tagen sie ein gemeine ney nadelen demen was / das ich selber ouch mit meinen ougen gesehen hab / vnd mich vast darab verwundert."

123. Ernest Wickersheimer, "Fries, Lorenz," *Neue Deutsche Biographie* 5 (1961): 609f, http://www.deutsche-biographie.de/pnd119451662.html.

124. Lorenz Fries, *Spiegel der Artzny des geleichen vormals nie von keinem doctor in tütsch vßgangen* [. . .] (Strasbourg: Johann Grüninger, 1518).

125. Fries, *Spiegel der Artzny*, 1518, fol. 9r: "mit grosem ernst zuo samen bracht / eins teils vß ler der alten berumpten meister / anders teils durch eigne müe vnn arbeit mit vil schönen erfarnen vnd bewerten stucken."

126. Fries, *Spiegel der Artzny*, 1518, fol. 13v: "spinnenfresser verlauffen henckerßbuoben / kesselringmacher / zanbrecher."

127. Fries, *Spiegel der Artzny*, 1518, fol. 13v: "irrend seind auch nit knecht der natur. Wan sie wissent was / aber nit warumb . . . erkennend nit ursach."

128. Fries, *Spiegel der Artzny*, 1518, fol. 14r: "Diascorides hat erfaren das der fenchel allweg guot sei den augen. Wann als die schlangen in der zeit des glentzen vß iren hölern kriechen / sach Diascorides das sie ire augen an den fenchel riben."

129. Fries, *Spiegel der Artzny*, 1518, fol. 14r: "scherer / hodenschneider / die zuo zeiten trenck geben / pflaster / salben vnn andere irer sachen gewisse gesellen"; "Etlich apotecker." The term *hodenschneider*, more literally a pig or sheep gelder, was also used for lithotomists, which seems more likely in this context. Cf. "hodenschneider" in Jacob and Wilhelm Grimm, *Deutsches Wörterbuch* (Leipzig, 1854–1961), vol. 10, col. 1654, http://www.woerterbuchnetz.de/DWB?lemma=hodenschneider.

130. Fries, *Spiegel der Artzny*, 1518, fol. 14r: "zwei schoener bücher."

131. Fries, *Spiegel der Artzny*, 1518, fol. 14r: "wie wol er im grund der ler vnn das latin verstanden / het wol gnuogge gethon / sovil er diser kunst von der artzney gewüßt / hat sich wol geübt."

132. Fries, *Spiegel der Artzny*, 1518, fol. 14r: "Es sei nit gut glauben geben der erfarung on vrsach vnn grundliche bewerung."

133. Lorenz Fries, *Uslegung der merkarthen* (Strasbourg: Grüninger, 1525), fol. 2v: "so haben die Marinal dise wind uß teglicher erfarung in solliche zal gesetzet." See also Kiening, "Erfahrung und Vermessung der Welt in der frühen Neuzeit."

134. Fries, *Uslegung der merkarthen*, fol. 2v: "die alten natürlichen meister / so vor alten zeiten in iren hütlin gesessen und ursach die natürlichen ding ersuchet."

135. Sachiko Kusukawa, *Picturing the Book of Nature: Image, Text, and Argument in Sixteenth-Century Human Anatomy and Medical Botany* (Chicago: University of Chicago Press, 2012), 8–19.

136. Lorenz Fries, *Spiegel der Artzny gemachet vnd widerumb mit ernst vbersehen* (Strasbourg: Johann Grüninger, 1529), fol. a2v: "welche wissenhaft weren /vnnd auch der anatomey / vnnd der natürlichen ding wol eruebet vnnd erfaren."

137. Fries, *Spiegel der Artzny*, 1529, fol. a2v: "Ein reformation machen / das die vngeschickten vßgemuster / vnn die gelerten vnd erfarnen herfür gezogen werden."

138. Fries, *Spiegel der Artzny*, 1529, fol. a2r: "Das wiewol er in teütscher sprach vertolmetschet / das es doch eigentlich alles in sich halt / was in den allerbesten autoribus der atzneyen erfunden / welchs ich auch selbs etwan nit geglaubt."

139. Michael Giesecke, "'Volkssprache' und 'Verschriftlichung des Lebens' im Spätmittelalter am Beispiel der Genese der gedruckten Fachprosa in Deutschland," in *Literatur in der Gesellschaft des Spätmittelalters*, ed. Hans Umbrecht (Heidelberg, 1980), 39–70.

140. Fries, *Spiegel der Artzny*, 1518, title page: "den gestreiffelten leyen."

Chapter Eight

1. Hieronymus Brunschwig, *Liber de arte Distillandi de Compositis* [Large book of distillation]. *Das buch der waren kunst zü distillieren die Composita vnn simplicia / vnd dz Buoch thesaurus pauperum / Ein schatz der armen* (Strasbourg: Johann Grüninger, 1512), fol. 283r: "schreigen vnnd rüffen ich hort / Wie gar / wie gar / wie gar hast du vnßer vergessen / Ich erwacht / Sprechende / Wer seind ir / Wir seint die armen / wer seint die armen / Die weder haller noch pfennig haben / Wer me / wir seind die bauleüt / auß den doerffern / Wer me / wir seind die von den schlossen vnnd vß den cleinen stettlin / Wan wir schon zimlich gelt hont / haben wir doch weder artzet noch die Ertzney / das wir vnß mügen behelffen."

2. Brunschwig, *Large Book*, fol. 283r: "schatz der armen Artzny / Vnd auch deren / die da vff den schlossen / vnd in den cleinen stetlicn / vnn dörffern wonen / Die da nit wol die hohen artzny erreichen mögen."

3. Brunschwig, *Large Book*, fol. 56v: "Darvmb diß buoch vnn ander bücher von mir gemacht / für den gemeinen vnn die da nit haben oder haben moegen einen gelerten vnd erfarenden artzet."

4. Alisha Rankin, "Germany," in *The Oxford History of Popular Print Culture*, ed. Joad Raymond, vol. 1. (New York: Oxford University Press, 2011), 205–13; Joachim Telle, "Medizin für den gemeinen Mann," in *Bibliotheca Palatina. Katalog zur Ausstellung vom 8. Juli bis 2. November 1986, Heiliggeistkirche Heidelberg*, ed. Elmar Mittler (Heidelberg: Braus, 1986), 335–40. On "popular" medical literature in England, see Mary Fissell, "Readers, Texts, and Contexts: Vernacular Medical Works in Early Mod-

ern England," in *The Popularization of Medicine, 1650–1850*, ed. Roy Porter (London: Routledge, 1992), 72–96; Mary Fissell, "The Marketplace of Print," in *Medicine and the Market in England and Its Colonies, c. 1450–c. 1850*, ed. Patrick Wallis and Mark Jenner (Basingstoke: Palgrave Macmillan, 2007), 108–32; Mary Fissell, "Popular Medical Books," in Raymond, *Oxford History of Popular Print Culture*, 417–31.

5. Lyndal Roper, "'The Common Man,' 'The Common Good,' 'Common Women': Gender and Meaning in the German Reformation Commune," *Social History* 12, no. 1 (1987): 1–21.

6. Verena Schmid Blumer, *Ikonographie und Sprachbild: Zur reformatorischen Flugschrift "Der gestryfft Schwitzer Baur"* (Tübingen: Niemeyer, 2004).

7. Hieronymus Brunschwig, *Liber pestilentialis de venenis epidimie. Das buoch der vergift der pestilentz das da genant ist der gemein sterbent der Trüsen Blatren* (Strasbourg: Johann Grüninger, 1500), fol. A3r: "angonden meistern vnn wund ertzt vnn scherern"; fol. 15r–v: "für den gemeinen man in schloesser doerffer."

8. Brunschwig, *Liber pestilentialis*, fol. 14v: "Etlich koestlich für die herren / etlich mittelmessig für die burger / etlich für dz gemein volck im land die / die apoteck nit erreichen mügen."

9. Brunschwig, *Liber pestilentialis*, fol. A3r: "den gelerten . . . wan sie vil anders groessers vnn merers zuo schaffen haben."

10. Hieronymus Brunschwig, *Liber de arte distillandi de simplicibus* [Small book of distillation]. *Das buch der rechten kunst zü distilieren die eintzigen ding* (Strasbourg: Johann Grüninger, 1500), fol. A2r–v: "die do begeren zuo leren die maß vnn kunst der distillierung diß wercks an zefahen."

11. Brunschwig, *Small Book*, fol. E1r: "für den gemeynen menschen / der weder artzet noch artzneney haben mag / auch nit zuo bezalen"; fol. D2r: "ist nit für den gemeynen man."

12. Hieronymus Brunschwig and Marsilio Ficino, *Liber de arte distulandi Simplicia et Composita* [Medicinarius]. *Das nüv buoch der rechten kunst zuo distillieren. Ouch von Marsilio Ficino vnn anderen hochberoempten Ertzte natürliche vnn guote künst* (Strasbourg: Johann Grüninger, 1509), fol. A1v: "vff den doerffern vnd schloessern die ein grosse stat mit einer apoteck nit wol erreichen moegen."

13. Brunschwig and Ficino, *Medicinarius*, 1509, fol. X3v: "vnnd das dritte buoch / sagt von dem leben von himmel herab . . . Das gar hoch zuo verston ist / deßhalb hie vß gelon."

14. Brunschwig, *Small Book*, fol. C1v: "dz groß buoch ich ouch sehen will lassen zuo siner zyt den gelerten diß wercks."

15. Brunschwig, *Large Book*, fol. 140r: "in der inern artzeny / weliche zuogehoerig seint dem phisico."

16. Christoph Friedrich and Wolf-Dieter Müller-Jahncke, *Geschichte der Pharmazie von der frühen Neuzeit bis zur Gegenwart* (Eschborn: Govi-Verlag, 2005), 79–92, 100–101; Joachim Telle, *Pharmazie und der gemeine Mann: Hausarznei und Apotheke der frühen Neuzeit* (Wolfenbüttel: Herzog August Bibliothek, 1988). On the common man as a reader of practical literature, see also William Eamon, *Science and the Secrets of Nature: Books of Secrets in Medieval and Early Modern Culture* (Princeton, NJ: Princeton University Press, 1994), 99–102.

17. Gundolf Keil, "Der Hausvater als Arzt," in *Haushalt und Familie in Mittelalter und früher Neuzeit*, ed. Trude Ehlert (Sigmaringen: Thorbecke, 1991), 219–43; Rudolf Schenda, "Der 'gemeine Mann' und sein medikales Verhalten im 16. und 17. Jahr-

hundert," in *Pharmazie und der gemeine Mann: Hausarznei und Apotheke der frühen Neuzeit*, ed. Joachim Telle (Braunschweig: Waisenhaus Verlag, 1982), 9–20; Joachim Telle, "Verzeichnis der Exponate: Bücher," in *Pharmazie und der gemeine Mann: Hausarznei und Apotheke der frühen Neuzeit*, ed. Joachim Telle (Wolfenbüttel: Herzog August Bibliothek, 1988), 108–9.

18. Roper, "Common Man."

19. Brunschwig, *Liber pestilentialis*, fol. 15r: Brunschwig, 1500b; fol. 15r: "die gemeinen burger."

20. Brunschwig, *Liber pestilentialis*, fol. 15r–v: "für den gemeinen man in schloesser doerffer auch für die rychen."

21. Brunschwig, *Liber pestilentialis*, fol. 14v: "die sindt nit als koestlich als die obern vnd sind für den gmeinen man der die apotecken erreichen vnd das bezalen mag."

22. Brunschwig, *Small Book*, fol. 210v: "ein anzeigung geben ist die weder schriben noch lesen kündent."

23. Brunschwig, *Large Book*, fol. 357r: "vff das dir nit geschehe wie mir / so ich lesen was die bücher der artzney / getruckt in lateinische zungen."

24. Brunschwig, *Large Book*, fol. 56v: "aber dem gemeinen . . . me schedlich dan nutzlich ist."

25. Michel Pastoureau, *The Devil's Cloth: A History of Stripes and Striped Fabric* (New York: Columbia University Press, 2001), 41–44; Ulinka Rublack, *Dressing Up: Cultural Identity in Renaissance Europe* (Oxford: Oxford University Press, 2010), 103–6; Schmid Blumer, *Ikonographie und Sprachbild*, 209.

26. Sebastian Brant, *Das Narren schyff* (Basel: Johann Bergmann, 1494), fol. o3v: "zerhacket / vnd gespreit // Mit aller varb." See also Schmid Blumer, *Ikonographie und Sprachbild*, 233.

27. Johann Geiler von Kaysersberg, *Des hochwirdigen doctor keiserspergs narrenschiff so er gepredigt hat zuo straßburg* (Strasbourg: Johann Grüninger, 1520), fol. 165v: "verlaßt das cleid seins vatters"; "becleidet sich vnd sein weib wie ein ritter."

28. Geiler von Kaysersberg, *Doctor keiserspergs narrenschiff*, fol. 28v: "von cleinen bletzlin sie zammen gestücket seint also dz sie me kosten zemachen den das thuoch wert ist."

29. Brant, *Das Narren schyff*, fol. i7r: "stryffecht vnd gelert."

30. Geiler von Kaysersberg, *Doctor keiserspergs narrenschiff*, fol. Z2r: "die gestreifletten leyen"; Johann Geiler von Kaysersberg, *Navicula sive speculum fatuorum* (Strasbourg: Matthias Schürer, 1510), fol. Z3v. Both cited in Schmid Blumer, *Ikonographie und Sprachbild*, 241.

31. Geiler von Kaysersberg, *Doctor keiserspergs narrenschiff*, fol. Z1v: Geiler von Kaysersberg, 1520, sig. Z1v: "irrungen vnd ketzereyen."

32. Thomas Murner, *Der Schelmen Zunfft* (Frankfurt: Beatus Murner, 1512), fol. p1v: "vier latynscher wort"; fol. p2r: "scharpffe question."

33. Schmid Blumer, *Ikonographie und Sprachbild*.

34. Brant, *Das Narren schyff*, fol. i6v; Sebastian Brant, *Stultifera navis* (Basel: Johann Bergmann, 1497), fol. 53v.

35. Tillmann Taape, "Common Medicine for the Common Man: Picturing the 'Striped Layman' in Early Vernacular Print," *Renaissance Quarterly* 74, no. 1 (2021): 1–58.

36. Melanie Panse, *Hans von Gersdorffs "Feldbuch der Wundarznei": Produktion, Präsentation und Rezeption von Wissen* (Wiesbaden: Reichert, 2012), 113–14.

37. Schmid Blumer, *Ikonographie und Sprachbild*, 215, 223–27.

38. Taape, "Common Medicine."

39. Brunschwig, *Liber pestilentialis*, fol. A2v: "wirdigen vnn hochgelerten doctores in den fryen künsten der artzeny deren diener vnd vnderthenig ich all zit bin / deren rath man billich pflegen soll ob man die haben mag."

40. Brunschwig, *Liber pestilentialis*, fol. D3r: "die weder zuo vil kalt noch zuo vil heiß sind sunder man sie alzyt bruchen mag."

41. Brunschwig, *Large Book*, fol. 56v: "ob du ein artzet haben magst / wil ich dir offenbaren sagen vnn ercleren / welcher zuo erwellen vnn die andern zuo vermeiden seint."

42. Roper, "Common Man."

43. Brunschwig and Ficino, *Medicinarius*, 1509, fol. 2v: "von manchen gelerten vnnd leyen man vnnd frowen"; the nuns are mentioned in Brunschwig, *Large Book*, fol. 56v.

44. Roper, "Common Man."

45. Brunschwig, *Liber pestilentialis*, fol. A2r: "mynen genedigen lieben herren der keiserlichen statt Straßburg"; fol. A3r: "in vnn der gantzen commun etwas darvon zuo schriben."

46. Brunschwig, *Large Book*, fol. 3r: "zuo wolgefallen allen iren burgern vnd ynwonenden / mit sampt der gantzen commun."

47. Lorenz Fries, *Spiegel der Artzny des geleichen vormals nie von keinem doctor in tütsch vßgangen* [. . .] (Strasbourg: Johann Grüninger, 1518), fol. 14r: "einer dem die armen lüt für sein hauß lauffen wenn sie wöllen fisch kauffen schryend / Ach lieber du hilffest allen menschen hilff vns auch / wir seind in den dörffern vnn schlössern / findent nit so vil fisch zuo kauffen als bei denem hauß Des halb er inen zwei schoener bücher gemacht hat / darin du findest vil hüpscher ler."

48. Fries, *Spiegel der Artzny*, 1518, fol. 20v: "Lieber meister du sagst mir von den complexionen / aber nit wie ich wol daruon hab gehoert sagen vnsern pfarrer / auch bin ich ein wenig ein gestreiffter ley hab dauon gelesen / dz nur vier seien."

49. Fries, *Spiegel der Artzny*, 1518, fol. 21r: "den gestreifften leyen muoß man mer vßlegung machen dann den gelerten."

50. Lorenz Fries, *Spiegel der Artzny* (Strasbourg: Balthasar Beck, 1532), fol. A2v: "yr minder schonen / dann Martinus Luther des Bapsts."

51. On Fries's stance on divulging medicine to the layman, see Schmid Blumer, *Ikonographie und Sprachbild*, 178–81, 250–55.

52. Schmid Blumer, *Ikonographie und Sprachbild*.

53. John L. Flood, "The Printed Book as a Commercial Commodity in the Fifteenth and Early Sixteenth Centuries," *Gutenberg-Jahrbuch* 76 (2001): 172–82.

Chapter Nine

1. Walther Hermann Ryff and Hieronymus Brunschwig, *New Vollkommen Distillierbuoch Wolgegründter kuenstlicher Distillation* (Frankfurt: Christian Egenolff, 1597), fol. π2r: "mit der Schweitzerischensprach vermengt / dz es also nit ein jeder zum besten verstehen moegen."

2. Hieronymus Brunschwig, *Boek der Wundenartzstedye* (Rostock: Ludwig Dietz, 1518). Chiara Benati, "The 1518 Low German Edition of Hieronymus Brunschwig's Buch der Cirurgia and its Terminology," in *Medieval German Textrelations: Trans-*

lations, Editions, and Studies, ed. S. Jefferis (Göppingen: Kümmerle Verlag, 2012), 189–245.

3. Staatsbibliothek zu Berlin (StaBi), Jg 3484.

4. Recent examples of this growing field include Tara Alberts, Sietske Fransen, and Elaine Leong, *Translating Medicine Across Premodern Worlds*, vol. 37, *Osiris* (Chicago: University of Chicago Press, 2022); Jaime Marroquin Arredondo and Ralph Bauer, *Translating Nature: Cross-Cultural Histories of Early Modern Science* (University of Pennsylvania Press, 2019); Sara K. Barker and Brenda M. Hosington, *Thresholds of Translation: Paratexts, Print, and Cultural Exchange in Early Modern Britain* (Basingstoke: Palgrave Macmillan, 2018).

5. Elaine Leong, "Transformative Itineraries and Communities of Knowledge in Early Modern Europe: The Case of Lazare Rivière's *The Practice of Physick*," in *Civic Medicine: Physician, Polity, and Pen in Early Modern Europe*, ed. J. Andrew Mendelsohn, Annemarie Kinzelbach, and Ruth Schilling (Routledge, 2020), 257–79.

6. Isabelle Pantin, "The Role of Translations in European Scientific Exchanges in the Sixteenth and Seventeenth Centuries," in *Cultural Translation in Early Modern Europe*, ed. Peter Burke and R. P. Hsia (Cambridge: Cambridge University Press, 2007), 163–79.

7. Marwa Elshakry and Carla Nappi, "Translations," in *A Companion to the History of Science*, ed. Bernard Lightman (Hoboken: John Wiley & Sons, 2019).

8. Harold Cook and Sven Dupré, "Introduction," in *Translating Knowledge in the Early Modern Low Countries*, ed. Harold J. Cook and Sven Dupré (Münster: LIT Verlag, 2012), 3–17.

9. Sietske Fransen, "Introduction: Translators and Translations of Early Modern Science," in *Translating Early Modern Science*, ed. Sietske Fransen, Niall Hodson, and Karl Enenkel (Leiden: Brill, 2017), 1–14; James A. Secord, "Knowledge in Transit," *Isis* 95, no. 4 (December 2004): 654–72, https://doi.org/10.1086/430657.

10. Cook and Dupré, "Introduction."

11. Peter Jones, *The Medicine of the Friars in Medieval England* (Woodbridge: Boydell & Brewer, 2024), 97–122.

12. Sietske Fransen, "Latin in a Time of Change: The Choice of Language as Signifier of a New Science?" *Isis* 108, no. 3 (September 2017): 629–3, https://doi.org/10.1086/694189.

13. Alix Cooper, "Latin Words, Vernacular Worlds: Language, Nature, and the 'Indigenous' in Early Modern Europe," *East Asian Science, Technology, and Medicine* 26 (2007): 17–39.

14. Shireen Hamza, "Vernacular Languages and Invisible Labor in Ṭibb," in Alberts, Fransen, and Leong, *Translating Medicine Across Premodern Worlds*, 115–38.

15. Alisha Rankin, "New World Drugs and the Archive of Practice: Translating Nicolás Monardes in Early Modern Europe," in Alberts, Fransen, and Leong, *Translating Medicine Across Premodern Worlds*, 67–88.

16. Anthony Grafton, *Inky Fingers: The Making of Books in Early Modern Europe* (Cambridge, MA: Harvard University Press, 2022), 29–55.

17. Thomas Sprat, *The History of the Royal Society* (1667), cited in Felicity Henderson, "Faithful Interpreters? Translation Theory and Practice at the Early Royal Society," *Notes and Records of the Royal Society* 67, no. 2 (2013): 103.

18. Felicity Henderson, "Making 'The Good Old Man' Speak English: The Reception of Antoni van Leeuwenhoek's Letters at the Royal Society, 1673–1723," in

Cook and Dupré, *Translating Knowledge in the Early Modern Low Countries*, 243–68; Henderson, "Faithful Interpreters?"; Felicity Henderson, "Translation in the Circle of Robert Hooke," in Fransen, Hodson, and Enenkel, *Translating Early Modern Science*, 17–40.

19. Belle and Hosington, *Thresholds of Translation*; Sara K. Barker and Brenda M. Hosington, *Renaissance Cultural Crossroads: Translation, Print and Culture in Britain, 1473–1640*, vol. 15 (Leiden: Brill, 2013).

20. Elaine Leong, "When the Tallamys Met John French: Translating, Printing, and Reading *The Art of Distillation*," in Alberts, Fransen, and Leong, *Translating Medicine Across Premodern Worlds*, 89–112, https://press.uchicago.edu/ucp/books/book/distributed/O/bo177851902.html; Maríaluz López-Terrada and José Pardo-Tomás, "Reading about New World Nature in the Low Countries: The Editions of *Crónicas De Indias*, 1493–1600," in Cook and Dupré, *Translating Knowledge in the Early Modern Low Countries*, 53–78.

21. Tara Alberts, Sietske Fransen, and Elaine Leong, "Translating Medicine, ca. 800–1900: Articulations and Disarticulations," in Alberts, Fransen, and Leong, *Translating Medicine Across Premodern Worlds*, 1–21.

22. Jane Tylus and Karen Newman, "Introduction," in *Early Modern Cultures of Translation*, ed. Jane Tylus and Karen Newman (Philadelphia: University of Pennsylvania Press, 2015), 1–24.

23. Theo Hermans, "Concepts and Theories of Translation in the European Renaissance," in *Übersetzung: ein internationales Handbuch zur Übersetzungsforschung* [An international encyclopedia of translation studies], vol. 3, ed. Harald Kittel, Juliane House, and Brigitte Schultze (Berlin: Walter de Gruyter, 2011), 1420–28.

24. Peter Burke, "Cultures of Translation in Early Modern Europe," in Burke and Hsia, *Cultural Translation in Early Modern Europe*, 7–38; Peter Burke, "Lost (and Found) in Translation: A Cultural History of Translators and Translating in Early Modern Europe," *European Review* 15, no. 1 (February 2007): 83–94, https://doi.org/10.1017/S1062798707000087.

25. Rita Copeland, *Rhetoric, Hermeneutics, and Translation in the Middle Ages: Academic Traditions and Vernacular Texts* (Cambridge: Cambridge University Press, 1995).

26. Hieronymus Brunschwig, *The Noble Experyence of the Vertuous Handy Warke of Surgeri* (London: Petrus Treveris, 1525), fol. A2r, T6v.

27. Piet J. A. Franssen, "Jan van Doesborch's Departure from Antwerp and His Influence on the Utrecht Printer Jan Berntsz," *Quaerendo* 18, no. 3 (January 1, 1988): 163–90, https://doi.org/10.1163/157006988X00097; Piet J. A. Franssen, *Tussen tekst en publiek: Jan van Doesborch, drukker-uitgever en literator te Antwerpen en Utrecht in de eerste helft van de zestiende eeuw* (Amsterdam: Rodopi, 1990), 27–39.

28. Hieronymus Brunschwig, *Dis ist das buch der cirurgia* [Cirurgia]. *Hantwirckung der wund artzny* (Strasbourg: Johann Grüninger, 1497), fol. 27v: "mit der scheeren vorgesprochen"; Brunschwig, *Hantwerck der cirurgien*, fol. D1r: "gelijke hier voor gefigureert staet."

29. Guy de Chauliac, *De Cyrurgie van meester Guid de Cauliaco* (Antwerp: Henrick Eckert van Homberch, 1507).

30. For example a passage stating that severe skull fractures had to be surgically widened to remove putrid matter. See Brunschwig, *Hantwerck der cirurgien*, fol. G2r–v: "Voort waer ennige excusacie . . . ist mogelijc van die cleyne contusien /

nochtans en ist als net van die groote." Cf. Chauliac, *De Cyrurgie van meester Guid de Cauliaco*, fol. 92r.

31. Brunschwig, *Cirurgia*, fol. 50v; Brunschwig, *Hantwerck der cirurgien*, fol. G2v; Chauliac, *De Cyrurgie van meester Guid de Cauliaco*.

32. Heidi Hausse, "The Locksmith, the Surgeon, and the Mechanical Hand: Communicating Technical Knowledge in Early Modern Europe," *Technology and Culture* 60, no. 1 (2019): 34–64; Heidi Hausse, *The Malleable Body: Surgeons, Artisans, and Amputees in Early Modern Germany* (Manchester: Manchester University Press, 2023), 155–209.

33. Brunschwig, *Cirurgia*, fol. 50r: "geresolviert mit hunig oder etlichen dinge als gemeldet ist in dem capitel von der zerknytschung."

34. Brunschwig, *Hantwerck der cirurgien*, fol. G2r: "resolveren met ghesouten wijn enn honich."

35. Brunschwig, *Noble Experyence*, fol. S4v; Brunschwig, *Hantwerck der cirurgien*, fol. G2r: "met wijn gekoect."

36. Brunschwig, *Cirurgia*, fol. 11v: "dy gewisten zeichen vor den anderen allen."

37. Brunschwig, *Noble Experyence*, fol. N2r.

38. Brunschwig, *Cirurgia*, t.p.

39. Brunschwig, *Noble Experyence*, t.p.

40. Brunschwig, *Noble Experyence*, t.p.

41. Brunschwig, *Cirurgia*, fol. 31v: "Item ich hab gehoert von einem warhafftigenn man"; Brunschwig, *Noble Experyence*, fol. C1v.

42. Brunschwig, *Cirurgia*, fol. 29v: "als er mir selbs mit sinem mund hat geseit wie er es her vß gethon het"; Brunschwig, *Hantwerck der cirurgien*, fol. C2v: "also hi mi selve geseyt heeft."

43. Brunschwig, *Cirurgia*, fol. 97v: "gesant von keiser friderichs wund artzet eim andren gütten wund artzet von dem ichs gelert hab."

44. Brunschwig, *Noble Experyence*, fol. R3r.

45. Brunschwig, *Noble Experyence*, fol. E1r.

46. Brunschwig, *Noble Experyence*, fol. R3r.

47. Brunschwig, *Noble Experyence*, fol. T6v.

48. Franssen, "Jan van Doesborch's Departure"; Franssen, *Tussen tekst en publiek*, 36, 76.

49. Henry Richard Tedder, "Andrewe, Laurence," *Dictionary of National Biography*, vol. 1 (Wikisource, 1885), https://en.wikisource.org/w/index.php?title=Dictionary_of_National_Biography,_1885-1900/Andrewe,_Laurence&oldid=10758860.

50. Piet J. A. Franssen, "Jan van Doesborch (?–1536), Printer of English Texts," *Quaerendo* 14 (1986): 259–80; Luuk Houwen, "The Noble Lyfe: An Early English Version of the Hortus Sanitatis," in *Schooling and Society*, ed. Alasdair MacDonald and Michael Twomey (2004).

51. Burke, "Cultures of Translation in Early Modern Europe"; Pantin, "Role of Translations in European Scientific Exchanges."

52. Hieronymus Brunschwig, *Vertuose Boke of Distyllacyon of the Waters of All Maner of Herbes / With the Fygures of the Styllatoryes*, trans. Laurence Andrew (London: Laurence Andrew, 1527), fol. ¶2r.

53. The index was improved as of the 1505 edition, but as Andrea Leerdam points out, the Dutch translation has the instructions printed inside the image of the brick mold, which was left blank in the 1505 edition. Andrea van Leerdam, *Woodcuts as*

Reading Guides: How Images Shaped Knowledge Transmission in Medical-Astrological Books in Dutch (1500–1550) (Amsterdam: Amsterdam University Press, 2023), 368–71.

54. Leerdam, *Woodcuts as Reading Guides*, 175–91.

55. Franssen, "Jan van Doesborch's Departure"; Franssen, *Tussen tekst en publiek*, 36, 92.

56. Hieronymus Brunschwig, *Die distellacien ende virtuyten der wateren* (Brussels: Thomas van der Noot, 1517), fol. 17r: "xiiii. tag oder lenger."; Brunschwig, *Die distellacien ende virtuyten der wateren*, fol. d2r: "veertig daghen oft langhere."

57. Brunschwig, *Small Book*, fol. 12v: "ein creütz von höltzlin also das es auch verspert sie mit eim hültzinn crützlin."; Brunschwig, *Die distellacien ende virtuyten der wateren*, fol. c3v: "een cruys van houtkens maken enn sperren dat selve met een houten cruysken inden muyl des ghelas."

58. Andrea van Leerdam, "Popularising and Personalising an Illustrated Herbal in Dutch," *Nuncius* 36, no. 2 (June 23, 2021): 356–93, https://doi.org/10.1163/18253911-03602006; Leerdam, *Woodcuts as Reading Guides*, 355–61.

59. Leerdam, "Popularising and Personalising."

60. Brunschwig, *Vertuose Boke*, fol. b1v. On Hessian crucibles, see Marcos Martinon-Torres, "The Tools of the Chymist: Archaeological and Scientific Analyses of Early Modern Laboratories," in *Chymists and Chymistry: Studies in the History of Alchemy and Early Modern Chemistry*, ed. Lawrence Principe (Philadelphia: Chemical Heritage Foundation, 2007).

61. Brunschwig, *Small Book*, fol. 7r: "weisse triangulierte filtz"; Brunschwig, *Vertuose Boke*, fol. a2r.

62. Brunschwig, *Vertuose Boke*, fol. b2r.

63. Brunschwig, *Small Book*, fol. 11r: "in ein backofen wan er durch warm ist"; Brunschwig, *Vertuose Boke*, fol. b1v.

64. Brunschwig, *Small Book*, fol. 10r: "also dz kein fuoge vff die ander gang"; Brunschwig, *Vertuose Boke*, fol. a6v.

65. Brunschwig, *Small Book*, fol. 16v: ".vi. oder .viii. tag morgens / mitags unn zuo nacht yedes mal uff .iii. oder .iiii. lot."

66. Brunschwig, *Vertuose Boke*, fol. A2r.

67. Brunschwig, *Vertuose Boke*, fol. a1r.

68. Brunschwig, *Vertuose Boke*, t.p.

69. Paul Slack, "Mirrors of Health and Treasures of Poor Men: The Uses of the Vernacular Medical Literature of Tudor England," in *Health, Medicine, and Mortality in the Sixteenth Century*, ed. Charles Webster (Cambridge: Cambridge University Press, 1979), 237–73.

70. D. Geoghegan, "A Licence of Henry VI to Practise Alchemy," *Ambix* 6, no. 1 (August 1, 1957): 10–17, https://doi.org/10.1179/amb.1957.6.1.10; Charles Webster, "Alchemical and Paracelsian Medicine," in *Health, Medicine, and Mortality in the Sixteenth Century*, 302–3.

71. Peter Jones, "Medical Libraries," in *A History of Libraries in Britain and Ireland*, ed. E. Leedham-Green and T. Webber (Cambridge: Cambridge University Press, 2006), 461–71.

72. Franssen, "Jan van Doesborch, Printer of English Texts"; Houwen, "The Noble Lyfe."

73. Mary Fissell, "Popular Medical Books," in *The Oxford History of Popular Print*

Culture, ed. Joad Raymond (New York: Oxford University Press, 2011), 417–31; Slack, "Mirrors of Health."

74. Anon., *A Newe Boke of Medecynes Intytulyd or Callyd the Treasure of Pore Men* (London: John Rastell for Richard Bankes, 1526), fol. 4r.

75. Lauren Kassell, "Secrets Revealed: Alchemical Books in Early Modern England," *History of Science* 49 (2011): 63–87.

76. Leonardo Fioravanti, *A Short Discours of the Excellent Doctour and Knight, Maister Leonardo Phiorauanti Bolognese Vppon Chirurgerie*, trans. John Hester (London, 1580), t.p.

77. Conrad Gessner, *The Newe Iewell of Health Wherein is Contayned the Most Excellent Secretes of Phisicke and Philosophie*, trans. George Baker (London: Denham, 1576), fol. 4r.

78. Deborah Harkness, *The Jewel House: Elizabethan London and the Scientific Revolution* (New Haven: Yale University Press, 2007), 90.

79. Isabelle Pantin, "John Hester's Translations of Leonardo Fioravanti: The Literary Career of a London Distiller," in *Renaissance Cultural Crossroads: Translation, Print and Culture in Britain 1473–1640*, ed. Sara Barker and Brenda Hosington (Leiden: Brill, 2013), 159–83.

80. Vivian Nutton, "Humanist Surgery," in *The Medical Renaissance of the Sixteenth Century*, ed. Andrew Wear, Roger Kenneth French, and Iain M. Lonie (Cambridge: Cambridge University Press, 1985), 75–99; Robert Ralley, "Medical Economies in Fifteenth Century England," in *Medicine and the Market in England and Its Colonies, c. 1450–c. 1850*, ed. Patrick Wallis and Mark Jenner (Basingstoke: Palgrave Macmillan, 2007), 24–46.

81. Celeste Chamberland, "Honor, Brotherhood, and the Corporate Ethos of London's Barber-Surgeons' Company, 1570–1640," *Journal of the History of Medicine and Allied Sciences* 64, no. 3 (2009): 300–332; Celeste Chamberland, "From Apprentice to Master: Social Disciplining and Surgical Education in Early Modern London, 1570–1640," *History of Education Quarterly* 53, no. 1 (2013): 21–44.

82. Thomas Gale, "An Enchiridion Conteining the Cure of Woundes, Fractures, and Dislocations," in *Certaine Workes of Chyrurgerie* (London: Rowland Hall, 1563), fol. A5r.

83. Thomas Gale, "An Excellent Treatise of Vvounds Made vvith Gonneshot," in *Certaine Workes of Chyrurgerie*, t.p.

84. Thomas Gale, *Certaine Workes of Galens, Called Methodus Medendi* (London: Thomas East, 1567), A6r–v.

85. Nutton, "Humanist Surgery."

86. Nutton, "Humanist Surgery," 96–98.

87. Gale, "Enchiridion," fol. A5r.

88. Guy de Chauliac, *Guydos Questyons*, trans. George Baker (London: Thomas East, 1579), fol. A2r.

89. Celeste Chamberland, "Between the Hall and the Market: William Clowes and Surgical Self-Fashioning in Elizabethan London," *Sixteenth Century Journal* 41, no. 1 (2010): 69–89.

90. Chauliac, *Guydos Questyons*, fol. A2r.

91. William Eamon, "Alchemy in Popular Culture: Leonardo Fioravanti and the Search for the Philosopher's Stone," *Early Science and Medicine* 5 (2000): 196–213;

William Eamon, "Pharmaceutical Self-Fashioning or How to Get Rich and Famous in the Renaissance Medical Marketplace," *Pharmacy in History* 45, no. 3 (2003): 123–29.

92. Gale, *Certaine Workes of Galens*, fol. A6r.

93. Gale, *Certaine Workes of Galens*, fol. A7v.

94. William Clowes, *A Short and Profitable Treatise Touching the Cure of the Disease Called Morbus Gallicus* (London: John Day, 1579), fol. A7v.

95. Chamberland, "Between the Hall and the Market"; Chamberland, "From Apprentice to Master"; Linda Payne, "'A Spedie Reformation:' Barber-Surgeons, Anatomization, and the Reformation of Medicine in Tudor London," in *Paracelsian Moments: Science, Medicine, and Astrology in Early Modern Europe*, ed. Gerhild Scholz Williams and Charles D. Gunnoe (Truman State University Press, 2002), 71–92.

96. Franciscus Arceus, *A Most Excellent and Compendious Method of Curing Woundes in the Head, and in Other Partes of the Body*, trans. John Read (London: Thomas East, 1588), fol. A3r.

97. Gale, *Certaine Workes of Galens*, fol. A4r.

98. Gessner, *Newe Iewell*, fol. *3v.

99. John Hester, *The First Part of the Key of Philosophie* (London: Richard Day, 1580), fol. D8v.

100. Hester, *First Part of the Key*, fol. 3v–4r.

101. Hester, *First Part of the Key*, fol. 4r: "so Experimentum experiendo confirmatur."

102. Leonardo Fioravanti, *A Compendium of the Rationall Secretes*, trans. John Hester (London: John Kingston, 1582), fol. 3r.

103. Joseph Du Chesne, *The Sclopotarie of Iosephus Quercetanus*, trans. John Hester (London: Roger Ward, 1590), fol. A3v.

104. Hester, *First Part of the Key*, fol. 4r.

105. Brunschwig, *Noble Experyence*, T6v.

106. Brunschwig, *Vertuose Boke*, t.p.

107. Brunschwig, *Vertuose Boke*, fol. ¶2r.

108. Helen Moore, "Gathering Fruit: The 'Profitable' Translations of Thomas Paynell," in *Tudor Translation*, ed. Fred Schurink (Basingstoke: Palgrave Macmillan, 2011), 39–57.

109. Chauliac, *Guydos Questyons*, t.p.

110. Gale, "Enchiridion," fol. A2r.

111. Leonardo Fioravanti, *A Ioyfull Iewell Contayning Aswell Such Excellent Orders, Preseruatiues and Precious Practises for the Plague, as Also Such Meruelous Medcins for Diuers Maladies, as Hitherto Haue Not Beene Published in the English Tung*, trans. John Hester (London: Wright, 1579), fol. A2r.

112. Arceus, *Most Excellent and Compendious Method*, fol. ¶1v, ¶3r.

113. Arceus, *Most Excellent and Compendious Method*, fol. A2r.

114. Gale, *Certaine Workes of Galens*, fol. A2v.

115. Fioravanti, *Ioyfull Iewell*, fol. A2r.

116. Gessner, *Newe Iewell*, fol. *3r.

117. Hugh Plat, *The Iewell House of Art and Nature Conteining Diuers Rare and Profitable Inuentions, Together with Sundry New Experimentes in the Art of Husbandry, Distillation, and Moulding* (London: Short, 1594), fol. B3v.

118. Harkness, *Jewel House*, 211–53.

119. Plat, *Iewell House of Art and Nature*, fol. B3r.

120. Ayesha Mukherjee, *Penury into Plenty: Dearth and the Making of Knowledge in Early Modern England* (London: Routledge, 2014).

Chapter Ten

1. Andrea van Leerdam, *Woodcuts as Reading Guides: How Images Shaped Knowledge Transmission in Medical-Astrological Books in Dutch (1500–1550)* (Amsterdam: Amsterdam University Press, 2023).

2. William H. Sherman, *Used Books: Marking Readers in Renaissance England. Philadelphia* (University of Pennsylvania Press, 2008), 8–10.

3. Dániel Margócsy, Mark Somos, and Stephen N. Joffe, *The Fabrica of Andreas Vesalius: A Worldwide Descriptive Census, Ownership, and Annotations of the 1543 and 1555 Editions* (Leiden: Brill, 2018).

4. The seminal article on Harvey was Lisa Jardine and Anthony Grafton, "'Studied for Action': How Gabriel Harvey Read His Livy," *Past & Present*, no. 129 (1990): 30–78. This was followed by elaborations such as Lisa Jardine, "'Studied for Action' Revisited," in *For the Sake of Learning: Essays in Honour of Anthony Grafton*, vol. 2, ed. Ann Blair and Anja-Silvia Goeing (Leiden: Brill, 2016), 999–1017. On Dee, see William H. Sherman, *John Dee: The Politics of Reading and Writing in the English Renaissance* (Amherst: University of Massachusetts Press, 1995). See also Kevin M. Sharpe, *Reading Revolutions: The Politics of Reading in Early Modern England* (New Haven: Yale University Press, 2000); and the Archaeology of Reading Project, (https://archaeologyofreading.org).

5. Jeroen Blaak, *Literacy in Everyday Life: Reading and Writing in Early Modern Dutch Diaries* (Leiden: Brill, 2009); Heidi Brayman Hackel, *Reading Material in Early Modern England: Print, Gender, and Literacy* (Cambridge: Cambridge University Press, 2005).

6. Sherman, *Used Books*.

7. Ann Blair, "Humanist Methods in Natural Philosophy: The Commonplace Book," *Journal of the History of Ideas* (1992): 541–51; Ann Blair, *Too Much to Know: Managing Scholarly Information Before the Modern Age* (New Haven: Yale University Press, 2010); Deborah Harkness, *The Jewel House: Elizabethan London and the Scientific Revolution* (New Haven: Yale University Press, 2007), 181–210; Elaine Leong, *Recipes and Everyday Knowledge: Medicine, Science, and the Household in Early Modern England* (Chicago: University of Chicago Press, 2018); Richard Yeo, *Notebooks, English Virtuosi, and Early Modern Science* (Chicago: University of Chicago Press, 2014).

8. Margócsy, Somos, and Joffe, *Fabrica of Andreas Vesalius*.

9. Sherman, *Used Books*, 20–24.

10. For an overview, see Leerdam, *Woodcuts as Reading Guides*, 240–43.

11. Sherman, *Used Books*, xi–xx; Cormack and Mazzio, *Book Use, Book Theory: 1500–1700*.

12. For a summary of different kinds of annotations and scholars' categories, see Sherman, *Used Books*, 15–20.

13. Henry E. Huntington Library (HEH), Pasadena, 61543. References to individual copies list the holding institution and the shelfmark.

14. Elaine Leong, "Collecting Knowledge for the Family: Recipes, Gender and

Practical Knowledge in the Early Modern English Household," *Centaurus* 55 (2013): 81–103; Sherman, *Used Books*, 53–67.

15. Staatsbibliothek zu Berlin (StaBi), Jl 8060.

16. Ada Palmer, *Reading Lucretius in the Renaissance* (Cambridge, MA: Harvard University Press, 2014), 43–98.

17. Margócsy, Somos, and Joffe, *Fabrica of Andreas Vesalius*, 114–20.

18. Cambridge University Library (CUL), Sel.3.149 no. 2.

19. StaBi, Jl 8060, 7v.

20. Humboldt University Library (HUL), Berlin, Io 53060:F4, fol. 3r.

21. Library of Congress Rosenwald Collection (LOCR), RS81.B673. See also Leerdam, *Woodcuts as Reading Guides*, 262–63.

22. HEH, 622050.

23. Herzog August Bibliothek (HAB), Wolfenbüttel, A: 56.11 Phys.

24. Wellcome Library (WL), London, EPB/D/6380/1.

25. Sherman, *John Dee*, 29–52, 62–65.

26. Sherman, *John Dee*, 65–75.

27. Margócsy, Somos, and Joffe, *Fabrica of Andreas Vesalius*, 91–106.

28. HAB, A: 21.2 Phys. 2°, 53r: "dis mag man ßo glauben ob man will."

29. British Library (BL), London, IB.1495, fol. 77v: "Stinckht grausa*m* ubell etc."

30. BL, IB.1495, fol. 99v: "Schweinbluot eingefast in derm und umb .10. ur vor mitag mit saltz und brodt \geessen/ (doch vor gesoten) ist guet für den hunger"; "mein gsell wust es sey war."

31. HAB, Xb 11131 (1).

32. HEH, 85771, back pastedown: "alda mayster werden"; "in der schuel gangen"; "bey gutten gesellenn."

33. HAB, A: 53 Med. 2°.

34. Brunschwig, *Cirurgia*, fol. 11v: "daz der siech dorecht wird."

35. HAB, A: 53 Med. 2°, fol. 21r: "Item die hefftung geschicht in funfferley weys Zum Ersten merck hernach."

36. BL, IB.1495.

37. BL, IB.1495: "das puch angehörtt dem closter Anndex auf dem heiligen perg."

38. HEH, 85769.

39. Bayerische Staatsbibliothek (BSB), Munich, 2 Inc.c.a. 3867: "emptus in sua ligatura anno 1501 Sub Henrico Abbate."

40. BSB, Rar. 2166 "emptus et inligatus."

41. WL, 1114_D, fol. 4v: "honig distillieren."

42. HEH, 61514, fol. A1r, X4r.

43. Marking one's place in the history of the printed book in this was a common practice, especially in Bibles. See Sherman, *Used Books*, 71–86.

44. HAB, 108.17 Med (2).

45. Johann Gottfried Biedermann, *Geschlechts-Register der Reichs Frey unmittelbaren Ritterschafft Landes zu Francken Löblichen Orts Ottenwald* (Culmbach: Spindler, 1751), Tabula 442 B, fol. Hhh4v.

46. Alisha Rankin, *Panaceia's Daughters: Noblewomen as Healers in Early Modern Germany* (Chicago: University of Chicago Press, 2013).

47. Hieronymus Brunschwig, *Liber de arte distillandi de simplicibus* [Small book of distillation]. *Das buch der rechten kunst zü distilieren die eintzigen ding* (Strasbourg:

Johann Grüninger, 1500), fol. 1v: "Das vor gedruckt distillier buoch hat gehebt ein hinder register / gar nahe so weit begriffen als dz buoch / vnn das zwei mal vßgelegt on not. Vnd hat dan wider gewisen in das recht buoch. Das selbig ist hie erspart vnn ein nüw register gemacht leichtlicher zuo verston vnn behender zuo finden."

48. Francisco Alonso-Almeida, "Genre Conventions in English Recipes, 1600–1800," in *Reading and Writing Recipe Books, 1550–1800*, ed. Michelle DiMeo and Sara Pennell (Manchester: Manchester University Press, 2013), 68–90.

49. Neil Kenny, *The Palace of Secrets: Béroalde de Verville and Renaissance Conceptions of Knowledge* (Clarendon, 1991).

50. Peter Stallybrass, "Books and Scrolls: Navigating the Bible," in *Books and Readers in Early Modern England*, ed. Jennifer Andersen (Philadelphia: University of Pennsylvania Press, 2002), 42–79.

51. Margócsy, Somos, and Joffe, *Fabrica of Andreas Vesalius*, 56–78.

52. BSB, Rar. 2166, fol. 17v–19v.

53. Harvard Botany Arnold Collection (HBA), GEN Ka B83.2 le.

54. WL, EPB/D/6380/1.

55. WL, EPB/D/6380/2.

56. BSB, Res/2 M.med.38.

57. WL, 1115/D.

58. WL, EPB/D/6382/1.

59. BSB, 2128.

60. WL, 1121_B, fol. 14r: "den frawen für ir kranckheyt vnd siechthumb."

61. BSB, 2 Inc.c.a. 3867, fol. 62r.

62. John Rylands Library (JRL), Manchester, 18946, fol. c4v.

63. Katharine Park, "Stones, Bones and Hernias: Surgical Specialists in Fourteenth- and Fifteenth-Century Italy," in *Medicine from the Black Death to the French Disease*, ed. Roger Kenneth French et al. (Aldershot: Ashgate, 1998), 110–30; Paolo Savoia, "Skills, Knowledge, and Status: The Career of an Early Modern Italian Surgeon," *Bulletin of the History of Medicine* 93, no. 1 (2019): 27–54.

64. Park, "Stones, Bones and Hernias."

65. Annemarie Kinzelbach, "Learning to Cut for the Stone: Training Specialised Surgeons in the Holy Roman Empire," in "Learning to Cut and Mend: Surgical Training and Practice, 1450–1800," ed. Maria Pia Donato, Elaine Leong, and Tillmann Taape (London: UCL Press, under review); H. Steinke, "Vom Schererlehrling zum Chirurgenmeister," in *Jakob Ruf, ein Zürcher Stadtchirurg und Theatermacher im 16. Jahrhundert*, ed. Hildegard Keller (Zürich: Chronos Verlag, 2006), 55–65.

66. Annemarie Kinzelbach and Monika Weber, *Steinreich: Das Schneidhaus der Fugger in Augsburg* (Ingolstadt: Deutschen Medizinhistorischen Museums, 2022).

67. Otto Winckelmann, *Das Fürsorgewesen der Stadt Strassburg* (Leipzig: Heinsius, 1922), 102.

68. Hieronymus Brunschwig, *Liber de arte Distillandi de Compositis* [Large book of distillation]. *Das buch der waren kunst zü distillieren die Composita vnn simplicia / vnd dz Buoch thesaurus pauperum / Ein schatz der armen* (Strasbourg: Johann Grüninger, 1512), fol. 332v: "aber mit schneiden kan man kum fol kumen darvon schreiben sunder von meistern geschnitten werden sol die dz vor gesehen vnnd gelert hant darumb so thuo ich hie kein ler daruon."

69. Brunschwig, *Small Book*, fol. 169r: "Aber in worheit selten hab ich gesehen

den gantzen herten stein zuo vertriben dann allein mit ußschnyden. Aber wan er nit volkumlich noch bestetiget oder befestet ist. so mag er wol verzert werden mit disen wasser."

70. Brunschwig, *Small Book*, fol. 168r: "grüen in den lenden oder der blasen oder in den nyeren"; fol. 169r: "zuo vertrihen unnd zuo verzeren."

71. Walter Pagel, *Paracelsus: An Introduction to Philosophical Medicine in the Era of the Renaissance* (Basel: Karger, 1958), 152–61.

72. Brunschwig, *Large Book*, fol. 332v.

73. Brunschwig, *Small Book*, fol. 169r: "dz man es in der harn kachlen wol sicht das vil sant von eim geet."

74. Brunschwig, *Small Book*, fol. 170v: "entpfach den harnn in ein glaß so vindest du in dem glase als wer es sandt / so sichstu das der stein vonn dem menschen geet."

75. Brunschwig, *Large Book*, fol. 333v: "die blaß durch loechert moecht werden von seiner scherfe."

76. Brunschwig, *Large Book*, fol. 264r: "das zuo versuochen so leg ein stein in diß wasser er zergeet / und wurt zuo sand in dreyen tagen"; "kan ich nit gantz bestetigen"; "wann ein cleiner vil ee dann ein grosser zertriben würt."

77. Michael Stolberg, *Experiencing Illness and the Sick Body in Early Modern Europe* (Springer, 2011), 95–142.

78. Brunschwig, *Small Book*, fol. 52v: "als guot als die ander."

79. Hieronymus Brunschwig and Marsilio Ficino, *Das buch der Gesuntheit* [Medicinarius]. *Liber de arte distillandi Simplicia et composita. Das nüv buoch der rechten kunst zuo distillieren. Ouch von Marsilio Ficino vnn andere hochberoepmter Artzte natürliche vnn gute künst* (Strasbourg: Johann Grüninger, 1505), fol. 59v: "Das hat man von eim hirten erfaren der ob .xx. iar vil kruter gekant hat / und die nature etlicher thier."

80. BSB, Rar. 2127, fol. 48r. On the place of the marvelous in Renaissance natural inquiry, see Lorraine Daston and Katharine Park, *Wonders and the Order of Nature, 1150–1750* (New York: Zone Books, 1998), 135–72.

81. CUL, Sel.3.149 no. 2, fol. N4r.

82. BL, IB.1495.

83. HAB, A: 51.2 Med., fol. 123v: "Ein gudtes Stücklein."

84. HAB, M: Mf 320. fol. 57r.

85. Leong, *Recipes and Everyday Knowledge*, 69–123.

86. Anna Dlabačová, Andrea van Leerdam, and John Thompson, *Identifying the Vernacular Reader in the Early Age of Print (c. 1450–1600)* (Leiden: Brill, 2023).

87. Miriam Usher Chrisman, *Lay Culture, Learned Culture: Books and Social Change in Strasbourg, 1480–1599* (New Haven: Yale University Press, 1982).

88. Leerdam, *Woodcuts as Reading Guides*, 262–63.

Conclusion

1. Lissa Roberts and Simon Schaffer, *The Mindful Hand: Inquiry and Invention from the Late Renaissance to Early Industrialisation* (Amsterdam: Royal Netherlands Academy of Arts and Sciences, 2007); Richard J. Oosterhoff, José Ramón Marcaida López, and Alexander Marr, *Ingenuity in the Making: Matter and Technique in Early Modern Europe* (Pittsburgh: University of Pittsburgh Press, 2021); Pamela H. Smith, *From Lived Experience to the Written Word: Reconstructing Practical Knowledge in the Early Modern World* (Chicago: University of Chicago Press, 2022), 43–64; Ursula

Klein, "Hybrid Experts," in *The Structures of Practical Knowledge*, ed. Matteo Valleriani (Cham: Springer, 2017), 287–306.

2. Pamela O. Long, "Trading Zones in Early Modern Europe," *Isis* 106, no. 4 (2015): 840–47, https://doi.org/10.1086/684652; Simon Schaffer et al., *The Brokered World: Go-Betweens and Global Intelligence, 1770–1820* (Sagamore Beach, MA: Science History Publications, 2009).

3. Miriam Usher Chrisman, *Lay Culture, Learned Culture: Books and Social Change in Strasbourg, 1480–1599* (New Haven: Yale University Press, 1982).

4. Projit Bihari Mukharji, *Doctoring Traditions: Ayurveda, Small Technologies, and Braided Sciences* (Chicago: University of Chicago Press, 2016).

5. Robin Wall Kimmerer, *Braiding Sweetgrass: Indigenous Wisdom, Scientific Knowledge and the Teachings of Plants* (Minneapolis, MN: Milkweed Editions, 2015).

6. Tillmann Taape, "Subtle Fire: Distillation as Low-Heat Technology and the Agency of Human Art," *Ambix*, forthcoming.

7. See, e.g., Alisha Rankin, *Panaceia's Daughters: Noblewomen as Healers in Early Modern Germany* (Chicago: University of Chicago Press, 2013), 59.

8. Tara Alberts, Sietske Fransen, and Elaine Leong, "Translating Medicine, ca. 800–1900: Articulations and Disarticulations," in *Translating Medicine Across Premodern Worlds*, ed. Tara Alberts, Sietske Fransen, and Elaine Leong, *Osiris* 37 (2022): 1–21, Chicago: University of Chicago Press.

9. See, e.g., Smith, *From Lived Experience to the Written Word*. For a full review of the literature, see chap. 5.

10. Mary Fissell, *Vernacular Bodies: The Politics of Reproduction in Early Modern England* (Oxford: Oxford University Press, 2004); Michael Stolberg, *Experiencing Illness and the Sick Body in Early Modern Europe* (Springer, 2011); Katharine Park, *Secrets of Women: Gender, Generation, and the Origins of Human Dissection* (New York: Zone Books, 2006); Barbara Duden, *The Woman Beneath the Skin: A Doctor's Patients in Eighteenth-Century Germany* (Cambridge, MA: Harvard University Press, 1991); Shigehisa Kuriyama and Natalie Köhle, *Fluid Matters: Flow and Transformation in the History of the Body* (Canberra: Australian National University Press, 2020).

Bibliography

Agricola, Georg. *Vom Bergkwerck XII Bücher darinn alle Empter, Instument, Gezeuge unnd alles zu disem handel gehörig, mitt schönen figuren, vorbilder und klärlich beschriben seindt*. Basel: Froben, 1557.

Alberts, Tara, Sietske Fransen, and Elaine Leong, eds. *Translating Medicine Across Premodern Worlds*. Vol. 37, *Osiris*. Chicago: University of Chicago Press, 2022.

Alberts, Tara, Sietske Fransen, and Elaine Leong. "Translating Medicine, ca. 800–1900: Articulations and Disarticulations." In *Translating Medicine Across Premodern Worlds*, edited by Tara Alberts, Sietske Fransen, and Elaine Leong. *Osiris* 37 (2022): 1–21. Chicago: University of Chicago Press.

Alioth, Martin. *Gruppen an der Macht: Zünfte und Patriziat in Strassburg im 14. und 15. Jahrhundert: Untersuchungen zu Verfassung, Wirtschaftsgefüge und Sozialstruktur*. Basel: Helbing & Lichtenhahn, 1988.

Alonso-Almeida, Francisco. "Genre Conventions in English Recipes, 1600–1800." In *Reading and Writing Recipe Books, 1550–1800*, edited by Michelle DiMeo and Sara Pennell, 68–90. Manchester: Manchester University Press, 2013.

Anon. *A Newe Boke of Medecynes Intytulyd or Callyd the Treasure of Pore Men*. London: John Rastell for Richard Bankes, 1526.

Arber, Agnes. *Herbals: Their Origin and Evolution*. Cambridge University Press, 1938.

Arceus, Franciscus. *A Most Excellent and Compendious Method of Curing Woundes in the Head, and in Other Partes of the Body*. Translated by John Read. London: Thomas East, 1588.

Arntz, Helmut. *Weinbrenner: Die Geschichte vom Geist des Weines*. Stuttgart: Seewald, 1975.

Arredondo, Jaime Marroquin, and Ralph Bauer, eds. *Translating Nature: Cross-Cultural Histories of Early Modern Science*. University of Pennsylvania Press, 2019.

Arrizabalaga, Jon, John Henderson, and Roger French. *The Great Pox: The French Disease in Renaissance Europe*. New Haven: Yale University Press, 1997.

Ashworth, William. "Emblematic Natural History of the Renaissance." In *Cultures of Natural History*, edited by Nicholas Jardine, James Secord, and Emma Spary, 17–37. Cambridge University Press, 1996.

Assion, Peter. *Altdeutsche Fachliteratur*. Berlin: Erich Schmidt, 1973.

Bachoffner, Pierre. "Du nouveau à propos de Jérôme Brunschwig." *Revue d'histoire de la pharmacie* 60, no. 212 (1972): 32–33.

Bachoffner, Pierre. "Jérôme Brunschwig, chirurgien et apothicaire strasbourgeois, portraituré en 1512." *Revue d'histoire de la pharmacie* 81, no. 298 (1993): 269–78.

Bachoffner, Pierre. "La pharmacie est réglementée." In *Histoire de la médecine à Strasbourg*, edited by Jacques Héran, 92–93. Strasbourg: La Nuée Bleue, 1997.

Bachoffner, Pierre. "Notes sur le Gaïac au XVIe et au début du XVIIe siècle." *Veröffentlichungen der internationalen Gesellschaft für Pharmazie* 36 (1970): 17–25.

Bachoffner, Pierre, and Théodore Vetter. "Brunschwig, Jérôme." In *Nouveau Dictionnaire de Biographie Alsacienne*, edited by Jean-Pierre Kintz, 393–94. Strasbourg: Fédération des sociétés d'histoire et d'archéologie d'Alsace, 1985.

Bakhtin, Mikhail. *Rabelais and His World*. Bloomington: Indiana University Press, 1984.

Barker, Sara K., and Brenda M. Hosington. *Renaissance Cultural Crossroads: Translation, Print and Culture in Britain, 1473–1640*. Vol. 15. Leiden: Brill, 2013.

Barnes, Jonathan. "Medicine, Experience and Logic." In *Science and Speculation*, edited by Jacques Brunschwig and Jonathan Barnes, 24–68. Cambridge University Press, 2005.

Baumann, Brigitte, and Helmut Baumann. *Die Mainzer Kräuterbuch-Inkunabeln "Herbarius Moguntinus" (1484), "Gart der Gesundheit" (1485), "Hortus sanitatis" (1491): wissenschaftshistorische Untersuchung der drei Prototypen botanisch-medizinischer Literatur des Spätmittelalters*. Stuttgart: Hiersemann, 2010.

Belle, Marie-Alice, and Brenda M. Hosington, eds. *Thresholds of Translation: Paratexts, Print, and Cultural Exchange in Early Modern Britain*. Basingstoke: Palgrave Macmillan, 2018.

Benati, Chiara. "The 1518 Low German Edition of Hieronymus Brunschwig's *Buch der Cirurgia* and its Terminology." In *Medieval German Textrelations: Translations, Editions, and Studies*, edited by S. Jefferis, 189–245. Göppingen: Kümmerle Verlag, 2012.

Benati, Chiara. *Dat boek der Wundenartzstedye und der niederdeutsche chirurgische Fachwortschatz*. Göppingen: Kümmerle, 2012.

Benzenhöfer, Udo. *Johannes de Rupescissa Liber de consideratione quintae essentiae omnium rerum, deutsch: Studien zur Alchemia medica des 15. bis 17. Jahrhunderts, mit kritischer Edition des Textes*. Stuttgart: Steiner Verlag Wiesbaden, 1989.

Bertoloni Meli, Domenico, and Cynthia Klestinec. "Renaissance Surgery Between Learning and Craft." *Journal of the History of Medicine and Allied Sciences* 72, no. 1 (2017): 1–5.

Bess, Reginald. "Hieronymus Brunschwig's 'De Arte Distillandi' and the Technical Language of Distillation." *Fifteenth-Century Studies* 11 (1985): 7–22.

Biedermann, Johann Gottfried. *Geschlechts-Register der Reichs Frey unmittelbaren Ritterschafft Landes zu Francken Löblichen Orts Ottenwald*. Culmbach: Spindler, 1751.

Blaak, Jeroen. *Literacy in Everyday Life: Reading and Writing in Early Modern Dutch Diaries*. Leiden: Brill, 2009.

Blair, Ann. "Humanist Methods in Natural Philosophy: The Commonplace Book." *Journal of the History of Ideas* (1992): 541–51.

Blair, Ann. *Too Much to Know: Managing Scholarly Information Before the Modern Age*. New Haven: Yale University Press, 2010.

Blumenberg, Hans. *Die Lesbarkeit der Welt*. Frankfurt: Suhrkamp, 1983.

Boehm, Laetitia. "Artes mechanicae und artes liberales im Mittelalter. Die praktischen Künste zwischen illiterater Bildungstradition und schriftlicher Wissenschaftskultur." In *Festschrift für Eduard Hlawitschka zum 65. Geburtstag*, edited by Karl Rudolf Schnith and Roland Pauler, 419–44. Kallmünz: Lassleben, 1993.

Brady, Thomas A. *The Politics of the Reformation in Germany: Jacob Sturm (1489–1553) of Strasbourg*. Atlantic Highlands: Humanities Press International, 1997.

Brady, Thomas A. "The Themes of Social Structure, Social Conflict, and Civic Harmony in Jakob Wimpheling's *Germania*." *Sixteenth Century Journal* 3, no. 2 (1972): 65–76.

Brant, Sebastian. *Das Narren schyff*. Basel: Johann Bergmann, 1494.

Brant, Sebastian. *Stultifera navis*. Basel: Johann Bergmann, 1497.

Brayman Hackel, Heidi. *Reading Material in Early Modern England: Print, Gender, and Literacy*. Cambridge University Press, 2005.

Brenner, Elma, and François-Olivier Touati, eds. *Leprosy and Identity in the Middle Ages*. Manchester University Press, 2021.

Brockliss, Laurence, and Colin Jones. *The Medical World of Early Modern France*. Oxford: Clarendon Press, 1997.

Brucker, Jean-Charles. *Strassburger Zunft-und Polizei-Verordnungen des 14. und 15. Jahrhunderts*. Strasbourg: Trübner, 1889.

Brunschwig, Hieronymus. *Boek der Wundenartzstedye*. Rostock: Ludwig Dietz, 1518.

Brunschwig, Hieronymus. *Das buch der wund Artzney. Handwirckung der Cirurgia von Jyeronimo brunschwick*. Strasbourg: Johann Grüninger, 1513.

Brunschwig, Hieronymus. *Die distellacien ende virtuyten der wateren*. Brussels: Thomas van der Noot, 1517.

Brunschwig, Hieronymus. *Dis ist das buch der cirurgia* [Cirurgia]. *Hantwirckung der wund artzny*. Strasbourg: Johann Grüninger, 1497.

Brunschwig, Hieronymus. *Dits dat hantwerck der cirurgien ende leert alle wonden, gehouden gesteken gheslaghen*. Utrecht: Jan Berntsz, 1535.

Brunschwig, Hieronymus. *Liber de arte Distillandi de Compositis* [Large book of distillation]. *Das buch der waren kunst zü distillieren die Composita vnn simplicia / vnd dz Buoch thesaurus pauperum / Ein schatz der armen*. Strasbourg: Johann Grüninger, 1512.

Brunschwig, Hieronymus. *Liber de arte distillandi de simplicibus* [Small book of distillation]. *Das buch der rechten kunst zü distilieren die eintzigen ding*. Strasbourg: Johann Grüninger, 1500.

Brunschwig, Hieronymus. *Liber pestilentialis de venenis epidimie. Das buoch der vergift der pestilentz das da genant ist der gemein sterbent der Trüsen Blatren*. Strasbourg: Johann Grüninger, 1500.

Brunschwig, Hieronymus. *The Noble Experyence of the Vertuous Handy Warke of Surgeri*. London: Petrus Treveris, 1525.

Brunschwig, Hieronymus. *Vertuose Boke of Distyllacyon of the Waters of All Maner of Herbes / With the Fygures of the Styllatoryes*. Translated by Laurence Andrew. London: Laurence Andrew, 1527.

Brunschwig, Hieronymus, and Marsilio Ficino. *Das buch der Gesuntheit* [Medicinarius]. *Liber de arte distillandi Simplicia et composita. Das nüv buoch der rechten kunst zuo distillieren. Ouch von Marsilio Ficino vnn andere hochberoepmter Artzte natürliche vnn gute künst*. Strasbourg: Johann Grüninger, 1505.

Brunschwig, Hieronymus, and Marsilio Ficino. *Liber de arte distulandi Simplicia et Composita* [Medicinarius]. *Das nüv buoch der rechten kunst zuo distillieren. Ouch von Marsilio Ficino vnn anderen hochberoempten Ertzte natürliche vnn guote künst*. Strasbourg: Johann Grüninger, 1509.

Burke, Jill. *How to Be a Renaissance Woman: An Untold History of Beauty and Female Creativity*. London: Profile Books, 2023.

Burke, Peter. "Cultures of Translation in Early Modern Europe." In *Cultural Translation in Early Modern Europe*, edited by Peter Burke and R. P. Hsia, 7–38. Cambridge University Press, 2007.

Burke, Peter. "Lost (and Found) in Translation: A Cultural History of Translators and Translating in Early Modern Europe." *European Review* 15, no. 1 (February 2007): 83–94. https://doi.org/10.1017/S1062798707000087.

Carey, Daniel. "The Problem of Credibility in Early Modern Travel." *Renaissance Studies* 33, no. 4 (2019): 524–47.

Cavallo, Sandra. *Artisans of the Body in Early Modern Italy: Identities, Families and Masculinities*. Manchester: Manchester University Press, 2007.

Chamberland, Celeste. "Between the Hall and the Market: William Clowes and Surgical Self-Fashioning in Elizabethan London." *Sixteenth Century Journal* 41, no. 1 (2010): 69–89.

Chamberland, Celeste. "From Apprentice to Master: Social Disciplining and Surgical Education in Early Modern London, 1570–1640." *History of Education Quarterly* 53, no. 1 (2013): 21–44.

Chamberland, Celeste. "Honor, Brotherhood, and the Corporate Ethos of London's Barber-Surgeons' Company, 1570–1640." *Journal of the History of Medicine and Allied Sciences* 64, no. 3 (2009): 300–332.

Chauliac, Guy de. *De Cyrurgie van meester Guid de Cauliaco*. Antwerp: Henrick Eckert van Homberch, 1507.

Chauliac, Guy de. *Guigonis de Caulhiaco (Guy de Chauliac), Inventarium sive Chirurgia magna. Volume One: Text*. Edited by Michael McVaugh. Leiden: Brill, 1997.

Chauliac, Guy de. *Guydos Questyons*. Translated by George Baker. London: Thomas East, 1579.

Chrisman, Miriam Usher. *Lay Culture, Learned Culture: Books and Social Change in Strasbourg, 1480–1599*. New Haven: Yale University Press, 1982.

Christie, John. "The Paracelsian Body." In *Paracelsus: The Man and His Reputation, His Ideas and Their Transformation*, edited by Ole Peter Grell, 269–91. Leiden: Brill, 1998.

Clarke, Mark, Bert De Munck, and Sven Dupré, eds. *Transmission of Artists' Knowledge*. Brussels: KVAB Press, 2012.

Clowes, William. *A Short and Profitable Treatise Touching the Cure of the Disease Called Morbus Gallicus*. London: John Day, 1579.

Conrad, Lawrence I., Michael Neve, Vivian Nutton, Roy Porter, and Andrew Wear. *The Western Medical Tradition: 800 BC to AD 1800, Volume 1*. Cambridge University Press, 1995.

Cook, Harold. *The Decline of the Old Medical Regime in Stuart London*. Ithaca, NY: Cornell University Press, 1986.

Cook, Harold. "Good Advice and Little Medicine: The Professional Authority of

Early Modern English Physicians." *Journal of British Studies* 33, no. 1 (1994): 1–31.

Cook, Harold. *Matters of Exchange: Commerce, Medicine, and Science in the Dutch Golden Age*. New Haven: Yale University Press, 2007.

Cook, Harold. "The New Philosophy and Medicine in Seventeenth-Century England." In *Reappraisals of the Scientific Revolution*, edited by David Lindberg and Robert Westman, 397–436. Cambridge University Press, 1990.

Cook, Harold. *Trials of an Ordinary Doctor: Joannes Groenevelt in Seventeenth-Century London*. Baltimore: Johns Hopkins University Press, 1994.

Cook, Harold, and Sven Dupré. "Introduction." In *Translating Knowledge in the Early Modern Low Countries*, edited by Harold J. Cook and Sven Dupré, 3–17. Münster: LIT Verlag, 2012.

Cooper, Alix. "Latin Words, Vernacular Worlds: Language, Nature, and the 'Indigenous' in Early Modern Europe." *East Asian Science, Technology, and Medicine* 26 (2007): 17–39.

Copeland, Rita. *Rhetoric, Hermeneutics, and Translation in the Middle Ages: Academic Traditions and Vernacular Texts*. Cambridge University Press, 1995.

Copenhaver, Brian. "Natural Magic, Hermetism, and Occultism in Early Modern Science." In *Reappraisals of the Scientific Revolution*, edited by Robert S. Westman and David C. Lindberg, 261–301. Cambridge University Press, 1990.

Cormack, Bradin, and Carla Mazzio. *Book Use, Book Theory: 1500–1700*. Chicago: University of Chicago Press, 2005.

Daston, Lorraine, and Elizabeth Lunbeck. *Histories of Scientific Observation*. Chicago: University of Chicago Press, 2011.

Daston, Lorraine, and Katharine Park. *Wonders and the Order of Nature, 1150–1750*. New York: Zone Books, 1998.

De Munck, Bert. "The Agency of Branding and the Location of Value. Hallmarks and Monograms in Early Modern Tableware Industries." *Business History* 54, no. 7 (2012): 1055–76.

De Munck, Bert. "Apprenticeship, Guilds, and Craft Knowledge." In *Encyclopedia of Early Modern Philosophy and the Sciences*, edited by Dana Jalobeanu and Charles T. Wolfe, 98–104. Cham: Springer, 2022.

De Munck, Bert. "Guilds, Product Quality and Intrinsic Value. Towards a History of Conventions?" *Historical Social Research / Historische Sozialforschung*, 2011, 103–24.

De Munck, Bert. *Technologies of Learning: Apprenticeship in Antwerp Guilds from the 15th Century to the End of the Ancien Regime*. Turnhout: Brepols, 2007.

De Munck, Bert, Steven L. Kaplan, and Hugo Soly, eds. *Learning on the Shop Floor: Historical Perspectives on Apprenticeship*. Berghahn Books, 2007.

Dear, Peter. *Discipline and Experience: The Mathematical Way in the Scientific Revolution*. Chicago: University of Chicago Press, 1995.

Dear, Peter. "The Meanings of Experience." In *The Cambridge History of Science*. Vol. 3, Early Modern Science, edited by Lorraine Daston and Katharine Park, 106–31. Cambridge University Press, 2006.

Deblock, Geneviève. *Le bâtiment des recettes: Présentation et annotation de l'édition Jean Ruelle, 1560*. Rennes: Presses Universitaires de Rennes, 2015.

Debus, Allen G. *The Chemical Philosophy: Paracelsian Science and Medicine in the Sixteenth and Seventeenth Centuries*. New York: Neale Watson, 1977.

Debus, Allen G. *Chemistry, Alchemy and the New Philosophy, 1550–1700: Studies in the History of Science and Medicine*. London: Variorum, 1987.

Demaitre, Luke. *Leprosy in Premodern Medicine: A Malady of the Whole Body*. Baltimore: Johns Hopkins University Press, 2007.

Demaitre, Luke. "Skin and the City: Cosmetic Medicine as an Urban Concern." In *Between Text and Patient: The Medical Enterprise in Medieval and Early Modern Europe*, edited by Florence Glaze and Brian Nance, 97–120. Florence: SISMEL, 2011.

DeVun, Leah. *Prophecy, Alchemy, and the End of Time: John of Rupescissa in the Late Middle Ages*. New York: Columbia University Press, 2009.

Dilg, Peter. "Die Strassburger Apothekerordnung (um 1500)." *Pharmazeutische Zeitung* 124 (1979): 2664–70.

Dlabačová, Anna, Andrea van Leerdam, and John Thompson, eds. *Identifying the Vernacular Reader in the Early Age of Print (c. 1450–1600)*. Leiden: Brill, 2023.

Dollinger, Philippe. "L'évolution politique des corporations strasbourgeoises à la fin du Moyen Âge." In *Artisans et ouvriers d'Alsace*, 127–35. Strasbourg: ISTRA, 1965.

Donato, Maria Pia, Elaine Leong, and Tillmann Taape, eds. "Learning to Cut and Mend: Surgical Training and Practice, 1450–1800." London: UCL Press, under review.

Draelants, Isabelle. "Expérience et autorités dans la philosophie naturelle d'Albert le Grand." In *Expertus sum: L'expérience par les sens dans la philosophie naturelle médiévale*, edited by Isabelle Draelants and Thomas Bénatouïl, 89–121. SISMEL, 2011.

Dross, Fritz. "Vom zuverlässigen Urteilen. Ärztliche Autorität, reichsstädtische Ordnung und der Verlust armer Glieder Christi in der Nürnberger Sondersiechenschau." *Medizin, Gesellschaft und Geschichte* 29 (2010): 9–46.

Du Chesne, Joseph. *The Sclopotarie of Iosephus Quercetanus*. Translated by John Hester. London: Roger Ward, 1590.

Duden, Barbara. *The Woman Beneath the Skin: A Doctor's Patients in Eighteenth-Century Germany*. Cambridge, MA: Harvard University Press, 1991.

Dumas, Geneviève. *Ymage de Vie*. Montpellier: Presses universitaires de la Méditerranée, 2019.

Dupré, Sven. "Doing It Wrong: The Translation of Artisanal Knowledge and the Codification of Error." In *The Structures of Practical Knowledge*, edited by Matteo Valleriani, 167–88. Cham: Springer International Publishing, 2017.

Dupré, Sven, ed. *Laboratories of Art: Alchemy and Art Technology from Antiquity to the 18th Century*. Cham: Springer, 2014.

Dupré, Sven. "The Value of Glass and the Translation of Artisanal Knowledge in Early Modern Antwerp." In *Trading Values in Early Modern Antwerp*, edited by Christine Göttler, Bart Ramakers, and Joanna Woodall, 138–61. Leiden: Brill, 2014.

Eagleton, Catherine. "Medieval Sundials and Manuscript Sources: The Transmission of Information About the Navicula and the *Organum Ptolomei* in Fifteenth-Century Europe." In *Transmitting Knowledge: Words, Images, and Instruments in Early Modern Europe*, edited by Sachiko Kusukawa and Ian Maclean, 41–71. Oxford: Oxford University Press, 2006.

Eamon, William. "Alchemy in Popular Culture: Leonardo Fioravanti and the Search for the Philosopher's Stone." *Early Science and Medicine* 5 (2000): 196–213.

Eamon, William. "From Secrets of Nature to Public Knowledge." In *Reappraisals of the Scientific Revolution*, edited by Robert Westman and David Lindberg, 333–66. Cambridge University Press, 1990.

Eamon, William. "Pharmaceutical Self-Fashioning or How to Get Rich and Famous in the Renaissance Medical Marketplace." *Pharmacy in History* 45, no. 3 (2003): 123–29.

Eamon, William. *The Professor of Secrets: Mystery, Medicine, and Alchemy in Renaissance Italy*. Washington, DC: National Geographic Society, 2010.

Eamon, William. *Science and the Secrets of Nature: Books of Secrets in Medieval and Early Modern Culture*. Princeton, NJ: Princeton University Press, 1994.

Eis, Gerhard. "Brunschwig, Hieronymus." *Neue Deutsche Biographie*, 688, 1955.

Elshakry, Marwa, and Carla Nappi. "Translations." In *A Companion to the History of Science*, edited by Bernard Lightman. Hoboken: John Wiley & Sons, 2019.

Erler, Mary C. "The First English Printing of Galen: The Formation of the Company of Barber-Surgeons." *Huntington Library Quarterly* 48, no. 2 (1985): 159–71. https://doi.org/10.2307/3817541.

Farr, James. *Hands of Honor: Artisans and Their World in Dijon, 1550–1650*. Ithaca, NY: Cornell University Press, 1988.

Fioravanti, Leonardo. *A Compendium of the Rationall Secretes*. Translated by John Hester. London: John Kingston, 1582.

Fioravanti, Leonardo. *A Ioyfull Iewell Contayning Aswell Such Excellent Orders, Preseruatiues and Precious Practises for the Plague, as Also Such Meruelous Medcins for Diuers Maladies, as Hitherto Haue Not Beene Published in the English Tung*. Translated by John Hester. London: Wright, 1579.

Fioravanti, Leonardo. *A Short Discours of the Excellent Doctour and Knight, Maister Leonardo Phiorauanti Bolognese Vppon Chirurgerie*. Translated by John Hester. London, 1580.

Fissell, Mary. "The Marketplace of Print." In *Medicine and the Market in England and Its Colonies, c. 1450–c. 1850*, edited by Patrick Wallis and Mark Jenner, 108–32. Basingstoke: Palgrave Macmillan, 2007.

Fissell, Mary. "Popular Medical Books." In *The Oxford History of Popular Print Culture*, edited by Joad Raymond, 417–31. New York: Oxford University Press, 2011.

Fissell, Mary. "Readers, Texts, and Contexts: Vernacular Medical Works in Early Modern England." In *The Popularization of Medicine, 1650–1850*, edited by Roy Porter, 72–96. London: Routledge, 1992.

Fissell, Mary. *Vernacular Bodies: The Politics of Reproduction in Early Modern England*. Oxford: Oxford University Press, 2004.

Flood, John L. "The Printed Book as a Commercial Commodity in the Fifteenth and Early Sixteenth Centuries." *Gutenberg-Jahrbuch* 76 (2001): 172–82.

Forbes, Robert. *A Short History of the Art of Distillation: From the Beginnings up to the Death of Cellier Blumenthal*. Leiden: Brill, 1948.

The Four Seasons of Human Life: Four Anonymous Engravings from the Trent Collection. Rotterdam: Erasmus, 2002.

Fransen, Sietske. "Introduction: Translators and Translations of Early Modern

Science." In *Translating Early Modern Science*, edited by Sietske Fransen, Niall Hodson, and Karl Enenkel, 1–14. Leiden: Brill, 2017.

Fransen, Sietske. "Latin in a Time of Change: The Choice of Language as Signifier of a New Science?" *Isis* 108, no. 3 (September 2017): 629–35. https://doi.org/10.1086/694189.

Franssen, Piet J. A. "Jan van Doesborch (?–1536), Printer of English Texts." *Quaerendo* 14 (1986): 259–80.

Franssen, Piet J. A. "Jan van Doesborch's Departure from Antwerp and His Influence on the Utrecht Printer Jan Berntsz." *Quaerendo* 18, no. 3 (January 1, 1988): 163–90. https://doi.org/10.1163/157006988X00097.

Franssen, Piet J. A. *Tussen tekst en publiek: Jan van Doesborch, drukker-uitgever en literator te Antwerpen en Utrecht in de eerste helft van de zestiende eeuw*. Amsterdam: Rodopi, 1990.

Frede, Michael. "The Method of the So-Called Methodical School of Medicine." In *Science and Speculation*, edited by Jaques Brunschwig and Jonathan Barnes, 1–23. Cambridge University Press, 2005.

Frederiksen, Jan. "Johannes (Hans) von Gersdorff." In *Die deutsche Literatur des Mittelalters: Verfasserlexikon*, edited by Wolfgang Stammler, Karl Langosch, and Kurt Ruh, 4:626–30. Berlin: Walter de Gruyter, 1978.

French, Roger Arrizabalaga. "Coping with the French Disease: University Practitioners' Strategies and Tactics in the Transition from the Fifteenth to the Sixteenth Century." In *Medicine from the Black Death to the French Disease*, edited by Roger Kenneth French, Andrew Cunningham, Luis García-Ballester, and Jon Arrizabalaga, 248–87. Aldershot: Ashgate, 1998.

Friedrich, Christoph, and Wolf-Dieter Müller-Jahncke. *Geschichte der Pharmazie von der frühen Neuzeit bis zur Gegenwart*. Eschborn: Govi-Verlag, 2005.

Fries, Lorenz. *Spiegel der Artzny des geleichen vormals nie von keinem doctor in tütsch vßgangen* [. . .]. Strasbourg: Johann Grüninger, 1518.

Fries, Lorenz. *Spiegel der Artzny gemachet vnd widerumb mit ernst vbersehen*. Strasbourg: Johann Grüninger, 1529.

Fries, Lorenz. *Uslegung der merkarthen*. Strasbourg: Grüninger, 1525.

Frisch, Andrea. *The Invention of the Eyewitness: Witnessing and Testimony in Early Modern France*. Chapel Hill: University of North Carolina Press, 2004.

Gale, Thomas. *Certaine Workes of Galens, Called Methodus Medendi*. London: Thomas East, 1567.

Gale, Thomas. "An Enchiridion Conteining the Cure of Woundes, Fractures, and Dislocations." In *Certaine Workes of Chyrurgerie*. London: Rowland Hall, 1563.

Gale, Thomas. "An Excellent Treatise of Vvounds Made vvith Gonneshot." In *Certaine Workes of Chyrurgerie*. London: Rowland Hall, 1563.

Geiler von Kaysersberg, Johann. *Des hochwirdigen doctor keiserspergs narrenschiff so er gepredigt hat zuo straßburg*. Strasbourg: Johann Grüninger, 1520.

Geiler von Kaysersberg, Johann. *Die Emeis*. Strasbourg: Johann Grüninger, 1516.

Geiler von Kaysersberg, Johann. *Navicula sive speculum fatuorum*. Strasbourg: Matthias Schürer, 1510.

Gelfand, Toby. *Professionalizing Modern Medicine: Paris Surgeons and Medical Science and Institutions in the 18th Century*. London: Bloomsbury Academic, 1980.

Geoghegan, D. "A Licence of Henry VI to Practise Alchemy." *Ambix* 6, no. 1 (August 1, 1957): 10–17. https://doi.org/10.1179/amb.1957.6.1.10.

Gersdorff, Hans von. *Feldtbuoch der wundtartzney*. Strasbourg: Johann Grüninger, 1517.

Gessner, Conrad. *The Newe Iewell of Health Wherein is Contayned the Most Excellent Secretes of Phisicke and Philosophie*. Translated by George Baker. London: Denham, 1576.

Gessner, Conrad. *Thesaurus Euonymi philiatri De remediis secretis, liber physicus, medicus et partim etiam chymicus, & oeconomicus*. Zurich: Gessner, 1552.

Giesecke, Michael. "'Volkssprache' und 'Verschriftlichung des Lebens' im Spätmittelalter am Beispiel der Genese der gedruckten Fachprosa in Deutschland." In *Literatur in der Gesellschaft des Spätmittelalters*, edited by Hans Umbrecht, 39–70. Heidelberg, 1980.

Goltz, D. "Aber erschrocklich ist es dem Menschenn. Beziehungen zwischen Chirurgie und Pharmazie im Mittelalter." *Pharmazeutische Zeitung* 123, no. 7 (1978): 278–84.

Grafton, Anthony. *Inky Fingers: The Making of Books in Early Modern Europe*. Harvard University Press. Cambridge, MA, 2022.

Gröber, Manfred. *Das wundärztliche Manual des Meisters Hans Seyff von Göppingen (ca. 1440–1518): der Cod. med. et phys. 20 8 der Württembergischen Landesbibliothek Stuttgart*. Göppingen: Kümmerle, 1998.

Hackett, Jeremiah. "Ego Expertus Sum: Roger Bacon's Science and the Origins of Empiricism." In *Expertus sum: L'expérience par les sens dans la philosophie naturelle médiévale*, edited by Isabelle Draelants and Thomas Bénatouïl, 145–73. SISMEL, 2011.

Hackett, Jeremiah. "Roger Bacons Concept of Experience: A New Beginning in Medieval Philosophy?" *Modern Schoolman* 86, no. 1/2 (2008): 123–46.

Hagendijk, Thijs. "Unpacking Recipes and Communicating Experience: The Ervarenissen of Simon Eikelenberg (1663–1738) and the Art of Painting." *Early Science and Medicine* 24, no. 3 (2019): 248–82.

Hagendijk, Thijs, Márcia Vilarigues, and Sven Dupré. "Materials, Furnaces, and Texts: How to Write About Making Glass Colours in the Seventeenth Century." *Ambix* 67, no. 4 (2020): 323–45.

Hamza, Shireen. "Vernacular Languages and Invisible Labor in Ṭibb." In *Translating Medicine Across Premodern Worlds*, edited by Tara Alberts, Sietske Fransen, and Elaine Leong. *Osiris* 37 (2022): 115–38. Chicago: University of Chicago Press.

Hankinson, R. Jim. "Art and Experience: Greek Philosophy and the Status of Medicine." *Quaestio* 4 (2004): 3–24.

Harkness, Deborah. *The Jewel House: Elizabethan London and the Scientific Revolution*. New Haven: Yale University Press, 2007.

Hausse, Heidi. "The Locksmith, the Surgeon, and the Mechanical Hand: Communicating Technical Knowledge in Early Modern Europe." *Technology and Culture* 60, no. 1 (2019): 34–64.

Hausse, Heidi. *The Malleable Body: Surgeons, Artisans, and Amputees in Early Modern Germany*. Manchester: Manchester University Press, 2023.

Henderson, Felicity. "Faithful Interpreters? Translation Theory and Practice at the Early Royal Society." *Notes and Records of the Royal Society* 67, no. 2 (2013): 101–22.

Henderson, Felicity. "Making 'The Good Old Man' Speak English: The Reception of Antoni van Leeuwenhoek's Letters at the Royal Society, 1673–1723." In *Trans-*

lating Knowledge in the Early Modern Low Countries, edited by Harold J. Cook and Sven Dupré, 243–68. Münster: LIT Verlag, 2012.

Henderson, Felicity. "Translation in the Circle of Robert Hooke." In *Translating Early Modern Science*, edited by Sietske Fransen, Niall Hodson, and Karl Enenkel, 17–40. Leiden: Brill, 2017.

Henry, John. "Doctors and Healers: Popular Culture and the Medical Profession." In *Science, Culture, and Popular Belief in Renaissance Europe*, edited by Stephen Pumfrey, Paolo Rossi, and Maurice Slawinski, 191–221. Manchester: Manchester University Press, 1991.

Héran, Jacques, ed. *Histoire de la médecine à Strasbourg*. Strasbourg: La Nuée Bleue, 1997.

Hermans, Theo. "Concepts and Theories of Translation in the European Renaissance." In *Übersetzung: ein internationales Handbuch zur Übersetzungsforschung* [An international encyclopedia of translation studies]. *Tbd.* 3, edited by Harald Kittel, Juliane House, and Brigitte Schultze, 1420–28. Berlin: Walter de Gruyter, 2011.

Hester, John. *The First Part of the Key of Philosophie*. London: Richard Day, 1580.

Heusinger, Sabine von. "Die Handwerksbruderschaften in Straßburg." In *Städtische Gesellschaft und Kirche im Spätmittelalter*, edited by Sigrid Schmitt and Sabine Klapp, 123–40. Vol. 62. Stuttgart: Steiner, 2008.

Heusinger, Sabine von. *Die Zunft im Mittelalter: zur Verflechtung von Politik, Wirtschaft und Gesellschaft in Straßburg*. Stuttgart: Steiner, 2009.

Heusinger, Sabine von. "The Topography of Sacred Space and the Representation of Social Groups—Confraternities in Strasbourg." In *Politics and Reformations: Communities, Polities, Nations, and Empires: Essays in Memory of Thomas A. Brady, Jr.*, edited by Thomas A. Brady and Christopher Ocker. Brill, 2007.

Heusinger, Sabine von. "Von 'antwerck' bis 'Zunft.' Methodische Überlegungen zu den Zünften im Mittelalter." *Zeitschrift für historische Forschung* 37 (2010): 37–71.

Heyd, Wilhelm von. "Widmann, Johann." In *Allgemeine Deutsche Biographie*, 355–57. Leipzig: Duncker und Humblot, 1897. https://www.deutsche-biographie.de/pnd102507368.html#adbcontent.

Houwen, Luuk. "The Noble Lyfe: An Early English Version of the Hortus Sanitatis." In *Schooling and Society*, edited by Alasdair MacDonald and Michael Twomey, 2004.

Ingold, Tim. *The Perception of the Environment: Essays on Livelihood, Dwelling and Skill*. New York: Routledge, 2000.

Jacquart, Danielle. "Theory, Everyday Practice, and Three Fifteenth-Century Physicians." *Osiris* 6 (1990): 140–60.

Jardine, Lisa. "'Studied for Action' Revisited." In *For the Sake of Learning: Essays in Honour of Anthony Grafton*, edited by Ann Blair and Anja-Silvia Goeing, vol. 2, 999–1017. Leiden: Brill, 2016.

Jardine, Lisa, and Anthony Grafton. "'Studied for Action': How Gabriel Harvey Read His Livy." *Past & Present*, no. 129 (1990): 30–78.

Jenner, Mark, and Patrick Wallis. *Medicine and the Market in England and Its Colonies, c. 1450–c. 1850*. Basingstoke: Palgrave Macmillan, 2007.

Jones, Peter. "Image, Word and Medicine in the Middle Ages." In *Visualizing Medieval Medicine and Natural History, 1200–1550*, edited by Jean Ann Givens, Karen Reeds, and Alain Touwaide, 1–24. Aldershot: Ashgate, 2006.

Jones, Peter. "Mediating Collective Experience: The *Tabula Medicine* (1416–1425) as a Handbook for Medical Practice." In *Between Text and Patient: The Medical Enterprise in Medieval and Early Modern Europe*, edited by Florence Glaze, Brian Nance, and Michael McVaugh, 279–307. Florence: Sismel, 2011.

Jones, Peter. "Medical Libraries." In *A History of Libraries in Britain and Ireland*, edited by E. Leedham-Green and T. Webber, 461–71. Cambridge University Press, 2006.

Jones, Peter Murray. *The Medicine of the Friars in Medieval England*. Woodbridge: Boydell & Brewer, 2024.

Jones, Peter. *Medieval Medicine in Illuminated Manuscripts*. London: The British Library, 1998.

Jütte, Robert. "Bader, Barbiere und Hebammen: Heilkundige als Randgruppen?" In *Randgruppen der spätmittelalterlichen Gesellschaft*, edited by Bernd-Ulrich Hergemöller, 90–121. Warendorf: Fahlbusch Verlag, 1994.

Jütte, Robert. "Zur Sozialgeschichte der Handwerkschirurgen im 16. Jahrhundert." In *Paracelsus und Salzburg*, edited by Heinz Dopsch and Peter Kramm, 45–60. Salzburg: Gesellschaft für Salzburger Landeskunde, 1994.

Kassell, Lauren. "Secrets Revealed: Alchemical Books in Early Modern England." *History of Science* 49 (2011): 63–87.

Kavey, Allison. *Books of Secrets Natural Philosophy in England, 1550–1600*. Urbana: University of Illinois Press, 2007.

Keil, Gundolf. "Brunschwig, Hieronymus." In *Die deutsche Literatur des Mittelalters: Verfasserlexikon*, edited by Wolfgang Stammler, Karl Langosch, and Kurt Ruh, 1074–75. Vol. 1. Berlin: Walter de Gruyter, 1978.

Keil, Gundolf. "Der Hausvater als Arzt." In *Haushalt und Familie in Mittelalter und früher Neuzeit*, edited by Trude Ehlert, 219–43. Sigmaringen: Thorbecke, 1991.

Keil, Gundolf. "Die medizinische Versorgung durch Bader und Wundärzte zur Zeit des Paracelsus." *Salzburger Beiträge zur Paracelsusforschung* 29 (1995): 50–74.

Keil, Gundolf. "Hieronymus Brunschwig, Buch der Cirurgia." *Beiträge zur Geschichte der deutschen Sprache und Literatur* 93 (1971): 475–77.

Keil, Gundolf, and Rolf Müller. "Deutsche Lanfrank-Übersetzungen des 14. und 15. Jahrhunderts." In *Medizingeschichte in unserer Zeit. Festgabe für Edith Heischkel-Artelt und Walter Artelt zum 65. Geburtstag*, edited by Hans-Heinz Eulner, Gunter Mann, Gert Preiser, Rolf Winau, and Otto Winckelmann, 90–110. Stutgart: Enke, 1971.

Kemp, Martin. "'The Mark of Truth': Looking and Learning in Some Anatomical Illustrations from the Renaissance and Eighteenth Century." In *Medicine and the Five Senses*, edited by William Bynum and Roy Porter, 85–121. Cambridge University Press, 1993.

Kenny, Neil. *The Palace of Secrets: Béroalde de Verville and Renaissance Conceptions of Knowledge*. Clarendon, 1991.

Keys, Thomas E. "The Earliest Medical Books Printed with Movable Type: A Review." *Library Quarterly* 10, no. 2 (1940): 220–30.

Kiening, Christian. "Erfahrung und Vermessung der Welt in der frühen Neuzeit." In *Text—Bild—Karte: Kartographien der Vormoderne*, edited by Christian Kiening and Jürg Glauser, 221–51. Freiburg im Breisgau: Rombach Verlag, 2007.

Kimmerer, Robin Wall. *Braiding Sweetgrass: Indigenous Wisdom, Scientific Knowledge and the Teachings of Plants*. Minneapolis, MN: Milkweed Editions, 2015.

Kintzinger, Martin. "Experientia lucrativa? Erfahrungswissen und Wissenserfahrung im europäischen Mittelalter." *Das Mittelalter. Perspektiven mediävistischer Forschung* 17, no. 2 (2012): 95–117.

Kintzinger, Martin. "Status medicorum. Mediziner in der städtischen Gesellschaft des 14. bis 16. Jahrhunderts." In *Städtisches Gesundheits- und Fürsorgewesen vor 1800*, edited by Peter Johanek, 63–91. Cologne: Böhlau, 2000.

Kinzelbach, Annemarie. "Arztsein, Handwerk und Politik. Die Wund-und Stadtärzte, Barbiere, Bader, Chirurgen in Nördlingen." *Jahrbuch des Historischen Vereins für Nördlingen* 35 (2017): 249–90.

Kinzelbach, Annemarie. "Erudite and Honoured Artisans? Performers of Body Care and Surgery in Early Modern German Towns." *Social History of Medicine* 27 (2014): 668–88.

Kinzelbach, Annemarie. "Learning to Cut for the Stone: Training Specialised Surgeons in the Holy Roman Empire." In "Learning to Cut and Mend: Surgical Training and Practice, 1450–1800," edited by Maria Pia Donato, Elaine Leong, and Tillmann Taape. London: UCL Press, under review.

Kinzelbach, Annemarie. "Zur Sozial-und Alltagsgeschichte eines Handwerks in der frühen Neuzeit: Wundärzte und ihre Patienten in Ulm." *Ulm und Oberschwaben* 49 (1994): 111–44.

Kinzelbach, Annemarie, and Monika Weber, eds. *Steinreich: Das Schneidhaus der Fugger in Augsburg*. Ingolstadt: Deutschen Medizinhistorischen Museums, 2022.

Klein, Gustav. *Das Buch der Cirurgia des Hieronymus Brunschwig*. Munich: Kuhn, 1911.

Klein, Ursula. "Hybrid Experts." In *The Structures of Practical Knowledge*, edited by Matteo Valleriani, 287–306. Cham: Springer, 2017.

Klein, Ursula, and Emma Spary, eds. *Materials and Expertise in Early Modern Europe: Between Market and Laboratory*. Chicago: University of Chicago Press, 2010.

Klestinec, Cynthia. "Renaissance Surgeons: Anatomy, Manual Skill and the Visual Arts." In *Early Modern Medicine and Natural Philosophy*, edited by Peter Distelzweig, Benjamin Goldberg, and Evan Ragland, 43–58. Springer, 2016.

Klestinec, Cynthia. "Translating Learned Surgery." *Journal of the History of Medicine and Allied Sciences* 72, no. 1 (2017): 34–50.

Kodera, Sergius. "The Art of the Distillation of 'Spirits' as a Technological Model for Human Physiology. The Cases of Marsilio Ficino, Joseph Duchesne and Francis Bacon." In *Blood, Sweat and Tears: The Changing Concepts of Physiology from Antiquity into Early Modern Europe*, edited by Manfred Horstmanshoff, Helen King, and Claus Zittel, 139–70. Leiden: Brill, 2012.

Kristeller, Paul Oskar. *Die Strassburger Bücherillustration im XV. und im Anfange des XVI. Jahrhunderts*. Leipzig: Seemann, 1888.

Kuriyama, Shigehisa, and Natalie Köhle. *Fluid Matters: Flow and Transformation in the History of the Body*. Canberra: Australian National University Press, 2020.

Kusukawa, Sachiko. *Picturing the Book of Nature: Image, Text, and Argument in Sixteenth-Century Human Anatomy and Medical Botany*. Chicago: University of Chicago Press, 2012.

Kusukawa, Sachiko. "The Uses of Pictures in the Formation of Learned Knowledge: The Cases of Leonhard Fuchs and Andreas Vesalius." In *Transmitting Knowledge:*

Words, Images, and Instruments in Early Modern Europe, edited by Sachiko Kusukawa and Ian Maclean, 73–96. Oxford: Oxford University Press, 2006.

Lanfranco di Milano. *Kleyne Wundartznei Des Hochberuemptten Lanfranci / Auß Fürbit Des Wol Erfarnen M. Gregorii Flüguß / Chyrurgen Vnn Wundartzt Zuo Straßburg / Durch Othonem Brunfels Verteutscht*. Translated by Otto Brunfels. Strasbourg: Christian Egenolff, 1528.

Lecouteux, Claude. *Le Livre des Grimoires*. Paris: Imago, 2002.

Leerdam, Andrea van. "Popularising and Personalising an Illustrated Herbal in Dutch." *Nuncius* 36, no. 2 (June 23, 2021): 356–93. https://doi.org/10.1163/18253911-03602006.

Leerdam, Andrea van. *Woodcuts as Reading Guides: How Images Shaped Knowledge Transmission in Medical-Astrological Books in Dutch (1500–1550)*. Amsterdam: Amsterdam University Press, 2023.

Leng, Rainer. "Social Character, Pictorial Style, and the Grammar of Technical Illustration in Craftsmen's Manuscripts in the Late Middle Ages." In *Picturing Machines 1400–1700*, edited by Wolfgang Lefèvre, 85–111. Cambridge, MA: MIT Press, 2004.

Leong, Elaine. "Collecting Knowledge for the Family: Recipes, Gender and Practical Knowledge in the Early Modern English Household." *Centaurus* 55 (2013): 81–103.

Leong, Elaine. "Making Medicines in the Early Modern Household." *Bulletin of the History of Medicine* 82 (2008): 145–68.

Leong, Elaine. *Recipes and Everyday Knowledge: Medicine, Science, and the Household in Early Modern England*. Chicago: University of Chicago Press, 2018.

Leong, Elaine. "Transformative Itineraries and Communities of Knowledge in Early Modern Europe: The Case of Lazare Rivière's *The Practice of Physick*." In *Civic Medicine: Physician, Polity, and Pen in Early Modern Europe*, edited by J. Andrew Mendelsohn, Annemarie Kinzelbach, and Ruth Schilling, 257–79. Routledge, 2020.

Leong, Elaine. "When the Tallamys Met John French: Translating, Printing, and Reading *The Art of Distillation*." In *Translating Medicine Across Premodern Worlds*, edited by Tara Alberts, Sietske Fransen, and Elaine Leong. *Osiris* 37 (2022): 89–112. Chicago: University of Chicago Press. https://press.uchicago.edu/ucp/books/book/distributed/O/bo177851902.html.

Leong, Elaine, and Alisha Rankin. *Secrets and Knowledge in Medicine and Science, 1500–1800*. Farnham: Ashgate Publishing, 2011.

Leong, Elaine, and Alisha Rankin. "Testing Drugs and Trying Cures: Experiment and Medicine in Medieval and Early Modern Europe." *Bulletin of the History of Medicine* 91, no. 2 (2017): 157–82.

Lloyd, Geoffrey Ernest Richard. *Magic, Reason, and Experience: Studies in the Origins and Development of Greek Science*. Indianapolis: Hackett Publishing Company, 1999.

Long, Pamela. *Artisan/Practitioners and the Rise of the New Sciences, 1400–1600*. Corvallis: Oregon State University Press, 2011.

Long, Pamela. *Openness, Secrecy, Authorship: Technical Arts and the Culture of Knowledge from Antiquity to the Renaissance*. Baltimore: Johns Hopkins University Press, 2001.

Long, Pamela O. "Trading Zones in Early Modern Europe." *Isis* 106, no. 4 (December 2015): 840–47. https://doi.org/10.1086/684652.

Lonicer, Adam. *Kreuterbuoch Künstliche Conterfeytunge der Baeume / Stauden / Hecken / Kreuter / Getreyde / Gewürtze*. Frankfurt: Egenolff heirs, 1564.

López-Terrada, Maríaluz, and José Pardo-Tomás. "Reading About New World Nature in the Low Countries: The Editions of *Crónicas De Indias*, 1493–1600." In *Translating Knowledge in the Early Modern Low Countries*, edited by Harold J. Cook and Sven Dupré, 53–78. Münster: LIT Verlag, 2012.

Maclean, Ian. *Logic, Signs and Nature: Learned Medicine in the Renaissance*. Cambridge University Press, 2002.

Margócsy, Dániel, Mark Somos, and Stephen N. Joffe. *The Fabrica of Andreas Vesalius: A Worldwide Descriptive Census, Ownership, and Annotations of the 1543 and 1555 Editions*. Leiden: Brill, 2018.

Mariotte, Jean-Yves. *Les Sources Manuscrites de l'histoire de Strasbourg*. Strasbourg: Archives municipales, 2000.

Martinon-Torres, Marcos. "The Tools of the Chymist: Archaeological and Scientific Analyses of Early Modern Laboratories." In *Chymists and Chymistry: Studies in the History of Alchemy and Early Modern Chemistry*, edited by Lawrence Principe. Philadelphia: Chemical Heritage Foundation, 2007.

Maurer, Rudolf. *Baden, schröpfen, amputieren: Die Geschichte der Bader in Baden bei Wien*. Vienna: Verlagshaus der Ärzte, 2004.

McVaugh, Michael. "Bedside Manners in the Middle Ages." *Bulletin of the History of Medicine* 71, no. 2 (1997): 201–23.

McVaugh, Michael. *The Rational Surgery of the Middle Ages*. Florence: Sismel, 2006.

McVaugh, Michael. "Surgical Education in the Middle Ages." *Dynamis* 20 (2000): 283–304.

Mendelsohn, J. Andrew, Annemarie Kinzelbach, and Ruth Schilling, eds. *Civic Medicine: Physician, Polity, and Pen in Early Modern Europe*. Abingdon: Routledge, 2020.

Meyer, Hans. "Die Strassburger Goldschmiedezunft von ihrem Entstehen bis 1681: Urkunden und Darstellung. Ein Beitrag zur Gewerbegeschichte des Mittelalters." *Staats- und socialwissenschaftliche Forschungen* 3, no. 2 (1881): 1–224.

Miller, Clyde Lee. *Reading Cusanus: Metaphor and Dialectic in a Conjectural Universe*. Washington, DC: Catholic University of America Press, 2003.

Moore, Helen. "Gathering Fruit: The 'Profitable' Translations of Thomas Paynell." In *Tudor Translation*, edited by Fred Schurink, 39–57. Basingstoke: Palgrave Macmillan, 2011.

Moran, Bruce. *Distilling Knowledge: Alchemy, Chemistry, and the Scientific Revolution*. Cambridge, MA: Harvard University Press, 2005.

Moran, Bruce T. *Paracelsus: An Alchemical Life*, 2019.

Mukharji, Projit Bihari. *Doctoring Traditions: Ayurveda, Small Technologies, and Braided Sciences*. Chicago: University of Chicago Press, 2016.

Mukherjee, Ayesha. *Penury into Plenty: Dearth and the Making of Knowledge in Early Modern England*. London: Routledge, 2014.

Müller, Jan-Dirk. "'Curiositas' und 'erfarung' der Welt im frühen deutschen Prosaroman." In *Literatur und Laienbildung im Spätmittelalter und in der Reforma-*

tionszeit, edited by Ludger Grenzmann and Karl Stackmann, 252–71. Stuttgart: Germanische Symposien, 1984.

Müller, Jan-Dirk. "Erfarung zwischen Heilssorge, Selbsterkenntnis und Entdeckung des Kosmos." *Daphnis* 15 (1986): 307–42.

Müller-Grzenda, Astrid. *Pflanzenwässer und gebrannter Wein als Arzneimittel zu Beginn der Neuzeit*. Stuttgart: Deutscher Apotheker Verlag, 1996.

Multhauf, Robert. "John of Rupescissa and the Origin of Medical Chemistry." *Isis* 45, no. 4 (1954): 359–67.

Multhauf, Robert. *The Origins of Chemistry*. London: Oldbourne, 1966.

Multhauf, Robert. "The Significance of Distillation in Renaissance Medical Chemistry." *Bulletin of the History of Medicine* 30, no. 4 (1956): 329–46.

Murner, Thomas. *Der Schelmen Zunfft*. Frankfurt: Beatus Murner, 1512.

Murphy, Hannah. *A New Order of Medicine: The Rise of Physicians in Reformation Nuremberg*. Pittsburgh: University of Pittsburgh Press, 2019.

Newman, William R. "From Alchemy to 'Chymistry.'" In *The Cambridge History of Science, vol. 3, Early Modern Science*, edited by L. Daston and K. Park, 497–515. Cambridge University Press, 2006.

Newman, William R., and Lawrence M. Principe. "Alchemy vs. Chemistry: The Etymological Origins of a Historiographic Mistake." *Early Science and Medicine* 3, no. 1 (1998): 32–65.

Nummedal, Tara. *Alchemy and Authority in the Holy Roman Empire*. Chicago: University of Chicago Press, 2007.

Nummedal, Tara. "Practical Alchemy and Commercial Exchange in the Holy Roman Empire". In *Merchants and Marvels: Commerce, Science, and Art in Early Modern Europe*, edited by Pamela Smith and Paula Findlen, 201–22. New York: Routledge, 2002.

Nummedal, Tara. *Anna Zieglerin and the Lion's Blood: Alchemy and End Times in Reformation Germany*. Philadelphia: University of Pennsylvania Press, 2019.

Nutton, Vivian. *Ancient Medicine*. London: Routledge, 2013.

Nutton, Vivian. "Humanist Surgery." In *The Medical Renaissance of the Sixteenth Century*, edited by Andrew Wear, Roger Kenneth French, and Iain M. Lonie, 75–99. Cambridge University Press, 1985.

Ogilvie, Brian. *The Science of Describing: Natural History in Renaissance Europe*. Chicago: University of Chicago Press, 2006.

Oltrogge, Doris. "Transmission of Artists' Knowledge in Germany." In *Transmission of Artists' Knowledge*, edited by Mark Clarke, Bert De Munck, and Sven Dupré, 25–30. Brussels: KVAB Press, 2012.

Oosterhoff, Richard J., José Ramón Marcaida López, and Alexander Marr, eds. *Ingenuity in the Making: Matter and Technique in Early Modern Europe*. Pittsburgh: University of Pittsburgh Press, 2021.

Pagel, Walter. *Das medizinische Weltbild des Paracelsus: Seine Zusammenhänge mit Neuplatonismus und Gnosis*. Wiesbaden: Steiner, 1962.

Pagel, Walter. *Paracelsus: An Introduction to Philosophical Medicine in the Era of the Renaissance*. Basel: Karger, 1958.

Palmer, Ada. *Reading Lucretius in the Renaissance*. Cambridge, MA: Harvard University Press, 2014.

Palmer, Richard. "Physicians and Surgeons in Sixteenth-Century Venice." *Medical History* 23, no. 4 (1979): 451–60.

Pangerl, Daniel Carlo. "Item als man dem kayser Fridrichen sin fuss abschnitt. Die Beinamputation an Kaiser Friedrich III. am 8. Juni 1493 in Linz." *Sudhoffs Archiv* 94, no. 2 (2010): 195–200.

Panse, Melanie. *Hans von Gersdorffs "Feldbuch der Wundarznei": Produktion, Präsentation und Rezeption von Wissen*. Wiesbaden: Reichert, 2012.

Pantin, Isabelle. "John Hester's Translations of Leonardo Fioravanti: The Literary Career of a London Distiller." In *Renaissance Cultural Crossroads: Translation, Print and Culture in Britain 1473–1640*, edited by Sara Barker and Brenda Hosington, 159–83. Leiden: Brill, 2013.

Pantin, Isabelle. "The Role of Translations in European Scientific Exchanges in the Sixteenth and Seventeenth Centuries." In *Cultural Translation in Early Modern Europe*, edited by Peter Burke and R. P. Hsia, 163–79. Cambridge University Press, 2007.

Paracelsus. *An Excellent Treatise Teaching Howe to Cure the French-Pockes*. Translated by John Hester. London: Charlwood, 1590.

Paracelsus. "On the Miners' Sickness and Other Miners' Diseases." In *Four Treatises of Theophrastus von Hohenheim Called Paracelsus*, edited by Henry Sigerist, translated by George Rosen, 43–126. Baltimore: Johns Hopkins University Press, 1941.

Park, Katharine. "Natural Particulars: Medical Epistemology, Practice, and the Literature of Healing Springs." In *Natural Particulars: Nature and the Disciplines in Renaissance Europe*, edited by Nancy Siraisi and Anthony Grafton, 347–67. Cambridge, MA: MIT Press, 1999.

Park, Katharine. "Observation in the Margins." In *Histories of Scientific Observation*, edited by Lorraine Daston and Elizabeth Lunbeck, 15–44. Chicago: University of Chicago Press, 2011.

Park, Katharine. *Secrets of Women: Gender, Generation, and the Origins of Human Dissection*. Zone Books New York, 2006.

Park, Katharine. "Stones, Bones and Hernias: Surgical Specialists in Fourteenth- and Fifteenth-Century Italy." In *Medicine from the Black Death to the French Disease*, edited by Roger Kenneth French, Jon Arrizabalaga, Andrew Cunningham, and Luis García-Ballester, 110–30. Aldershot: Ashgate, 1998.

Pastoureau, Michel. *The Devil's Cloth: A History of Stripes and Striped Fabric*. New York: Columbia University Press, 2001.

Payne, Linda. "'A Spedie Reformation:' Barber-Surgeons, Anatomization, and the Reformation of Medicine in Tudor London." In *Paracelsian Moments: Science, Medicine, and Astrology in Early Modern Europe*, edited by Gerhild Scholz Williams and Charles D. Gunnoe, 71–92. Truman State University Press, 2002.

Pelling, Margaret. "Appearance and Reality: Barber-Surgeons, the Body and Disease." In *London 1500–1700: The Making of the Metropolis*, edited by A. L. Beier and Roger Finlay, 82–112. London: Longman, 1986.

Pelling, Margaret. *Medical Conflicts in Early Modern London: Patronage, Physicians, and Irregular Practitioners, 1550–1640*. Oxford: Oxford University Press, 2003.

Pelling, Margaret, and Charles Webster. "Medical Practitioners." In *Health, Medicine, and Mortality in the Sixteenth Century*, edited by Charles Webster, 165–236. Cambridge University Press, 1979.

Pennell, Sara, and Michelle DiMeo, eds. *Reading and Writing Recipe Books*. Manchester: Manchester University Press, 2013.

Pereira, Michela. "Heavens on Earth: From the Tabula Smaragdina to the Alchemical Fifth Essence." *Early Science and Medicine* 5 (2000): 131–44.

Pereira, Michela. "Mater Medicinarum: English Physicians and the Alchemical Elixir in the Fifteenth Century." In *Medicine from the Black Death to the French Disease*, edited by Roger Kenneth French, Andrew Cunningham, Luis García-Ballester, and Jon Arrizabalaga, 26–52. Aldershot: Ashgate, 1998.

Pfeiffer, Gerhard. "Technologische Entwicklung von Destilliergeräten vom Spätmittelalter bis zur Neuzeit." 1986.

Plat, Hugh. *The Iewell House of Art and Nature Conteining Diuers Rare and Profitable Inuentions, Together with Sundry New Experimentes in the Art of Husbandry, Distillation, and Moulding*. London: Short, 1594.

Pogliani, Annarita. "Il successo editoriale del *Büchlein von den ausgebrannten Wässern* di Michael Puff aus Schrick." In *La letteratura tecnico-scientifica nel Medioevo germanico: Fachliteratur e Gebrauchstexte*, edited by Letizia Vezzosi, 193–212. Alessandria: Edizioni dell'Orso, 2009.

Pomata, Gianna. *Contracting a Cure: Patients, Healers, and the Law in Early Modern Bologna*. Baltimore: Johns Hopkins University Press, 1998.

Pomata, Gianna. "Observation Rising: Birth of an Epistemic Genre, 1500–1650." In *Histories of Scientific Observation*, edited by Lorraine Daston and Elizabeth Lunbeck, 45–80. Chicago: University of Chicago Press, 2011.

Pomata, Gianna. "*Praxis Historialis*: The Uses of *Historia* in Early Modern Medicine." In *Historia: Empiricism and Erudition in Early Modern Europe*, edited by Gianna Pomata and Nancy G. Siraisi, 105–46. Cambridge, MA: MIT Press, 2005.

Pomata, Gianna. "A Word of the Empirics: The Ancient Concept of Observation and Its Recovery in Early Modern Medicine." *Annals of Science* 68, no. 1 (2011): 1–25.

Pomata, Gianna, and Nancy Siraisi. *Historia: Empiricism and Erudition in Early Modern Europe*. Cambridge, MA: MIT Press, 2005.

Porter, Roy. *The Greatest Benefit to Mankind: A Medical History of Humanity*. London: Harper Collins, 1997.

Principe, Lawrence M. *The Secrets of Alchemy*. Chicago: University of Chicago Press, 2013.

Ragland, Evan. "Chymistry and Taste in the Seventeenth Century: Franciscus Dele Boe Sylvius as a Chymical Physician Between Galenism and Cartesianism." *Ambix* 59, no. 1 (2012): 1–21.

Ralley, Robert. "Medical Economies in Fifteenth Century England." In *Medicine and the Market in England and Its Colonies, c. 1450–c. 1850*, edited by Patrick Wallis and Mark Jenner, 24–46. Basingstoke: Palgrave Macmillan, 2007.

Rampling, Jennifer M. "Depicting the Medieval Alchemical Cosmos: George Ripley's Wheel of Inferior Astronomy." *Early Science and Medicine* 18, no. 1–2 (2013): 45–86.

Rampling, Jennifer M. *The Experimental Fire: Inventing English Alchemy, 1300–1700*. University of Chicago Press, 2020.

Rampling, Jennifer M. "John Dee and the Sciences: Early Modern Networks of Knowledge." *Studies in History and Philosophy of Science* 43, no. 3 (2012).

Rankin, Alisha. "Duchess, Heal Thyself: Elisabeth of Rochlitz and the Patient's Perspective in Early Modern Germany." *Bulletin of the History of Medicine* 82, no. 1 (2008): 109–44.

Rankin, Alisha. "Germany." In *The Oxford History of Popular Print Culture*, edited by Joad Raymond, 205–13. Vol. 1. New York: Oxford University Press, 2011.

Rankin, Alisha. "How to Cure the Golden Vein: Medical Remedies as Wissenschaft in Early Modern Germany." In *Ways of Making and Knowing*, edited by Pamela H. Smith, Harold Cook, and Amy Meyers, 109–33. Ann Arbor: University of Michigan Press, 2014.

Rankin, Alisha. "New World Drugs and the Archive of Practice: Translating Nicolás Monardes in Early Modern Europe." In *Translating Medicine Across Premodern Worlds*, edited by Tara Alberts, Sietske Fransen, and Elaine Leong. *Osiris* 37 (2022): 67–88. Chicago: University of Chicago Press.

Rankin, Alisha. *Panaceia's Daughters: Noblewomen as Healers in Early Modern Germany*. Chicago: University of Chicago Press, 2013.

Rankin, Alisha. *The Poison Trials: Wonder Drugs, Experiment, and the Battle for Authority in Renaissance Science*. Chicago: University of Chicago Press, 2021.

Rawcliffe, Carole. *Leprosy in Medieval England*. Woodbridge: Boydell Press, 2006.

Reith, Reinhold. "Know-How, Technologietransfer und die Arcana Artis im Mitteleuropa der Frühen Neuzeit." *Early Science and Medicine*, 2005, 349–77.

Roberts, Lissa, and Simon Schaffer. *The Mindful Hand: Inquiry and Invention from the Late Renaissance to Early Industrialisation*. Amsterdam: Royal Netherlands Academy of Arts and Sciences, 2007.

Roper, Lyndal. "'The Common Man,' 'The Common Good,' 'Common Women': Gender and Meaning in the German Reformation Commune." *Social History* 12, no. 1 (1987): 1–21.

Röslin, Eucharius. *Kreutterbuoch von Allem Erdtgewaechs*. Frankfurt: Egenolff, 1533.

Roxo Beltran, Maria. "Sixteenth Century Books of Distillation." In *Between the Natural and the Artificial: Dyestuffs and Medicines*, edited by Gérard Emptoz and Patricia Aceves Pastrana, 145–55. Turnhout: Brepols, 2000.

Rublack, Ulinka. *Dressing Up: Cultural Identity in Renaissance Europe*. Oxford: Oxford University Press, 2010.

Rublack, Ulinka. "Fluxes: The Early Modern Body and the Emotions." *History Workshop Journal* 53, no. 1 (2002): 1–16.

Rupescissa, John of. *Ioannis de Rupescissa . . . de consideratione quintae essentiae rerum omnium*. Basel: anon., 1561.

Ryff, Walther Hermann. *Das New groß Distillier Buoch Wolgegründter Künstlicher Distillation*. Frankfurt: Christian Egenolff, 1545.

Ryff, Walther Hermann. *Die kleyner Chirurgi D. Gualtheri H. Ryff. Das ist / der grund vnnd kern gemeyner eynleytung der gantzen wundartzney*. Strasbourg: Balthasar Beck, 1542.

Ryff, Walther Hermann, and Hieronymus Brunschwig. *New Vollkommen Distillierbuoch Wolgegründter kuenstlicher Distillation*. Frankfurt: Christian Egenolff, 1597.

Sachs, Hans. *Eygentliche Beschreibung Aller Staende auff Erden / Hoher vnd Nidriger / Geistlicher vnd Weltlicher / Aller Künsten / Handwercken vnd Haendeln / etc. vom groesten biß zum kleinesten / Auch von irem Vrsprung / Erfindung vnd gebreuchen*. Frankfurt: Georg Raben for Sigmund Feyerabend, 1568.

Sander, Sabine. *Handwerkschirurgen*. Göttingen: Vandenhoeck & Ruprecht, 1989.

Savoia, Paolo. *Gaspare Tagliacozzi and Early Modern Surgery: Faces, Men, and Pain*. Routledge, 2020.

Savoia, Paolo. "Skills, Knowledge, and Status: The Career of an Early Modern Italian Surgeon." *Bulletin of the History of Medicine* 93, no. 1 (2019): 27–54.

Schaffer, Simon, Lissa Roberts, Kapil Raj, and James Delbourgo, eds. *The Brokered World: Go-Betweens and Global Intelligence, 1770–1820*. Science History Publications Sagamore Beach, MA, 2009.

Schenda, Rudolf. "Der 'gemeine Mann' und sein medikales Verhalten im 16. und 17. Jahrhundert." In *Pharmazie und der gemeine Mann: Hausarznei und Apotheke der frühen Neuzeit*, edited by Joachim Telle, 9–20. Braunschweig: Waisenhaus Verlag, 1982.

Schlegelmilch, Sabine. "Promoting a Good Physician: Letters of Application to German Civic Authorities, 1500–1700." In *Civic Medicine: Physician, Polity, and Pen in Early Modern Europe*, edited by J. Andrew Mendelsohn, Annemarie Kinzelbach, and Ruth Schilling, 88–109. Abingdon: Routledge, 2020.

Schlich, Thomas. "Introduction: What Is Special About the History of Surgery?" In *The Palgrave Handbook of the History of Surgery*, edited by Thomas Schlich, 1–24. Cham: Springer, 2018.

Schmid Blumer, Verena. *Ikonographie und Sprachbild: Zur reformatorischen Flugschrift "Der gestrygfft Schwitzer Baur."* Tübingen: Niemeyer, 2004.

Schmidt, Peter. "Beschrieben, bemalt, zerschnitten: Tegernseer Mönche interpretieren einen Holzschnitt." In *Einblattdrucke des 15. und frühen 16. Jahrhunderts: Probleme, Perspektiven, Fallstudien*, edited by Volker Honemann, Sabine Griese, Falk Eisermann, and Marcus Ostermann, 245–76. Tübingen: Max Niemeyer, 2000.

Schreiner, Klaus. "Laienbildung als Herausforderung für Kirche und Gesellschaft. Religiöse Vorbehalte und soziale Widerstände gegen die Verbreitung von Wissen im späten Mittelalter und in der Reformation." *Zeitschrift für historische Forschung* 11, no. 3 (1984): 257–354.

Schulz, Knut. *Handwerksgesellen und Lohnarbeiter: Untersuchungen zur oberrheinischen und oberdeutschen Stadtgeschichte des 14. bis 17. Jahrhunderts*. Sigmaringen: Thorbecke, 1985.

Schulz-Grobert, Jürgen. *Das Straßburger Eulenspiegelbuch*. Max Niemeyer Verlag, 1999.

Schütt, Hans-Werner. "Die frühe Alchemie und die Technik der Distillation." In *Brückenschläge: 25 Jahre Lehrstuhl für Geschichte der exakten Wissenschaften und der Technik an der Technischen Universität Berlin: 1969–1994*, edited by Hans-Werner Schütt and Burghard Weiss, 293–309. Berlin: Engel, 1995.

Secord, James A. "Knowledge in Transit." *Isis* 95, no. 4 (December 2004): 654–72. https://doi.org/10.1086/430657.

Shapin, Steven. "The Invisible Technician." *American Scientist* 77, no. 6 (1989): 554–63.

Shapin, Steven. *A Social History of Truth: Civility and Science in Seventeenth-Century England*. Chicago: University of Chicago Press, 1994.

Shapin, Steven, and Simon Schaffer. *Leviathan and the Air-Pump: Hobbes, Boyle, and the Experimental Life*. Princeton, NJ: Princeton University Press, 1985.

Sharpe, Kevin M. *Reading Revolutions: The Politics of Reading in Early Modern England*. New Haven: Yale University Press, 2000.

Sherman, William H. *Used Books: Marking Readers in Renaissance England*. Philadelphia: University of Pennsylvania Press, 2008.

Sherman, William H. *John Dee: The Politics of Reading and Writing in the English Renaissance*. University of Massachusetts Press, 1995.

Sigerist, Henry. *Hieronymus Brunschwig and His Work: A Fifteenth Century Surgeon*. New York: Ben Abrahamson, 1946.

Siraisi, Nancy. *Medieval and Early Renaissance Medicine*. Chicago: University of Chicago Press, 1990.

Slack, Paul. "Mirrors of Health and Treasures of Poor Men: The Uses of the Vernacular Medical Literature of Tudor England." In *Health, Medicine, and Mortality in the Sixteenth Century*, edited by Charles Webster, 237–73. Cambridge University Press, 1979.

Smith, Pamela H. *The Body of the Artisan: Art and Experience in the Scientific Revolution*. Chicago: University of Chicago Press, 2004.

Smith, Pamela H. "Craft Techniques and How-To Books." In *Transmission of Artists' Knowledge*, edited by Mark Clarke, Bert De Munck, and Sven Dupré, 75–84. Brussels: KVAB Press, 2012.

Smith, Pamela H. *From Lived Experience to the Written Word: Reconstructing Practical Knowledge in the Early Modern World*. Chicago: University of Chicago Press, 2022.

Smith, Pamela H. "In a Sixteenth-Century Goldsmith's Workshop." In *The Mindful Hand: Inquiry and Invention from the Late Renaissance to Early Industrialisation*, edited by Lissa Roberts and Simon Schaffer, 33–57. Amsterdam: Royal Netherlands Academy of Arts and Sciences, 2007.

Smith, Pamela H. *Secrets of Craft and Nature in Renaissance France. A Digital Critical Edition and English Translation of BnF Ms. Fr. 640*. New York: Making and Knowing Project, 2020. https://edition640.makingandknowing.org/.

Smith, Pamela H. "Vermilion, Mercury, Blood, and Lizards: Matter and Meaning in Metalworking." In *Materials and Expertise in Early Modern Europe: Between Market and Laboratory*, edited by Ursula Klein and Emma Spary, 29–48. Chicago: University of Chicago Press, 2010.

Smith, Pamela H. "What Is a Secret? Secrets and Craft Knowledge in Early Modern Europe." In *Secrets and Knowledge in Medicine and Science, 1500–1800*, edited by Elaine Leong and Alisha Rankin, 47–66. Farnham: Ashgate Publishing, 2011.

Smith, Pamela H., Harold Cook, and Amy Meyers. *Ways of Making and Knowing*. Ann Arbor: University of Michigan Press, 2014.

Spruit, Leen. "Albert the Great on the Epistemology of Natural Science." In *Erfahrung und Beweis. Die Wissenschaften von der Natur im 13. und 14. Jahrhundert*, edited by Alexander Fidora and Matthias Lutz-Bachmann, 14:61–75. Oldenbourg Verlag, 2007.

Stallybrass, Peter. "Books and Scrolls: Navigating the Bible." In *Books and Readers in Early Modern England*, edited by Jennifer Andersen, 42–79. Philadelphia: University of Pennsylvania Press, 2002.

Stammler, Wolfgang, Karl Langosch, and Kurt Ruh, eds. *Die deutsche Literatur des Mittelalters: Verfasserlexikon*. Vol. 1. Berlin: Walter de Gruyter, 1978.

Stein, Claudia. "The Meaning of Signs: Diagnosing the French Pox in Early Modern Augsburg." *Bulletin of the History of Medicine* 80, no. 4 (2006): 617–48.

Stein, Claudia. *Negotiating the French Pox in Early Modern Germany*. Aldershot: Ashgate, 2009.

Steinke, H. "Vom Schererlehrling zum Chirurgenmeister." In *Jakob Ruf, ein Zürcher*

Stadtchirurg und Theatermacher im 16. Jahrhundert, edited by Hildegard Keller, 55–65. Zurich: Chronos Verlag, 2006.

Stolberg, Michael. "Der gesunde Leib: Zur Geschichtlichkeit frühneuzeitlicher Körpererfahrung." In *"Erfahrung" als Kategorie der Frühneuzeitgeschichte*, edited by Paul Münch, 37–57. München: Oldenbourg, 2001.

Stolberg, Michael. *Experiencing Illness and the Sick Body in Early Modern Europe*. Springer, 2011.

Stolberg, Michael. "Learning from the Common Folks: Academic Physicians and Medical Lay Culture in the Sixteenth Century." *Social History of Medicine* 27 (2014): 649–67.

Stolberg, Michael. *Uroscopy in Early Modern Europe*. Farnham: Routledge, 2016.

Stolberg, Michael. "You Have No Good Blood in Your Body: Oral Communication in Sixteenth-Century Physicians' Medical Practice." *Medical History* 59 (2015): 63–82.

Stuart, Kathy. *Defiled Trades and Social Outcasts: Honor and Ritual Pollution in Early Modern Germany*. Cambridge University Press, 2006.

Sudhoff, Karl. *Deutsche medizinische Inkunabeln. Bibliographisch-literarische Untersuchungen*. Leipzig: Barth, 1908.

Swan, Claudia. "The Uses of Realism in Early Modern Illustrated Botany." In *Visualizing Medieval Medicine and Natural History, 1200–1550*, edited by Jean Ann Givens, Karen Reeds, and Alain Touwaide, 239–49. Aldershot: Ashgate, 2006.

Taape, Tillmann. "The Body and the Senses in Ms. Fr. 640: Towards a 'Material Sensorium.'" In *Secrets of Craft and Nature in Renaissance France. A Digital Critical Edition and English Translation of BnF Ms. Fr. 640*, edited by Pamela H. Smith Making and Knowing Project. New York: Making and Knowing Project, 2020. https://edition640.makingandknowing.org/#/essays/ann_302_ie_19.

Taape, Tillmann. "A Commerce of Secrets: Digital and Performative Approaches to an Early Modern How-to Manuscript at the Making and Knowing Project." In *Tracts of Action: Material, Visual, and Practical Dimensions of Early Modern How-to Books*, edited by Stefan Laube, 126–51. Leiden: Brill, 2024.

Taape, Tillmann. "Common Medicine for the Common Man: Picturing the 'Striped Layman' in Early Vernacular Print." *Renaissance Quarterly* 74, no. 1 (2021): 1–58.

Taape, Tillmann. "Distilling Reliable Remedies: Hieronymus Brunschwig's *Liber de Arte Distillandi* (1500) Between Alchemical Learning and Craft Practice." *Ambix* 61, no. 3 (2014): 236–56.

Taape, Tillmann. "Subtle Fire: Distillation as Low-Heat Technology and the Agency of Human Art." *Ambix*, forthcoming.

Taape, Tillmann. "Trade and Industry: Chemical Economies and the Business of Distillation." In *A Cultural History of Chemistry in the Early Modern Age*, edited by Bruce Moran. London: Bloomsbury, 2021.

Tedder, Henry Richard. "Andrewe, Laurence." *Dictionary of National Biography*, Vol. 1. Wikisource, 1885. https://en.wikisource.org/w/index.php?title=Dictionary_of_National_Biography,_1885-1900/Andrewe,_Laurence&oldid=10758860.

Telle, Joachim. "Das Rezept als literarische Form. Zum multifunktionalen Gebrauch des Rezepts in der Deutschen Literatur." *Berichte zur Wissenschaftsgeschichte* 26, no. 4 (2003): 251–74.

Telle, Joachim. "Medizin für den gemeinen Mann." In *Bibliotheca Palatina. Katalog zur Ausstellung vom 8. Juli bis 2. November 1986, Heiliggeistkirche Heidelberg,* edited by Elmar Mittler, 335–40. Heidelberg: Braus, 1986.

Telle, Joachim, ed. *Pharmazie und der gemeine Mann: Hausarznei und Apotheke der frühen Neuzeit.* Wolfenbüttel: Herzog August Bibliothek, 1988.

Telle, Joachim. "Verzeichnis der Exponate: Bücher." In *Pharmazie und der gemeine Mann: Hausarznei und Apotheke der frühen Neuzeit,* edited by Joachim Telle, 51–133. Wolfenbüttel: Herzog August Bibliothek, 1988.

Temkin, Owsei. "The Role of Surgery in the Rise of Modern Medical Thought." *Bulletin of the History of Medicine* 25, no. 3 (1951): 248–59.

Thomas, Nicolas. "L'alambic dans la cuisine?" In *La cuisine et la table dans la France de la fin du Moyen-Âge. Contenus et contenants du XIVe au XVIe siècle,* edited by Fabienne Ravoir and Anne Dietrich, 35–50. Caen: CRAHM, 2009.

Thorndike, Lynn. *A History of Magic and Experimental Science.* Vol. 2. New York: Columbia University Press, 1923.

Tylus, Jane, and Karen Newman. "Introduction." In *Early Modern Cultures of Translation,* edited by Jane Tylus and Karen Newman, 1–24. Philadelphia: University of Pennsylvania Press, 2015.

Valleriani, Matteo, ed. *The Structures of Practical Knowledge.* Cham: Springer, 2017.

Vollmuth, Ralf. "Wann einer mit einer puchssen geschossen wirt. Eine Mehrschritt-Therapie für Schußwunden aus der zweiten Hälfte des 15. Jahrhunderts." *Sudhoffs Archiv,* 1998, 102–4.

Vollmuth, Ralf. "War Klaus von Matrei der Lehrer Hans von Gersdorffs?" *Sudhoffs Archiv,* 1996, 109–17.

Wall, Wendy. "Distillation: Transformations in and out of the Kitchen." In *Renaissance Food from Rabelais to Shakespeare: Culinary Readings and Culinary Histories,* edited by Joan Fitzpatrick, 89–104. Farnham: Ashgate, 2010.

Wallis, Faith. "Pre-Modern Surgery: Wounds, Words, and the Paradox of 'Tradition.'" In *The Palgrave Handbook of the History of Surgery,* edited by Thomas Schlich, 49–70. Cham: Springer, 2018.

Wallis, Patrick, and Catherine Wright. "Evidence, Artisan Experience, and Authority in Early Modern England." In *Ways of Making and Knowing,* edited by Pamela Smith, Harold Cook, and Amy Meyers, 138–63. Ann Arbor: University of Michigan Press, 2014.

Wear, Andrew. "Early Modern Europe." In *The Western Medical Tradition: 800 BC to AD 1800, Volume 1,* edited by Lawrence I. Conrad, Michael Neve, Vivian Nutton, Roy Porter, and Andrew Wear, 215–361. Cambridge University Press, 1995.

Webster, Charles. "Alchemical and Paracelsian Medicine." In *Health, Medicine, and Mortality in the Sixteenth Century,* edited by Charles Webster, 301–34. Cambridge University Press, 1979.

Webster, Charles. *Paracelsus: Medicine, Magic and Mission at the End of Time.* New Haven: Yale University Press, 2008.

Weisser, Olivia. "Grieved and Disordered: Gender and Emotion in Early Modern Patient Narratives." *Journal of Medieval and Early Modern Studies* 43, no. 2 (2013): 247–73.

Werrett, Simon. *Fireworks: Pyrotechnic Arts and Sciences in European History.* Chicago: University of Chicago Press, 2010.

Wickersheimer, Ernest. "Fries, Lorenz." *Neue Deutsche Biographie* 5 (1961): 609f. http://www.deutsche-biographie.de/pnd119451662.html.

Wickersheimer, Ernest. "Le gaïac à Strasbourg au XVIe siècle." *Analecta Medico-Historica* 1 (1966): 55–66.

Widmann, Johann. *Tractatus Clarissimi Medicinarum Doctoris Joannis Joannis Widman*. Strasbourg: Grüninger, 1497.

Wiegand, Wilhelm, ed. Urkundenbuch der Stadt Strassburg. Vol. 4. Strasbourg: Karl Trübner, 1898.

Wieger, Friedrich. *Geschichte der Medicin und ihrer Lehranstalten in Strassburg vom Jahre 1497 bis zum Jahre 1872*. Strasbourg: Karl Trübner, 1885.

Winckelmann, Otto. *Das Fürsorgewesen der Stadt Strassburg*. Leipzig: Heinsius, 1922.

Yeo, Richard. *Notebooks, English Virtuosi, and Early Modern Science*. Chicago: University of Chicago Press, 2014.

Zech, Kristin. "Le 'Schwörbrief' de 1482: L'origine et les conséquences de l'exclusion du Grand conseil pour les baigneurs de Strasbourg." *Revue d'Alsace*, no. 140 (2014): 59–78.

Zech, Kristin. "Zunftauflösungen als Spiegel politischer Partizipationschancen und -grenzen sozialer Gruppen in der Stadt: Straßburg, Colmar, Schlettstadt." In *Politische Partizipation in spätmittelalterlichen Städten am Oberrhein*, edited by Olivier Richard and Gabriel Zeilinger, 205–40. Berlin: Erich Schmidt, 2017.

Zilsel, Edgar. "The Sociological Roots of Science." *American Journal of Sociology* 47 (1942): 544–62.

Index

Page numbers in italics refer to figures.